Medical Practice, 1600–1900

Clio Medica

PERSPECTIVES IN MEDICAL HUMANITIES

VOLUME 96

The titles published in this series are listed at *brill.com/clio*

Medical Practice, 1600–1900

Physicians and Their Patients

Edited by

Martin Dinges
Kay Peter Jankrift
Sabine Schlegelmilch
Michael Stolberg

Translated by

Margot Saar

BRILL
RODOPI

LEIDEN | BOSTON

Cover illustration: Frans van Mieris (1635–1681), *The Doctor's Visit* (Los Angeles, J. Paul Getty Museum).

Library of Congress Cataloging-in-Publication Data

Medical practice, 1600-1900 : physicians and their patients / edited by Martin Dinges, Kay Peter Jankrift, Sabine Schlegelmilch, Michael Stolberg.
p. ; cm. -- (Clio medica: perspectives in medical humanities, ISSN 0045-7183 ; volume 96)
Includes bibliographical references and index.
ISBN 978-90-04-30329-4 (hardback : alk. paper) -- ISBN 978-90-04-30332-4 (e-book)
I. Dinges, Martin, 1953- , editor. II. Jankrift, Kay Peter, editor. III. Schlegelmilch, Sabine, editor. IV. Stolberg, Michael, 1957- , editor. V. Series: Clio medica (Amsterdam, Netherlands) ; 96. 0045-7183
[DNLM: 1. General Practice--history--Europe. 2. General Practice--history--Germany. 3. History, 17th Century--Europe. 4. History, 17th Century--Germany. 5. History, 18th Century--Europe. 6. History, 18th Century--Germany. 7. History, 19th Century--Europe. 8. History, 19th Century--Germany. 9. Physicians--Europe. 10. Physicians--Germany. W1 CL933 v.96 2016 / WZ 70 GA1]

R729.5.G4
610.68--dc23

2015032411

This publication has been typeset in the multilingual 'Brill' typeface. With over 5,100 characters covering Latin, IPA, Greek, and Cyrillic, this typeface is especially suitable for use in the humanities. For more information, please see www.brill.com/brill-typeface.

ISSN 0045-7183
ISBN 978-90-04-30329-4 (hardback)
ISBN 978-90-04-30332-4 (e-book)

This book is printed on acid-free paper.

Contents

List of Illustrations

About the Contributors

Marion Baschin

Dr. phil., researcher at the Institut für Geschichte der Medizin der Robert Bosch Stiftung, Stuttgart. Academic training: history, political science and sociology at the Universität Stuttgart, PhD in history (Stuttgart, 2010).

Research interests: Social history of medicine; patient history; history of homeopathy and historical demography.

Martin Dinges

Prof. Dr. phil., deputy director of the Institut für Geschichte der Medizin der Robert Bosch Stiftung, Stuttgart, teaches modern history at the Universität Mannheim. Academic training: law, history and political science in Cologne, Mainz, Bonn, Berlin and Bordeaux, PhD in history (Freie Universität Berlin, 1986), *habilitation* in modern history (Mannheim, 1991).

Research interests: History of health (modern age); gender history.

Elisabeth Dietrich-Daum

Prof. Dr. phil., professor for economic and social history at the Universität Innsbruck, PhD in economic and social history (Innsbruck, 1986).

Research interests: Social history of medicine; history of psychiatry; regional history and gender studies.

Lina Gafner

Dr. des., Universität Zürich. Academic training: history, philosophy and German literature at the Universität Bern. PhD in history (Berne, 2015). Since 2010 research fellowship (Schweizer Nationalfonds).

Research interests: History of knowledge; administration; medicine and sexual-reform-movements.

Susanne Grosser

Dr. phil. Academic training: history and political science at the Universität Erlangen-Nürnberg (2006–2011).

Research interests: Early modern medicine and science, esp. epistolary networks, letter editions, museology.

Volker Hess

Prof. Dr. med., director of the Institut für Geschichte der Medizin, Berlin-Charité and affiliated professor at the history department of the Humboldt-Universität, Berlin. Academic training: medicine, philosophy, and history in

Konstanz and Berlin, MD-thesis in the history of medicine (Berlin, 1992), *habilitation* (Berlin, 1999).

Research interests: History of medical practices.

Marina Hilber

Dr. phil., 2008–2015 researcher at the Institut für Geschichtswissenschaften und Europäische Ethnologie, Universität Innsbruck. Academic training: history, English and European ethnology in Innsbruck, PhD in history (Innsbruck, 2011).

Research interests: Social history of medicine (19th and 20th century); history of maternity hospitals and midwifery; social history of mining and miners (17th-19th century).

Kay Peter Jankrift

Prof. Dr. phil., teaches medieval history at the Universität Münster. Academic training: history, semitic languages and islamic science at the universities of Münster and Tel Aviv, PhD in medieval history (Münster, 1995), *habilitation* in medieval history (Münster, 2002).

Research interests: Cultural and social history of the middle ages and the early modern period; history of the Mediterranean; encounters between Christians, Jews and Muslims; historical peace and conflict research; history of diplomacy; history of medicine and science (middle ages/early modern period).

Annemarie Kinzelbach

Dr. phil., researcher. Academic training: history, linguistics, history of German language and literature, and art history at the universities in Munich (LMU) and Heidelberg, PhD in history (Heidelberg, 1994).

Research interests: History of medicine and health; early modern administration and culture in the cities; knowledge-transfer in pre-industrial societies; gender studies.

Philipp Klaas

MA, research assistant at the Institut für Medizingeschichte, Universität Bern. Academic training: early modern and modern history, German philology and philosophy.

Research interests: Social history of medicine (18th-19th century); social and cultural history; history of psychiatry.

Stephanie Neuner

Dr. phil., postdoc at the Institut für Geschichte der Medizin, Universität Würzburg, and curator at the Berliner Medizinhistorisches Museum der Charité. Academic training: history and politics at universities in Munich (LMU) and Edinburgh. PhD in history (Munich, 2009).

Research interests: Social history of medicine, history of psychiatry, history of scientific knowledge production, history of social politics, museum studies.

Karen Nolte

PD Dr. phil., assistant professor at the Institut für Geschichte der Medizin, Universität Würzburg. Academic training: history, sociology and cultural anthropology in Göttingen. PhD in history (Göttingen, 2002), *habilitation* (Würzburg, 2010).

Research interests: History of psychiatry (19th century); gender history (18th–20th century); body history (18th–20th century); history of medical ethics (18th–20th century); nursing history (18th/19th century).

Iris Ritzmann

Prof. Dr. med., lecturer for medical history at the Universität Zürich. Academic training: medicine (1981–1988), history, medical history and social history at the Universität Zürich, MD (Zurich, 1991), *habilitation* (Zurich, 2005).

Research interests: history of pediatrics 18th-20th century; history of patients; medicalization of marginalized groups.

Marion Maria Ruisinger

Prof. Dr. phil., director of the Deutsches Medizinhistorisches Museum Ingolstadt. Academic training: medicine at the Universität of Erlangen-Nürnberg (1983–1989), MD-thesis in the history of medicine (Erlangen, 1997), diploma as specialist in general medicine (2000), *habilitation* in the history of medicine (Erlangen, 2005).

Research interests: History of surgery; patients' history; museology; Greek health-care system in the 19th century.

Ruth Schilling

Prof. Dr. phil., Juniorprofessorin at the Universität Bremen and research co-ordinator at the Deutsche Schifffahrtsmuseum. Academic training: modern and contemporary history, ancient history and ancient Greek at the Humboldt-Universität Berlin, PhD in modern and contemporary history (Berlin, 2008).

Research Interests: Early modern political culture; history of medicine and science (early modern period); concepts of space; memorial practices in interdisciplinary perspective.

Sabine Schlegelmilch

Dr. phil., assistant professor at the Institut für Geschichte der Medizin, Universität Würzburg. Academic training: classics and German literature in Würzburg and London. Interdisciplinary PhD in classics, archeology, Egyptology (Würzburg, 2008).

Research interests: medical theory and practice of the 16th & 17th century; medicine in film.

Hubert Steinke

Prof. Dr. med. Dr phil., professor for the history of medicine at the Universität Bern. Academic training: MD-thesis in the history of medicine (Bern, 1995), PhD in history (Oxford, 2003).

Research interests: Medical theory and practice from the 16th to the 19th century; history of the republic of letters.

Michael Stolberg

Prof. Dr. med. Dr. phil., director of the Institut für Geschichte der Medizin, Universität Würzburg. Academic training: MD-thesis in history of medicine (Munich, 1986), *habilitation* in medical history and medical sociology (Munich 1992), PhD in history (Munich, 1994).

Research interests: early modern medicine; history of medical ethics.

Alois Unterkircher

Dr. phil., research assistant at the Medizinhistorische Objektsammlung, Universität Zürich Academic training: history, European ethnology and German literature PhD in history (Innsbruck, 2012).

Research interests: social history of medicine; history of masculinities; museology.

Eberhard Wolff

PD Dr. rer. soc., cultural anthropologist and medical historian. Academic training: PhD (Dr. rer. soc.) in cultural anthropology (Tübingen, 1995), *habilitation* (Basel, 2009).

Research interests: science communication and museology; medicine and culture; doctor-patient-interaction.

Introduction

Martin Dinges and Michael Stolberg

'Practices' and 'praxeological' approaches have acquired a prominent place in recent historical writing and, for that matter, in the humanities and the social sciences in general.[1] For medical historians, these approaches would seem particularly appropriate and promising. Not only did this 'practice turn' receive a major impetus from the work of Bruno Latour, Andrew Pickering and other writers with a particular interest in the history of medicine and the life sciences.[2] Historically, 'practice' has always had a prominent place in medicine. Medicine was an applied science *par excellence*, *techne* or *ars* as well as *episteme* or *scientia*. In some modern European languages, the term 'practice', viz. its respective equivalent, refers especially – or, as in colloquial modern German, even primarily – to the 'practice' of private physicians (and dentists). It indicates the specific site where patients are seen, i.e. the 'office' or 'surgery', as well as, in a more abstract sense, the sum of an individual physician's medical activities, i.e. an organisational and economic unit within the health care system.

Half a century ago, Erwin H. Ackerknecht already called for a 'bevariorist approach' to medicine and demanded a 'more extensive and more critical analysis of what doctors did in addition to what they thought and wrote'.[3] To this day, we know remarkably little, however, about the daily lives and routines of typical, ordinary physicians. Who came to see them or asked for their advice? How many of their patients were men and how many women? To what degree did they treat children or old people? Which complaints, which diseases were they consulted for? How often did they see their patients? How did they diagnose and treat them? How did actual practice relate to the established textbook knowledge of the time? How expensive were their services compared to those of other, non-academic healers? Did they usually visit the patients in their homes – or did the patients come to see them in their own houses or in a pharmacy? Did physicians work on their own or could they rely on the support of wives, children or *famuli*? How did they keep record of their patients and

---

1 Theodore R. Schatzki, Karin Knorr Cetina and Eike von Savigny (eds), *The Practice Turn in Contemporary Theory* (London/New York 2001).

2 Bruno Latour and Steve Woolgar, *Laboratory Life. The Social Construction of Scientific Facts* (Beverly Hills/London 1979); Andrew Pickering, *The Mangle of Practice: Time, Agency and Science* (Chicago 1995).

3 Erwin H. Ackerknecht, 'A Plea for a "Behaviorist" Approach in Writing the History of Medicine', *Journal of the History of Medicine and Allied Sciences* 22 (1967), 211–14.

 | DOI 10.1163/9789004303324_002

their diseases, of the treatment they prescribed and the money owed to them? These are just some of the questions which are essential for an adequate understanding of physicians' lives and careers and of their (changing) place in ordinary health care – and they have largely remained unanswered.

A major reason for this ignorance is, of course, the relative scarcity of sources. Printed sources, such as medical textbooks, learned treatises and journal articles tend to be of very limited value in this respect. It was not until fairly recently that historians have started to pay more attention to other genres and to unpublished, archival documents that can throw some light on these issues. At first, with the rise of the new social history of medicine and linked to debates on the 'birth of the clinic' (Michel Foucault), hospital journals and other serial clinical sources took the limelight.[4] The individual patient tends to appear essentially as a case in these sources rather than a suffering individual, but the patients' perspective on medical practices soon also gained more prominence. Published case-histories – an increasingly popular genre since the 1550s – were found to offer valuable insights not only into the application of theoretical knowledge to the individual case but also into the patients' experience.[5] For the more educated classes, letters in which patients or their relatives asked physicians for epistolary advice as well as private correspondences dealing with health matters in general, were found to offer a wealth of information.[6] Autobiographies and diaries have been surveyed for passages describing illness episodes and the patients' interactions with physicians and other healers.[7]

4 Guenter B. Risse, *Hospital Life in Enlightenment Scotland* (Cambridge 1986), 119–76; Johanna Bleker, Eva Brinkschulte and Pascal Grosse (eds), *Kranke und Krankheiten im Juliusspital zu Würzburg 1819–1829. Zur frühen Geschichte des allgemeinen Krankenhauses in Deutschland* (Husum 1995).

5 Barbara Duden, *The Woman Beneath the Skin: A Doctor's Patients in Eighteenth-Century Germany* (Cambridge 1991).

6 Michael Stolberg, *Experiencing Illness and the Sick Body in Early Modern Europe* (Basingstoke 2011); Martin Dinges and Vincent Barras (eds), *Krankheit in Briefen im deutschen und französischen Sprachraum 17.–21. Jahrhundert* (Stuttgart 2007); Nicole Schweig, *Gesundheitsverhalten von Männern: Gesundheit und Krankheit in Briefen 1800–1950* (Stuttgart 2009); Judith Oxfort, *Meine Nerven tanzen. Die Krankheiten der Madame de Graffigny (1695–1758)* (Köln 2010).

7 Martin Dinges, 'Soldatenkörper in der Frühen Neuzeit. Erfahrungen mit einem unzureichend geschützten, formierten und verletzten Körper in Selbstzeugnissen' in Richard van Dülmen (ed.), *Körpergeschichten* (Frankfurt am Main 1996), 71–98; *idem*, 'Schmerzerfahrung und Männlichkeit. Der russische Gutsbesitzer und Offizier Andrej Bolotow (1738–1795)', *Medizin, Gesellschaft und Geschichte* 15 (1997), 55–78; Susanne Hoffmann, *Gesunder Alltag im 20. Jahrhundert? Geschlechterspezifische Diskurse und gesundheitsrelevante Verhaltensstile in deutschsprachigen Ländern* (Stuttgart 2010).

Physicians' correspondences came to be studied also for what they had to say about their authors' medical practices.[8]

Comprehensive and representative insights into physicians' everyday practice, historians have come to realise, however, could be gained above all from casebooks, practice journals and similar practice records that originate from and directly reflect the physician's daily business.[9] Before 1990, such sources had occasionally been analysed in studies on individual physicians but no systematic survey of these sources had ever been undertaken and the evidence for the individual physicians was mostly fragmentary or second-hand only. With the exception of the fifty-four volumes of patient journals of Samuel Hahnemann, the founder of homeopathy, larger series of casebooks from the practice of an individual physician were virtually unknown.[10] It was nothing less than a

8 Thomas Schnalke, *Medizin im Brief: der städtische Arzt des 18. Jahrhunderts im Spiegel seiner Korrespondenz* (Stuttgart 1997); Marion Maria Ruisinger, *Patientenwege. Die Konsiliarkorrespondenz Lorenz Heisters (1683–1758) in der Trew-Sammlung Erlangen* (Stuttgart 2008).

9 For an overview see Martin Dinges, 'Arztpraxen 1500–1900. Zum Stand der Forschung' in Elisabeth Dietrich-Daum, Martin Dinges, Robert Jütte and Christine Roilo (eds), *Arztpraxen im Vergleich. 18.–20. Jahrhundert* (Innsbruck 2008), 23–61; Jacalyn Duffin, *Langstaff. A Nineteenth-Century Medical Life* (Toronto 1993); Philip G. Perneel, *Het Beroepsjournaal van Dr. J.F.Ph. Hers, Arts te Oud-Beijerland (1881–1915). Een reconstructie van een plattelandspraktijk omtreks 1900* (Rotterdam 2000); Anne Hilde van Baal, *In Search of a Cure. The Patients of the Ghent Homoeopathic Physician Gustave A. van den Berghe (1837–1902)* (Rotterdam 2008); cf. John Harley Warner, 'The Uses of Patient Records by Historians – Patterns, Possibilities and Perplexities', *Health and History* 1 (1999), 101–11; Brian K. Nance, 'Medical Casebooks in Early Modern Europe. A Survey of Recent Research and Strategies for Teaching' in Chester R. Burns *et al.* (eds), The Proceedings of the 37th International Congress for the History of Medicine (Galveston 2002), 214–17; the title of David Shepard, 'The Casebook, the Daybook, and the Diary as Sources in Medical Historiography', *Canadian Bulletin of Medical History* 17 (2000), 245–55, is somewhat misleading; the author discusses the surviving records of a couple of Canadian practices in the nineteenth century. On France, see the various references in Laurence W.B. Brockliss and Colin Jones, *The Medical World of Early Modern France* (Oxford 1997).

10 Archives of the Institut für Geschichte der Medizin der Robert Bosch-Stiftung in Stuttgart; these journals are of paramount interest to historians as well as to practitioners of homeopathy today and a number of them have been edited, starting with Heinz Henne (ed.), *Hahnemanns Krankenjournale Nr. 2 und 3* (Stuttgart 1963); critical editions started with Samuel Hahnemann, *Krankenjournal D 5: (1803–1806)*, ed. by Helene Varady (Heidelberg 1991); more recent publications are Samuel Hahnemann, *Krankenjournal D 16 (1817–1818)*, ed. with transcription and commentary by Ulrich Schuricht, 2 vols. (Stuttgart 2004); Samuel Hahnemann, *Krankenjournal D 38 (1833–1835)*, ed. with transcription and commentary by Monika Papsch (Stuttgart 2007); Samuel Hahnemann, *Krankenjournal D 22*

sensation therefore, when, in the late 1990s, the 244 patient journals of one single physician were found which contained entries on some 87.000 patients over a period of fifty-two years (1847–1899). More remarkably even, that physician, Franz von Ottenthal, was not a particularly famous man and he practised in a fairly remote Alpine area. Christine Roilo of the *Landesarchiv* in Bozen, who discovered this collection in the attic of the Ottenthal family mansion, and Elisabeth Dietrich-Daum, a social historian at the University of Innsbruck, successfully applied for EU-funding which enabled them, from 2002 to 2007, to establish a data-base of all entries in these journals[11] and to start analysing the data from different perspectives.[12]

The discovery of von Ottenthal's patient journals gave a new impetus, in turn, to the search for similar sources. A loose network of scholars emerged with related interests in medical practices in general and in these types of sources in particular. Workshops and conferences were organised. In 2006, another extensive series of practice journals, fifty-five volumes covering some thirty years of practice, was found among the papers of the Swiss physician Caesar Adolf Bloesch.[13] A collective volume which assembled the contributions to an international conference on physicians' practices from the eighteenth to the twentieth centuries offered, for the first time, a series of case studies on individual physicians' practices in that period.[14]

The conversation began to widen. Other historians were asked about their acquaintance with similar sources and, as it turned out, a considerable body of such sources could be identified, reaching back to the fifteenth century.[15]

(1821), ed. with transcription and commentary by Markus Mortsch, 2 vols. (Stuttgart 2008); for Hahnemann's French journals see Samuel Hahnemann, *Krankenjournal DF2: (1836–1842)*, ed. with transcription and commentary by Arnold Michalowski (Stuttgart, 2003); for further editions see http://www.igm-bosch.de/content/language1/html/11855.asp.

11 URL: http://www.uibk.ac.at/ottenthal/english/.

12 On Ottenthal see Elena Taddei, 'Historiae morborum: Le anamnesi del medico generico Franz von Ottenthal', *Medicina e storia* 9 (2005), 89–98; Elena Taddei, *Franz von Ottenthal. Arzt und Tiroler Landtagsabgeordneter (1818–1899)* (Wien 2010) and the contribution by Elisabeth Dietrich-Daum, Marina Hilber and Eberhard Wolff to this volume.

13 Nadine Boucherin, 'Die Krankengeschichten von C.A. Bloesch. (1804–1863)' in Dietrich-Daum *et al.*, *Arztpraxen*, 2008, 147–65; see the contribution by Lina Gafner to this volume.

14 Dietrich-Daum *et al.*, *Arztpraxen*, 2008.

15 The earliest sources of this type which have been identified to this point are the *Rezeptbücher* of Hartmann Schedel (1440–1514) in Nördlingen, Amberg and Nürnberg. They start in 1470 and list hundreds of patients for whom Schedel prescribed medicines (see the forthcoming work by Katharina Beck, *Dr. Hartmann Schedel (1440–1514) und seine Patienten*, PhD thesis, Universität Würzburg); cf. Klaus G. König, *Der Nürnberger Stadtarzt*

Plans for more systematic, organized collaborative research emerged. In 2009, an Austrian-German-Swiss research network, funded by the *Deutsche Forschungsgemeinschaft* (DFG), took up its work, with altogether eight projects, in Berlin, Bern, Ingolstadt, Innsbruck/Bozen, Stuttgart, Würzburg and Zürich.[16]

This volume comes out of that collective enterprise and presents its principal results. Drawing primarily on casebooks and similar serial practice records, the eight projects span the period from the Thirty-Years-War to the late nineteenth century, which has made it possible to trace long-term developments and to study the impact of new medical theories and practices as well as the effects of socioeconomic and cultural change.[17] The choice of the individual practices was guided above all by the available sources but it also reflects an attempt to include different kinds of localities as well as different types of practice. Six research groups dealt with the private practices of university-trained physicians. They range from the physician-astrologer J. Magirus in seventeenth-century Berlin and Zerbst, the eighteenth-century Nuremberg physician J.C. Götz and J.F. Glaser, town physician of Suhl and the son of a hangman, to C.A. Bloesch in the Swiss town of Biel, the physician-homeopath F. von Bönninghausen in Münster and the above-mentioned F. von Ottenthal in the nineteenth century. Two further practices, both in the nineteenth century, do not quite fit into this pattern, namely that of G. Wachter, a healer in a rural area south of Lake Constance, who had no formal academic training, and the policlinics which C.H. Fuchs ran in Göttingen and Würzburg and which provided medical care for poor patients in their own homes. They were chosen in order to broaden the range of our analysis and to highlight the specifics of academic physicians' private practice.

Dr. Georg Palma (1543–1591) (Stuttgart 1961); Peter Assion and Joachim Telle, 'Der Nürnberger Stadtarzt Johannes Magenbuch. Zu Leben und Werk eines Mediziners der Reformationszeit', *Sudhoffs Archiv* 56 (1972), 353–421; for the early seventeenth century see Franziska Ofenhitzer, *Das Diarium practicum des Breslauer Arztes Petrus Kirstenius, 1612–1616*, MD thesis, Universität Würzburg [forthcoming].

16 The network's common activities and five German projects were funded by a grant from the *Deutsche Forschungsgemeinschaft*; the two projects in Bern and Zurich were funded by the *Schweizer Nationalfonds*, the project in Innsbruck/Bozen by the *Tiroler Kulturstiftung* and the *Robert Bosch-Stiftung*, Stuttgart.

17 Late nineteenth- and twentieth-century practices were deliberately excluded. New medical theories, findings (e.g. in bacteriology) and practices, the growing importance of medical technology, the rise of hospital medicine and the massive expansion of health insurance schemes brought dramatic changes which would have made any consistent effort at comparative analysis across the centuries very difficult. Also in view of the huge body of relevant sources which has survived, especially from the twentieth century, the more recent history of physicians' practices would warrant a major research initiative of its own.

In geographical terms, work thus focused on the German speaking territories. It was the explicit intention, from the start, however, to encourage scholars in other countries to embark on similar studies. The members of the network met at regular intervals to exchange findings and experiences and to help turn the enterprise into a truly cooperative venture. In 2012, the results were presented to a larger audience at an international conference in Stuttgart, with outside commentators and with invited speakers such as Lauren Kassell and Philipp Rieder presenting the results of their own work on other geographical contexts.[18]

Reflecting these years of collective research – some of which is still on going –, this volume is divided into two parts. The first part consists of five papers that analyse, from a comparative and diachronic perspective, fundamental aspects of medical practice(s) from 1600 to 1900: patients and their reasons for consulting a particular healer; diagnostic and therapeutic practices and their relationship to the dominant theories of the day; the 'daily business' of running a practice; the social, religious and cultural context of medical practice; and medical record keeping. This part of the book reflects a novel and, we believe, particularly fruitful approach to collaborative historical research and writing: each of the five chapters was penned by two or three authors working in different research groups. The authors could draw on the respective findings of all eight projects, however. At a time, when evaluationism is rampant and impact factors and the length of publication lists are crucial in the competition for positions and third-party funding, it is worth underlining what this meant: those working on the individual projects generously supplied whatever data and information the authors of these chapters needed, usually including, to a very substantial degree, unpublished results of their own research.

This is not the place to even try and summarize the most important findings. Suffice it to say that these chapters cast a new light even on some very basic, widely shared assumptions about the history of medical practice. The proportion of infants and children, for instance, was remarkably high, throughout the period under consideration, as was that of old and very old patients – both groups have been widely perceived as receiving professional medical treatment only in exceptional cases. The assembled data for the private practices of physicians also leaves no doubt, to cite another important finding, that

18 Cf. Lauren Kassell, 'Casebooks in Early Modern England: Medicine, Astrology, and Written Records', *Bulletin of the History of Medicine* 88 (2014), 595–625; Philip Rieder, 'Médecins et patients à Genève: offre et consommations thérapeutiques à l'époque moderne', *Revue d'histoire moderne et contemporaine* 52 (2005), 39–63; idem, *La figure du patient au XVIIIe siècle* (Geneva 2010).

the physicians, throughout the period under consideration, treated patients from more or less all walks of society rather than focusing largely on upper-class patients; in fact, in our samples, the latter usually constituted only a small minority. Looking at the typical site of the medical consultation led to another surprise: premodern physicians have been described as seeing and treating their sick almost exclusively in the patients' homes. By contrast, for two of the three seventeenth- and eighteenth-century practices we examined we found a very considerable number of patients who came to see the physician in his own home.

Each of the eight (briefer) chapters in the second part of this volume is devoted to an individual practice. The authors were encouraged to put a particular focus on what was remarkable or unique about that specific practice and the physician or healer who ran it, for example his characteristic approach to keeping patient records, his preferred diagnostic and therapeutic approach, the kinds of patients that sought his help and his other medical activities, such as learned publications, and his place in the rural or urban society in which he lived and worked.

Some of the results which are laid out in this volume, have also been presented to the wider public in a very different format, namely by means of an exhibition. Designed by Thomas Schnalke, Berlin, and Maria Ruisinger, Ingolstadt, and their teams, and with the financial support of the *Deutsche Forschungsgemeinschaft*, the exhibition presents the individual practices and their respective contexts as well as the historical sources and the methods used. It is accompanied by a catalogue written, again, in close cooperation with the scholars working on the individual projects. The exhibition opened at the *Berliner Medizinhistorische Museum der Charité* in 2013 and will be shown in different museums over the next years.[19]

Acknowledgements

We gratefully acknowledge funding by the *Deutsche Forschungsgemeinschaft*. Special thanks to Margot Saar for her careful translation of the chapters and her copy-editing of this introduction, to Leila Al-Deri and Theresa Sanzenbacher for their editorial assistance and to Brian Dolan and the editorial board of *Clio medica* for making this book possible.

19 Isabel Atzl, Roland Helms, Stephanie Neuner and Ruth Schilling, *Praxiswelten. Zur Geschichte der Begegnung zwischen Arzt und Patient* (Ingolstadt 2013); Isabel Atzl and Philipp Klaas (eds), Das kranke Biel: auf Arztvisite um 1850 (Biel 2015).

PART 1

CHAPTER 1

Cornucopia Officinae Medicae: Medical Practice Records and Their Origin

Volker Hess and Sabine Schlegelmilch

Introduction

> When the noises of the day have died down and the stillness of the evening invites to quiet reflection, [the physician] should devote a few hours to calm contemplation of his patients….Here, in the still of the night, matters may appear in a different light; ideas and inspirations will arise in him that were impossible with the distractions of the day. It is only now, when the inner life awakens, that these impressions can find entry to the inner life and meet with genuine interest and attention.[1]

In this idyll Christoph Wilhelm Hufeland (1762–1836) has captured the legacy of fifty years of experience as a medical practitioner and scholar:[2] the regular nightly recording of the observations of the day. One should 'write down the most important points of the patient's history, the changes that have occurred, one's own comments and ideas about the origin and treatment of a disease, the remedies applied, and one should duly reflect on them'. The regular keeping of a 'patient journal' (*Kranken-Journal*), neglected by the young physicians, was for Hufeland 'not only the main means of, but also the indispensable prerequisite for attaining greatness and perfection not only in practice, but also in the art as such'.[3]

* We thank the following other authors of the individual projects in the DFG Research Network 'Medical practices (1600–1900)', on whose partly still unpublished results this contribution is largely founded: *Kay Peter Jankrift/Annemarie Kinzelbach/Susanne Grosser/Marion Ruisinger* (*Chr. Götz*), *Ruth Schilling* (*J.F. Glaser*), *Stephanie Neuner/Karen Nolte* (*Ch. Fuchs*), *Katharina Weikl/Alois Unterkircher/Iris Ritzmann* (*G. Wachter*), *Lina Gafner/Philipp Klaas* (*C.A. Bloesch*), *Marion Baschin* (*F.P. von Bönninghausen*), *Marina Hilber/Eberhard Wolff/Alois Unterkircher* (*F. von Ottenthal*).

1 Christoph Wilhelm Hufeland, *Enchiridion medicum oder Anleitung zur medicinischen Praxis. Vermächtnis einer fünfzehnjährigen Erfahrung* (Herisau 1837), 500.

2 Rüdiger Campe: 'Die Schreibszene, Schreiben', in Sandro Zanetti (ed.), *Schreiben als Kulturtechnik* (Berlin 2012), 269–82.

3 *Ibid.*, quotation abbreviated.

 | DOI 10.1163/9789004303324_003

The 'patient journal', as Hufeland called it, or 'practice record' as we refer to these medical practice notes, so as to distinguish them from the institutional format, occupied a central place in the self-reflection of physicians in the past. Practice records not only documented daily consultations and justified the accounts, they were also of particular significance for the physician's thinking and actions. Over and above that, they shaped – as the idyll presented above shows – the self-image of the learned physician. This is what makes this genre of sources so attractive. Scarcely any other source material is as much an inherent part of the daily practice routine, linking this routine with medical science and academic teachers such as Herman Boerhaave (1668–1738), Friedrich Hoffmann (1660–1742) or Maximilian Stoll (1742–1787) who were – not only for Hufeland – role-models for good recording and documenting practice.[4]

In view of the number of studies presented in recent years – in particular on the eighteenth and nineteenth centuries (see reading list) – we will avoid the often-expressed criticism that medical practice records belong to the type of source that has been neglected by research.[5] Since the late 1990s the growing endeavour to find historical access to the everyday medical practice has made it increasingly clear that the body of records evaluated so far is hardly homogeneous.[6] Even if the studies that only convey indirect insights into the physician's activities are excluded,[7] the comprehensive survey of Martin Dinges alone reveals a variety of note-taking formats that elude immediate categorization. It begins with the ambiguous designations for the sources:[8] The research

4 *Ibid.* On the bedside-practice of the role-models mentioned cf. Christian Probst, *Der Weg des ärztlichen Erkennens am Krankenbett. Herman Boerhaave und die ältere Wiener medizinische Schule, vol. 1: 1707–1787* (Wiesbaden 1972); Johanna Geyer-Kordesch, 'Medizinische Fallbeschreibungen und ihre Bedeutung in der Wissensreform des 17. und 18. Jahrhunderts', *Medizin in Geschichte und Gesellschaft* 9 (1990), 7–19.

5 For instance Paul Weindling, 'Medical Practice in Imperial Germany. The Casebook of Alfred Grotjahn', *Bulletin of the History of Medicine* 61 (1987), 391–410: 392; Katharina Ernst, 'Fragments and Gaps. The Form of Illness Narratives in 18th Century Wurttemberg Pietist Diaries', *Medizin in Gesellschaft und Geschichte* 24 (2005), 33–45: 34.

6 Martin Dinges, 'Arztpraxen 1500–1900. Zum Stand der Forschung', in Elisabeth Dietrich-Daum, Martin Dinges, Robert Jütte and Christine Roilo (eds), *Arztpraxen im Vergleich. 18.–20. Jahrhundert* (Innsbruck 2008), 23–61.

7 There is an example in the study of Duden 1987 based on the analysis of patient observations which an almost seventy-year-old physician, in looking back over a rich working life, had compiled (mostly from previously published case histories) and published for teaching purposes.

8 We rarely find exact descriptions of the source as given, for instance, by Edmund Hobhouse (ed.), *The Diary of a West Country Physician, A.D. 1684–1726* (London 1934), 50 or Brian K. Nance, *Turquet de Mayerne as Baroque Physician. The Art of Medical Portraiture* (Amsterdam 2001), 30–56. It is interesting that particularly in socio-historical publications we find nothing

literature differentiates between 'daybooks',[9] 'medical diaries',[10] 'patient records',[11] 'memorandum books'[12] or simply 'ledgers'[13] even though the sources designated in this way barely justify such differentiation.[14] Even the reports of medical activity one finds in the private diary of a physician strewn among visits to relatives, household expenses and family stories, are often used as documentation of medical practice.[15] The most commonly used term

on the make-up of the source. Cf. for instance Lucinda McCray Beier, *Sufferers and Healers. The Experience of Illness in Seventeenth-Century England* (London 1987).

9 Jacalyn Duffin, *Langstaff. A Nineteenth-Century Medical Life* (Toronto 1993, repr. 1999).

10 Peter Assion and Joachim Telle, 'Der Nürnberger Stadtarzt Johannes Magenbuch. Zu Leben und Werk eines Mediziners der Reformationszeit', *Sudhoffs Archiv* 56 (1972), 353–421; David Shephard, 'The Casebook, the Daybook, and the Diary as Sources in Medical Historiography', *Canadian Bulletin of Medical History* 17 (2000), 245–55; Steven M. Stowe, *Doctoring the South. Southern Physicians and Everyday Medicine in the Mid-Nineteenth Century* (Chapel Hill 2004).

11 Henrik R. Wulff and Kirsten Jungersen, 'A Danish Provincial Physician and His Patients. The Patient Records from the Pratice of Christopher Detlev Hahn in Aarhus around 1800', *Medizinhistorisches Journal* 40 (2005), 321–45.

12 Rajendra Persaud, 'The Reporting of Psychiatric Symptoms in History. The Memorandum Book of Samuel Coates, 1785–1825', *History of Psychiatry* 4 (1993), 499–510.

13 Irvine Loudon, 'The Nature of Provincial Medical Practice in Eighteenth-Century England', *Medical History* 29 (1985), 1–32; E.M. Sigsworth and P. Swan, 'An Eighteenth-Century Surgeon and Apothecary. William Elmhirst (1721–1773)', *Medical History* 26 (1982), 191–98.

14 This also applies to the German-language research literature that mentions *Besuchsbuch* (Wolfgang Balster, 'Medizinische Wissenschaft und ärztliche Praxis im Leben des Bochumer Arztes Karl Arnold Kortum (1745–1824). Medizinhistorische Analyse seines Patiententagebuches' (Med. diss., University of Bochum, 1990); Franz Dumont, 'Nicht nur Hölderlin. Das ärztliche Besuchsbuch Soemmerings als Quelle für sein soziales Umfeld in Frankfurt am Main', *Medizinhistorisches Journal* 28 (1993), 123–54), *Ordinationsbuch* (Eugen Wehren, 'Das medizinische Werk des Wundarztes Michel Schüppach (1707–1781) an Hand seiner Rezept- und Ordinationsbücher', *Berner Zeitschrift für Geschichte und Heimatkunde* 47 (1985), 87–166), *Praxisjournal* or *Krankenjournal* (Regina Engel, 'Das Patientengut eines praktischen Arztes um die Mitte des 19. Jahrhunderts, gezeigt am Beispiel des Patientenjournals von Dr. Heinrich Grotjahn in Schladen am Harz' (Med. diss., University of East-Berlin, 1978); Robert Jütte (ed.), *Die Krankenjournale von Samuel Hahnemann; D 2: (1801–1802) / nach der Ed. von Heinz Henne. bearb. von Arnold Michalowski*, 2 (Heidelberg 1993) without justifying such differentiation. The account book (*Rechnungsbuch*) (see e.g. Willem F. Daems, *Johann Anton Grass von Portein, 1684–1770. Arzt, Chirurg, Zahnarzt, Harndiagnostiker, Pharmazeut, Viehdoktor und Dorfpolitiker. Ein Beitrag zur Kultur- und Medizingeschichte des Domleschgs und Heinzenbergs im 18. Jahrhundert* (Chur 1985)) is very different in form from the practice records (Andrea Thümmler, 'Rekonstruktion des Alltags eines thüringischen Arztes im 18. Jahrhundert anhand seines Praxistagebuchs 1750–1763' (Med. diss., University of Berlin, 2004)).

15 Cf. Hobhouse, 1934.

'casebook' figures as an example for the ambiguity of this genre as it is applied to entirely different types of sources[16] – from early modern collections of *observationes* to bound and printed journal sheets.[17] While research inclines to include early modern – hand-written or printed – *observationes* in the category of practice records,[18] it assigns the more recent practice documentation to the same category,[19] as if the doctor's practice constituted a transhistorical entity that can be defined in general, abstract terms. But – to summarize this outline of the recent research literature – there is no consensus as to what the term 'practice records' actually entails.

The question therefore is: can these historically very dissimilar forms be defined as a type of source in their own right? If the answer is yes, a second question follows: do the 'practice records', as a type of source, constitute a separate literary genre comparable to the early modern *observationes* or the 'case history'?[20] Despite the progress made by historical research with this kind of material, these questions have so far seldom been addressed.

16 Nance, 2001; Anne Hilde van Baal, *In Search of a Cure. The Patients of the Ghent Homoeopathic Physician Gustave A. van den Berghe (1837–1902)* (Rotterdam 2008); Amalie M. Kass, 'The Obstetrical Casebook of Walter Channing, 1811–22', *Bulletin of the History of Medicine* 67 (1993), 494–523.

17 Cf. Michael Stolberg, 'Formen und Funktionen medizinischer Fallberichte in der Frühen Neuzeit (1500–1800)', in Johannes Süßmann, Susanne Scholza and Gisela Engel (eds), *Fallstudien. Theorie – Geschichte – Methode* (Berlin 2007), 81–95; Volker Hess, 'Formalisierte Beobachtung. Die Genese der modernen Krankenakte am Beispiel der Berliner und Pariser Medizin (1725–1830)', *Medizinhistorisches Journal* 45 (2010), 293–340.

18 An impressive example for this is Joan Lane, *John Hall and His Patients. The Medical Practice of Shakespeare's Son in Law. With Medical Commentary by Melvin Earles* (Stratford-upon-Avon 1996).

19 Weindling, 1987.

20 On the genre of the *Observatio* cf. Gianna Pomata, 'Observation Rising. Birth of an Epistemic Genre, 1500–1650', in Lorraine Daston and Elizabeth Lunbeck (eds), *Histories of Scientific Observation* (Chicago 2011), 45–80; on *case history* cf. Harriet Nowell-Smith, 'Nineteenth-Century Narrative Case Histories. An Inquiry into Stylistics and History', *Canadian Bulletin of Medical History* 12 (1995), 47–67. The case history is generally seen as a separate genre among the 'simple forms' (André Jolles). Rudolf Behrens and Carsten Zelle conclude, however, in the introduction to their anthology, with a view to the research literature available that 'the medical case history in its textual structure and – relatedly – in its historical-epidemic foundations as well with regard to its cultural and medial variance (textbook, magazine, collection of cases etc.) has only been researched rudimentarily'. (Rudolf Behrens and Carsten Zelle (eds), *Der ärztliche Fallbericht. Epistemische Grundlagen und textuelle Strukturen dargestellter Beobachtung* (Wiesbaden 2012), viii).

This deficiency is essentially due to the fact that the preserved records are mostly seen as mere sources of historical information. The medium as such is rarely examined: How were the records composed? Which patterns did they follow? What purpose did they serve? What were the intentions of the writer? Such questions are only asked when the handwritten material fails to be 'readable' as a historical source. Despite their material proximity to the medical practice, practice records do not provide unhindered access to the past. Even if they – as can be assumed (see below) – lay on the physician's desk, the fundamental reservation, put forward by Foucault in his well-known statement, still applies to these hand-written records: despite the propinquity of the preserved records they are unable to capture the lives of physicians and their patients in the past.[21] It is therefore the more important for any historical analysis to thoroughly examine the material limitations of this 'window' to the past and to establish its qualities, material inconsistencies and literary 'construction principles'.

Our contribution focuses on the practice records analysed by the various projects that collaborated in the Research Network 'Medical Practices (1600–1900)'. The sources underlying this cooperative research project are highly representative at least in one respect: they are very heterogeneous in terms of scale, historical usage, presumed purpose, temporal origin and material form. As regards scale: the *Diarium* that Johannes Magirus (1615–1697) kept over a period of nine years only has 540 pages, while the fifty-five preserved volumes of the 800-page *Tagebuch* of Caesar Adolph Bloesch (1804–1863) fill half a bookcase. Their purpose also varied: the *Register* of Johann Friedrich Glaser (1707–1789) served primarily as proof of treatment and billing, while the carefully copied out *Krankheitsgeschichten* in Conrad Fuchs' (1803–1855) record were mostly used for research and instruction. Friedrich von Bönninghausen (1828–1910) wrote down his practice records during the consultations so as to document the complaints of his patients in as much detail as possible. There are however, examples of a more negligent documentation of the practitioners' activities: the Swiss lay-healer Gottfried Wachter (1776–1861) was often late in conforming to the duty of documentation imposed on him by the Thurgau health officer. The use of Latin is also not a sufficient criterion for differentiating the various purposes of these records. While seventeenth- and eighteenth-century practice records tend to be written in Latin, our sample shows that the German language crept in very early in the eighteenth century. Magirus and Götz wrote most of their notes in Latin, but Glaser

21 Michel Foucault, 'La vie des hommes infâmes (1977)', in Daniel Defert and François Ewald, in collaboration with Jacques Lagrange (eds), *Dits et Ecrits*, vol. 3 (Paris 1994), 237–53.

wrote his in German. Yet even in the early nineteenth century some physicians remained faithful to the *lingua franca* of humanism. Bloesch tended to switch from German to French, especially in consultations that were conducted by letter. Ottenthal's choice of using the old language of scholars in his practice records seems at variance with the modern form of documentation he applied (see below). We cannot derive an obvious rule: physicians in other countries also wrote their practice notes in their own language, even in the early seventeenth century, as we can see from the numerous Anglo-Saxon examples.[22]

The differences in form are equally pronounced: Magirus followed the humanist tradition of common-placing in writing down his observations. Despite the traditional name he gave to his records – *Historia morborum* – the Tyrolese physician Franz von Ottenthal (1818–1899) entered his notes in standardized text fields similar to those of modern-day patient files.[23]

How, then, can this type of source be adequately defined? Considering the heterogeneity of the source materials, a purely content-based definition of our material as 'records of medical practice' does not sufficiently characterize the purpose and use of these records. A definition based on function alone is similarly problematic because we find such a variety of functions and purposes: by defining them too narrowly one settles for particular functions (which first need to be investigated) and risks prematurely excluding other, related records. If the definition is too broad (such as: 'serves as record of medical activity') it will not provide sufficient differentiation from other genres such as 'diaries' or 'account books'. All practice records have a definite memory function. Human beings are not, unlike God, able to remember everything, as the Kolberg physician Balthasar Timaeus von Güldenklee (1600–1667) conceded in the preface to his collection of case histories. Many physicians will therefore have introduced record-keeping as a means of saving 'useful information' from falling into oblivion, even if they did not say it quite so explicitly.[24] In retrospect, the

22 Cf. the *medical case-book* (1565–1573) of 'magistro Polewheele' (NL MS. 28023), the *Entry-book* (1619–1666) of the physician Thomas Arthur from Limerick (NL MS 31855), the *Casebook* (1619–1622) of a 'provincial physician' (NL MS Sloane 1112) or the *Casebook* (1676–1696) of Sir Edmund King (NL MS Sloane 1589). François Boissier Sauvages de la Croix recorded his patient observations (ca. 1740–1760) mostly in French and occasionally in Latin (Archives départementales de l'Hérault, 10 F 51–53).

23 On *Historia* cf. Gianna Pomata, 'Praxis Historialis. The Uses of Historia in Early Modern Medicine', in Gianna Pomata and Nancy G. Siraisi (eds), *Historia. Empiricism and Erudition in Early Modern Europe* (Cambridge 2005), 105–46.

24 '*Ego cum praxin medicam successu, quod praefiscini dixerim, non infelici aliquandiu exercuissem & multa quae aegris profuerant a me repeterentur, oblivioni iam tradita, dolore coepi de neglecta eorundem annotatione: Memoriam enim omnium tenere divinitatis potius*

uses in each individual case are difficult to reconstruct from the context and remain the subject of historical interpretation. Safeguarding the recorded information from being forgotten is, however, a function that all records have in common and that distinguishes them from written-down rules or instructions (signs). In order to adequately assess the value and significance of this genre of sources we must therefore look at other aspects. A definition of 'practice record' must primarily consider the material and textual constitution of these sources – two different but interdependent dimensions of record-keeping:

1. From the point of view of literary science, practice records could be seen as factual narratives because they do not grant the same artistic license as fictional writing. The form of recording is not arbitrary, as we pointed out before: it depends on the (historically contingent) needs and constraints of medical practice. These constraints limit poetic expression as well as artistic freedom in the – thematic and formal – composition of the patient reports. For instance, records that are listed by date reveal temporal gaps in the observation, while the pre-printed tabular listing of 'prescriptions' and 'new signs' facilitates semiotic observation.[25]
2. As regards the material medium we can conclude that the form of the practice record is not determined by its function (see item 3 below); it is in itself a central element of the practice organization. Whatever the original intention behind the documentation, it structures essential steps in the daily medical practice that go far beyond a purely memory-related function: the form determines what is documented as 'medical service' and it therefore also determines the fee that is to be charged. The form justifies actions as medically required and ultimately sets, as an information management tool, the parameters for the verbalization of medical knowledge. It establishes who belongs to the 'main clientele' and provides more or less detailed information on each member of this clientele, contributing considerably, in doing so, to the shaping of the physician-patient-contact.
3. It needs to be borne in mind that the mutual dependence of form and content is not independent of other factors: the practice of documenting is embedded in notions of how physicians ought to record their observations and conduct their business. We can therefore assume that the

esse quam humanitatis iam olim agnovit antiquitas, nec eadem semper in inveniendis & componendis medicamentis animo obvenit cogitatio' (Balthasar Timaeus von Güldenklee, *Serenissimi Electoris Brandenburgici Archiatri Casus Medicinales Praxi Triginta Sex Annorum Observati* (Leipzig 1667), 2Ar).

25 Cf. for instance the profiles of Ottenthal's or Bönninghausen's records.

different recording techniques in our sample were not 'newly' invented in each case, but acquired at school, university or during training – as a learned method,[26] in the personal teacher-student relationship (see below), as a subject in clinical training[27] or as the outcome of other registration or recording practices.[28]

The interdependence of form and content becomes evident when one examines the practice of note-taking as such. We will therefore try to find an answer to our question of definition by looking at the sources evaluated by our Research Network (see the overview in the appendix of this volume). As well as information from the recent research literature, we will include a number of practice records that are easily accessible in public facilities.

Recording Practices

Despite their diversity, practice records can be distinguished from other genres of medical case descriptions if the actual recording process is included in the

26 Ann Blair, 'The Practices of Erudition According to Morhof', in Françoise Waquet (ed.), *Mapping the World of Learning. The Polyhistor of Daniel Georg Morhof* (Wiesbaden 2000), 59–74; Helmut Zedelmaier, 'De ratione excerpendi. Daniel Georg Morhof und das Exzerpieren', in Waquet, 2000, 75–92.

27 Guenter B. Risse, 'Clinical Instruction in Hospitals. The Boerhaavian Tradition in Leyden, Edinburgh, Vienna and Padua', in Harm Beukers and John Michael Henderson Molls (eds), *Clinical Teaching, Past and Present* (Amsterdam 1989), 1–19; Sabina Verena Griesel, 'Medizinische und chirurgische Observationen, 1785–1787' (Med. diss., University of Basel, 1984); Silvia Stolz, 'Ärztliche Theorie und Praxis im Spiegel von 61 Falldarstellungen eines Züricher Medizinschülers, 1785–1787' (Med. diss., University of Basel, 1984); Volker Hess, 'Das Material einer guten Geschichte. Register, Reglements und Formulare', in Sheila Dickson, Stefan Goldmann and Christof Wingertszahn (eds), *Fakta, und kein moralisches Geschwätz. Zu den Fallgeschichten im 'Magazin zur Erfahrungsseelenkunde' (1783–1793)* (Göttingen 2011), 115–39.

28 Why budding physicians in the second half of the nineteenth century received training in laboratory and experimentation practices is a question that has so far not been sufficiently answered, since the skills necessary for assessing data of laboratory experimentation were not relevant to medical practice. It can be assumed that, in addition to the symbolic value of such training, it was expected that the benefits of scientific documentation would come to bear in other fields too (cf. Volker Hess, 'Medizin zwischen Sammeln und Experimentieren', in Rüdiger vom Bruch and Heinz-Elmar Tenorth (eds), *Geschichte der Universität zu Berlin, 1810–2010. Biographie einer Institution, Praxis ihrer Disziplinen*, vol. 4 (Berlin 2010), 489–566: 530.

historical analysis. One defining property of the practice record is that it has a known author: while entries in hospital records, ward journals and *cahiers de visite* were usually made by diverse writers, practice records were, as a rule, kept by one person. Third-party entries as we know them from Götz' practice were carried out as delegated tasks, authorized and agreed by the author. Interestingly, the only source in our sample that has several authors comes from a policlinic, that is to say an institutional context.[29] It is more difficult to delineate the practice records against other forms of chronological documentation, such as diaries in general: the transition from diary to medical diary is seamless as is apparent from the octavo book of the British surgeon Richard Kay of Baldingstone (1715/16–1751) which served both purposes.[30] But there are other forms of patient-related documentation composed by one author that cannot always be clearly differentiated from practice records: the documentation (1619–1622) of James Rant started as a casebook and ended up as a collection of prescriptions.[31]

We therefore think that, in order to define the practice record, we need to focus on the way it is produced. For both the function and content of the practice record can only be reconstructed in a historical sense, if we look at the written record as a material residue of the medical activity.[32] Practice records were picked up, consulted, amended and even revised on a daily basis. They were not always kept in the doctor's house but – to some extent – carried around and probably examined if not studied by others. They provide the documentation of a physician's practice of which they are at the same time a constituent part. This documentation increased through continued use and lacks therefore the linearity of a pre-determined registration. The material aspect of the source is consequently a central factor in the overall context. The evaluation of our source sample from this perspective yields three fundamental observations.

29 The journal of the Zurich clinician Burckhardt can count as a practice record because it was not product of any institutional note-taking practice.

30 William Brockbank and Marjory Lilian Kay, 'Extracts from the Diary of Richard Kay of Baldingstone, Bury, Surgeon (1737–50)', *Medical History* 3 (1959), 58–68.

31 British Library MS Sloane 1112. The notebook of François Boissier de Sauvages, which is known as *Livre de Raison*, is another example. It included patient observations among other things, cf. Volker Hess and Andrew Mendelsohn, 'Fallgeschichte, Historia, Klassifikation. François Boissier de Sauvages bei der Schreibarbeit', NTM. *Zeitschrift für Geschichte der Wissenschaften, Technik und Medizin* 21 (2013), 61–92.

32 Cf. the argumentation of Marc Berg and Geoffrey Bowker, 'The Multiple Bodies of the Medical Record. Toward a Sociology of an Artifact', *The Sociological Quarterly* 38 (1997), 513–37 for the example of the patient file.

The Recording Technique

Unfortunately, not many reports are available on how physicians produced their records. Normative instructions, usually imposed by an authority, are more frequent. Early modern physicians, for instance, tended to justify the methods they used by referring to the unquestionable authority of Hippocrates:

> 'He transferred everything that occurred in the patients under his quiet observation onto wax tablets, the entire patient history, with details on what happened with the patient after how many days and on which day, in which hour and in which moment. The results in each case, whether favourable or not, he wrote down, listing each individual patient by name, as is sufficiently shown by the 'epidemics'.[33]

In the seventeenth century the Baconian canon of scientific virtue was added to this, recommending that physicians ought to observe like Hippocrates and describe as diligently as painters.[34] The keeping of a record was therefore presumably taught directly by academic teachers.[35] Bloesch already kept a journal-like record as a student in Göttingen.[36] Glaser, in assembling his journal, the *Medizinisches Register*, probably copied his father who had also recorded his services, such as drastic treatments and 'embarrassing interrogations', in a 'register' himself.[37] Bönninghausen junior also adopted his father's form and technique of record-keeping.[38] For the sixteenth and seventeenth centuries no special instructions are known so far regarding the composition of medical

33 François Valleriola, *Observationum medicinalium libri sex* (Lyon 1573), Praefatio: '*qui cum ea omnia, quae aegris accidere conspicabatur, tacitus observaret, ea postea in tabulas referebat, enarrata aegritudinis historia omni, et quid quoto quoque die, hora, atque momentis aegro accidisset: et quid item ab iis casibus vel boni vel mali secutum fuisset, de scripto protulit: nominatis etiam singulatim aegris, quod Epidemiorum libri abunde testantur*'.

34 '[*E*]*xquisitam* Pictorum *industriam imitando, qui vel naevos et levissimas maculas in imagine exprimunt*': Thomas Sydenham, *Observationes Medicae Circa Morborum Acutorum Historiam Et Curationem* (London 1676); *praefatio*, s.p., emphasis by original author.

35 Cf. note 27.

36 See the contribution by Lina Gafner, 'Administrative and Epistemic Aspects of Medical Practice: Caesar Adolf Bloesch (1804–1863)' (in this volume).

37 Thüringisches Staatsarchiv Meiningen, 499002-Zinck Mattenberg-Sammlung/Akten 1327, 9. Personengeschichtliche und genealogische Quellen, Register oder Aufzeichnungs-Büchlein auf Dreißigacker. (Tagebuch des Scharfrichters Joh. Jeremias Glaser, geb. 1653, gest. 1725): entry of 10 June 1686 and 11 January 1714 (with thanks to Ruth Schilling).

38 Cf. note 77.

records. They were probably unnecessary on two accounts: commonplacing was part of the methodical competence that academic doctors acquired in school and at university[39] and the physician's observation did not constitute a separate genre within the humanist text genre of the *observatio*. Physicians transferred the humanist technique of knowledge management to the note-taking that arose from their own practice (see below). That these learned observation and recording techniques were not considered common practice outside academic circles is apparent from the curricula of eighteenth century training institutions for surgeons. Since army surgeons in particular could not be expected to have any academic training these institutions tended to teach the basic competences of a higher education.[40]

While there are hardly any indications as to whether or how the documentation of the medical activity was taught and practised, the writing down of patient observations was part of the canon of practical training. We know of Giovanni Battista da Monte (1498–1551) that his students had to write their own patient histories.[41] The proper presentation of the course of illness was seen as evidence for a physician's skill and experience: the 'exact history of the illness' was regarded 'extremely necessary even indispensable' not only for young physicians. For experienced doctors, too, the regular recording of patient observations in writing was 'always very useful', but 'very difficult' and could only 'be mastered with plenty of practice'. A patient history would therefore always most convincingly reveal whether someone was 'a bungler or a master of his art'.[42] As a consequence, students had to practise writing patient histories and, from the early eighteenth century, this was included in the by then institutionalized training. Whether in Edinburgh, Paris, Göttingen, Würzburg, Jena, Pavia or Berlin:[43] trainee physicians proved their practical skills by writing a learned patient history which, in its comprehensiveness and literary form, can hardly be compared with the notes written down in the practice records. This clinical ideal was even adopted in the training of surgeons. The students of the *Medizinische-Chirurgische Institut* in Zurich, which was

39 Ann Moss, *Printed Commonplace-Books and the Structuring of Renaissance Thought* (Oxford 1996).

40 Otto Schickert, *Die Militärärztlichen Bildungsanstalten von ihrer Gründung bis zur Gegenwart. Festschrift zur Feier des hundertjährigen Bestehens des medicinisch-chirurgischen Friedrich-Wilhelms-Instituts* (Berlin 1895), 34–5.

41 Cf. Pomata, 'Praxis', 2005, 128.

42 Johann Friedrich Rahn: *Fieberlehre. Ein nachgelassenes Bruchstück des Handbuchs der praktischen Heilkunde* (Zürich 1814), Aph. 11. and Aph. 5 and 6.

43 Toby Gelfand, *Professionalizing Modern Medicine. Paris Surgeons and Medical Science and Institutions in the 18th Century* (Westport [Conn] 1980); Hess, 'Beobachtung', 2010.

founded in 1782, learned how to compose case histories in the style of learned observations.[44] The medico-surgical schools in Paris and Berlin had similar objectives.

There were probably few opportunities in daily practice for the time-consuming writing of a *historia morbi* or patient history in the humanist tradition. With the gradual relocation of clinical training to hospitals the daily observations were increasingly noted down in tabular forms, which could be copied or obtained as preprints.[45] Such tables served as templates for the composition of 'exact patient histories' – at the Göttingen Policlinic for instance[46]– but the technique also established itself in general practice. From the end of the eighteenth century preprints could be purchased from booksellers, and by the mid-nineteenth century medical practitioners were able to choose from a wide range of tabular patient forms 'in order to optimize the recording of their observations and their ways of healing'.[47] This 'formularization' of records is also apparent in our sample: Bönninghausen as well as Ottenthal used a recording technique based on such templates. But while Bönninghausen used pre-printed forms which offered a defined space for comments (clinical picture) but fixed table elements to enter information on individual consultations, Ottenthal worked with his own handwritten forms, which he divided into sections according to his own requirements and therefore did not need to name them.

Additional instructions on writing are rare. Those that exist all focus on one question: how can bedside observations be recorded in a time-saving and efficient manner? We find frequent advice – from Jean Razoux (1791–1821) or

---

44 Griesel, 1984, 50–2. The *observationes* that have been preserved suggest that the writer struggled to meet the request of the training institute to pronounce the Latin words 'without insulting people's ears'. (*ibid.*, 41).

45 Instruction at the first Leiden clinic already entailed that the two physicians who worked in the St Caecilia Hospital had to keep reports on the clinical cases 'written down in books and had to mention the name of the patient, his disease and the date when certain treatments were carried out or prescribed' (Tim Huisman, *The Finger of God. Anatomical Practice in 17th-Century Leiden* (Leiden 2009), 131).

46 *Die ärztliche Besuchs-Anstalt der erkrankten Armen und die ambulante Klinik in Würzburg* (Würzburg 1820), 54.

47 'Rezension: Medicinisch-praktischer Geschäfts- und Address-Kalender auf das Jahr 1809. Für praktische Aerzte, Chirurgen und Apotheker. Herausgegeben von Dr. Carl Heinrich Ludwig Schulz. Nebst 12 Monatstafeln. Leipzig, bey Heinrich Gräff', *Heidelbergische Jahrbücher der Literatur* 3 (1810), 221–2: 222; 'Schema zu zweckmässigen, in Folio anzulegenden Krankentabellen für die Praxis (das aber noch jeder Arzt beliebig modificiren kann)', *Medicinischer Almanach für das Jahr 1836* 1 (1836), 72.

Heinrich Küchler (1811–1873) for instance – to use special signs, letters or numbers so that all that needed documenting would fit into one line.[48]

The Time Delay between Observation and Recording

Hufeland's idyll suggests that observation and recording were not simultaneous events. The instructions for the proper composition of patient histories discussed above also suggest that there was a delay in time between sickbed and documentation which was a cause for concern: as early as the eighteenth century, physicians were admonished to 'keep their medical diaries precise and convenient'. It was recommended to trainee doctors in particular to 'record the most necessary information day by day' in their records.[49] Some advisers recommended binding the daily notes – on a monthly or quarterly basis – and asking an assistant to 'establish an index for each volume and…an alphabetic ledger' that could also serve as account book.[50]

The practice of recording is, however, not one that belongs only to modern times. In illustrations of early modern physicians one sometimes sees a practice record lying open on the table during consultations so the previous treatment could be quickly recapitulated (see Figure 1.1). The following description is by a patient of the physician Turquet de Mayerne (1573–1655):

> Usually I went in a morning for his advice, about 7 of ye clock, where I us'd to find him in his study, which was a large room furnish'd with books & Pictures; and as one of ye chiefest he had ye picture of ye head of Hyppocrates that great physitian; & upon his table he had ye proportion of a man in wax, to set forth ye ordure & composure of every part; before his table he had a frame with shelves, whereon he set some books, and

48 Jean Razoux, *Tables nosologiques et météorologiques très-étendues dressées à l'Hôtel-Dieu des Nîmes depuis le premier juin 1757 jusques au premier janvier 1762* (Basel 1767); Hermann Küchler, *Bemerkungen über die Erkenntniss der Krankheiten nebst einer Anleitung zur Führung einer geordneten Krankencontrole* (Darmstadt 1844).

49 'Anzeige neu entworfener Krankentabellen, zum Gebrauche für Ärzte und Wundärzte zu genauer und bequemer Führung ihrer medizinischen Tagebücher', *Medicinische Nationalzeitung für Deutschland und die mit selbigem zunächst verbundenen Staaten: Intelligenz-Blatt* 10 (1799), 39–40, see also Küchler, 1844.

50 Hermann E. Richter, '169. Bemerkungen über die Erkenntniss der Krankheiten, nebst einer Anleitung zur Führung einer geordneten Krankencontrole. Küchler, Arzt zu Darmstadt', *Carl Christian Schmidt's Jahrbücher der in- und ausländischen gesammten Medicin* 44 (1844), 128–9.

FIGURE 1.1 *Frontispiece of Johannes Magirus'* Old and New Writing Calender *(1653)*
STADTVERWALTUNG ALTENBURG, STADTARCHIV HISTORISCHE HAUS-UND SCHREIBKALENDER.

> behind this he sat to receive those that came for his advise. His custom is to record in a book the diseases and remedies of all his patients if they be of difficulties, so that sending for his book he finds what he had done to her formerly, & thereupon prescribes ye same.[51]

The image of the physician enthroned among the insignia of medical science is hardly realistic and was probably used to lend a dramatic note to the report.[52] The practice records that have been preserved reveal that the physicians only rarely wrote down their observation during or immediately after the treatment session.[53] In our sample, too, the regular writing and the even saturation of the ink show that the notes of one or more days were written down in one sitting, most probably in the evening. Johannes Magirus, for instance, complained

51 Nance, 2001, 24.

52 Cf. Bloesch's portrait (in this volume) and the painting by Johann Georg Platzer 'Der Besuch beim Arzt' that shows the physician at his desk, disrupted while writing. When calling on patients the physician is however usually shown without a notebook.

53 As the *Casebook* (1565–1573) of Magister Polewheele (British Library (hereafter BL), MS 28023), the copied *Diarum medicum* (1680–1688) of Johannes Munnick (BL MS Sloane 1256), the *Daybook* (1676–1696, 1595–1605) of Sir Edmund King (BL MS Sloane 1589).

bitterly that the noisy students kept him from his 'good thoughts and *laboribus*' at night, allowing him less peace and quiet than 'weary horses and oxen' had.[54] Even Johann Christoph Götz, who would add to his practice records several times during the day, most likely did this only on returning from calling on patients. Others, such as the non-licensed surgeon Wachter, wrote down their case descriptions in the form we know them today only weeks or even, as in the case of the London surgeon Joesph Binns (died 1664) of St Bartholomews Hospital, years later – presumably as collections of selected cases.[55] Large, heavy folios were usually too valuable to be taken on patient visits. Physicians preferred to use small notebooks for jotting down some notes which they could later transfer to their larger-format fascicles or books. 'Every morning the practitioner takes with him a writing pad in octavo or duodecimo format to note down the most important points, either during the examination or just after, using the pad at the same time as forms for prescriptions, certificates etc'.[56] Such memory aids simply consisted of a small notepad with a list of patients to be visited (as in Soemmerring's case), or in a multiply-folded sheet of paper as recommended by Hermann Eberhard Richter (1808–1876).[57] No matter how the physicians kept their 'notebooks', whether on house calls or in their homes: conscientiously maintained practice records are not usually minuted accounts of the physician-patient encounter, but well-reflected documentations of aspects of that interaction that seemed particularly important to the writer.

The Practice Record as Part of a Multistage Recording System

The interval between the bedside visit and the entry into the practice record was often bridged through the use of a 'notebook' or similar formats suitable for swift note-taking (loose leaves, wastebooks etc.),[58] which would then serve as the basis for the nocturnal reflection. Often several formats were used. In the late eighteenth century the *Medizinische Nationalzeitung* suggested the use of two parallel records: one for day-to-day chronological documentation, and a

54 Sabine Schlegelmilch, 'Johannes Magirus. Stadtarzt in Zerbst (1651–1656)', *Mitteilungen des Vereins für Anhaltische Geschichte* 20 (2011), 9–30: 24.

55 *Surgical Casebook* (1633–1663) of Joseph Binns (BL MS Sloane 153). The total number of 671 case histories over a recording period of more than thirty years (1633–1663) suggests selective copying from former records, as do the regular handwriting and the saturation of the ink in case descriptions that are far apart in time.

56 Richter, 1844.

57 Dumont, 1993; Richter, 1844.

58 Küchler, 1844.

patient-oriented one to summarize the course of treatment.[59] The practice records, in turn, served as the basis for billing. Alphabetical indices of the patients are a sign of this kind of usage. As early as in the seventeenth century we find exemplary registers subdivided in alphabetical sections (similar to document files), some with alternating red and black labelling to allow for a better overview over the individual patients.[60] Whether the records were kept in chronological order or as collections of case histories, many physicians had separate account books in which they entered the fees charged, settled payments, deferred payments or payment in kind. Our sample also provides evidence of note-keeping practices that involved, at least temporarily, the parallel use of an account book: we have a separate account book for Götz from 1720, a documentation of treatment costs from the time around 1840 and Bönninghausen's separate '*Conto-Buch*' (cash-book).[61]

Authors with scientific aspirations drew on their practice record for research purposes. This is apparent from selections of case histories, subsequently added corrections of spelling or grammatical errors or the addition of literary references, as illustrated by the *Surgical Casebook* of the London surgeon Joseph Binns.[62] The introduction of occasional indices into the practice records also strongly suggests such aspirations. In some cases one can even track the journey of such notes right up to their publication. Götz used his practice notes on Duke Ernst von Metternich (1656–1727), who was afflicted with bladder stones, for an article in his newly-founded journal 'The sincere physician' (*Der Aufrichtige Medicus*) which was published from 1726 to 1727.[63] Magirus had other ways of making his records public: he annotated them extensively with excerpts from scientific literature and used individual cases to illustrate the theoretical content of his teaching. Even though it is hardly possible to reconstruct the in-between stages in each case, one can deduct from markings, deletions

59 Anzeige, 1799.

60 Cf. MS Sloane 1589 (Sir Edmund Kings *Daybook* (1676–1696)) and Sloane 1256, 1258 (Christopher Love Morley's copy of Johannes Munnicks *Diarium medicum* (1680–1688)).

61 Institute for the History of Medicine of the Robert Bosch Foundation, P156.

62 As note 55. The *Diarium medicum practicum* of Christopher Detlev Hahn from Aarhus is another example for indexing for scientific purposes (Wulff *et al.*, 2005).

63 Susan Splinter, '"Der Aufrichtige Medicus". Eine Zeitschrift des Nürnberger Arztes Johann Christoph Götz (1688–1733) als Vorläufer des "Commercium Litterarium"', *Jahrbuch für Kommunikationsgeschichte* 13 (2011), 5–15: 10; Kay-Peter Jankrift, Annemarie Kinzelbach and Marion Ruisinger, 'Ernst von Metternich (1656–1727). Ein patientenzentrierter Einblick in den medizinischen Markt um 1720', in Lina Gafner, Iris Ritzmann and Katharina Weikl (eds), *Penning Patient's Histories – Doctors' Records and the Medical Market in the 18th and 19th Century* [special issue: *Gesnerus* 69, no. 1 (2012)], 12–35.

and additions, as we find them with Götz for instance, that the notes were revised subsequently in several stages. The case histories of the Würzburg policlinic are a particularly impressive example of this multistage process: the neatly hand-written and literary *observationes* were composed from the printed forms which were mandatory in Bavaria for poor-doctors[64] and which were presumably filled in by the trainee doctors on the basis of notes they had jotted down previously in their notebooks. This step-by-step processing of observations – from the sick-bed to the note-book, from the notebook to the journal, from the journal to the policlinical 'patient histories' – constituted the methodological basis of a clinical science that manifested, for instance, in the casuistic publications of 'Annual Reports'. This particular case even goes a step further in that Fuchs used the case histories composed by his students for a textbook he published: *Lehrbuch der speciellen Nosologie und Therapie* (2 vols, 1845–1848).[65] Here, we can pursue in detail how, in a multistage process that involved revising, transferring, copying, excerpting, condensing and generalizing, hastily jotted notes were transformed into textbook knowledge.

Attempting a Typology

Despite the great variety, we suggest to distinguish three types of practice records. This distinction is based on the way the records were produced and takes, as will be discussed in detail, the actual note-taking practice as the basis for a definition. This allows us to establish formal criteria that refer to the organization and technique of the record-keeping. As with any typology, this proposal does not aim at an accurate distinction, but focuses on ideal-typical forms of record-keeping.

Scholarly Records of the 'Casebook' Type

The first type of practice records emerges as the result of scholarly penmanship and is demonstrated by Gianna Pomata using the example of the early

64 Besuchs-Anstalt, 1820, appendix with tables.

65 Fuchs' patient histories (see the contribution by Stephanie Neuner and Karen Nolte: 'Bedside Training and Healthcare for the Poor in the Würzburg and Göttingen Policlinics in the First Half of the Nineteenth Century' with the manuscripts and publications mentioned there). Partly also published in Conrad Heinrich Fuchs, *Bericht über die Vorgänge an der Poliklinik zu Göttingen im Jahre 1838 bis 39* (Hannover 1840).

modern *observatio*.[66] In our sample, this type is represented by the *Diarium medicum* (Magirus) where the recording technique is based on common-placing.[67] The *locus* is the patient's name (or occupation) which is noted at the top of the left-hand column. All observations are entered to the right, in chronological order. A space would be kept for these entries, usually it was the whole page which was folded to create two columns. Once the right column was full, the notes started again at the top of the left column and continued by date. This way of recording has its limitations, as Michael Stolberg showed recently: some pages remain blank if only little information is gathered on one *locus*.[68] In Magirus' *Diarium* not all pages are full. When the one page set aside for a patient turned out to not be enough in the course of a long-drawn-out or recurrent treatment, the notes were continued on a new page which was not necessarily the next following one since that might have been allocated to another patient in the meantime. This form of recording consequently required a register or index to provide information on the various *loci* of a case history.[69] Fuchs' patient notes can also be seen as following the tradition of the scholarly commonplace books. Their contents, once revised by the author, served as indices in that they conflated the individual entries into an *observatio* (see below).[70]

Journal-Type Records

This case-oriented recording practice has to be distinguished from the strictly chronological journal. In the journal the date constitutes the first order of

66 Pomata, 'Praxis', 2005; *eadem*, 'Sharing Cases. The Observationes in Early Modern Medicine', *Early Science and Medicine* 15 (2010), 193–36.

67 Magirus kept, as well as a practice diary, commonplace books in thematic order, for which he used the same technique (Sabine Schlegelmilch, 'Vom Nutzen des Nebensächlichen. Paratexte in den Kalendern des Johannes Magirus', in Klaus-Dieter Herbst (ed.), *Astronomie – Literatur – Volksaufklärung. Der Schreibkalender der Frühen Neuzeit mit seinen Text- und Bildbeigaben* (Bremen 2012), 393–411).

68 Michael Stolberg, 'Medizinische *Loci communes*. Formen und Funktionen einer ärztlichen Aufzeichnungspraxis im 16. und 17. Jahrhundert', *NTM. Zeitschrift für Geschichte der Wissenschaften, Technik und Medizin* 21 (2013), 37–60.

69 Further examples for the learned recording practice include the casebooks of: Georg Palma (1543–1591), cf. Klaus G. König, *Der Nürnberger Stadtarzt Dr. Georg Palma (1543–1591)* (Stuttgart 1961); Mayerne (1573–1655), cf. Nance, 2001; Albrecht Haller (1730s), cf. Urs Boschung, 'Albrecht Haller's Patient Records (Berne 1731–1736)', *Gesnerus* 53 (1996), 5–14; Christopher Hahn (1744–1822), cf. Wulff *et al.*, 2005.

70 On such a process of subsequent assembly cf. Hess, 2011.

entries: all observations of that day are recorded consecutively. Long-term treatments are always listed under the date of each patient contact, connected by cross-references (*vid. datum*). Reconstruction of the course of an illness, and therefore assembly of a case history, is difficult and requires much browsing through the records. Johann Friedrich Glaser's *Medizinisches Register* is a prototype of this form of recording. It shows all the patients treated each day over a period of fourteen years. That he chose to call it a 'register' might be an indication of its central function: it assesses and registers – similar to a hospital admission book – the services rendered. We propose that chronological records of this kind, as a rule, also served as account books (as template or explication): it lists the fees for each service rendered (cf. Bloesch), the medicines prescribed and (occasionally) their settlement (cf. Götz) or the dispensation of medicines (cf. Glaser). The forerunners of the examples presented here were probably the prescription journals of the sixteenth and seventeenth centuries, which were used by physicians to record – on a day-to-day basis and in a form that had been adapted from the pharmacy books – their prescriptions (and what they charged for them) without referring to the symptoms.[71] With this type we rarely find observations of several consecutive days subsumed under one date, as with Götz or the British surgeon and private physician Sir Edmund King (1629–1709).[72] The date does not, in these cases, signify the day of treatment but that of the recording.

The two types described so far are relatively easy to distinguish: while the learned recording practice of the casebooks reveals the history of a patient (almost) immediately, this is not the case with the journals. The entire patient history can often only be made visible through electronic data capture[73] – and this might be difficult if patients are not clearly identified. Journals, on the other hand, reveal immediately what the physician did on a particular day. We see at a glance whether he saw three, four or even more patients – or on which days he had no patients, an aspect that remains obscure or can only be reconstructed through electronic data processing with the scholarly records.

71 Franziska Ofenhitzer, Würzburg, is just completing a dissertation on the prescription diary of Petrus Kirstenius in Breslau, 1612–1616.

72 Ruth Schilling, Sabine Schlegelmilch and Susan Splinter, 'Stadtarzt oder Arzt in der Stadt? Drei Ärzte der Frühen Neuzeit und ihr Verständnis des städtischen Amtes', *Medizinhistorisches Journal* 46 (2011), 99–133: 113 (Familie Monath); British Library Sloane MS 1589.

73 Jacalyn Duffin, 'A Rural Pratice in Nineteenth-Century Ontario. The Continuing Medical Education of James Mile Langstaff', *Canadian Bulletin of Medical History* 5 (1988), 3–28: 5.

(An interesting solution is the *numerus currens* introduced by Götz.[74]) While the scholarly records culminate in the case history, the journal mostly aims at registration; while the forum of the former is the academic world of the *res publica litteraria*, the latter aims, as will be discussed below, at the *res publica* of offices and registries.[75]

The Printed-form Type of Record

The third type is more difficult to delineate. If one looks at the logic of recording it is a combination of the previous two: as with the casebook type, the records are case-oriented; but the services rendered (treatment and prescription) are recorded and billed strictly according to date. This recording practice constitutes a separate type because of its formularization. Cloze text and columns direct the writing and give the entries a different narrative structure. Such 'schemes' bring uniformity to the entries and facilitate the operational linking of individual cases.

This linking facilitates, as in the case of Clemens von Bönninghausen, comparison or the 'necessary individualization of all characteristic signs of each case of illness'.[76] It also supports, as with Ottenthal, the accounting process because the services rendered on individual days (journal function) can be totalled per patient and invoiced (case-based documentation). In either case, the book-keeping aspect is prioritized be it for economic reasons as in the case of Ottenthal or out of moral consideration, as with the lay healer Clemens von

74 Each case was allocated a number in Roman numerals, with letters being added in ascending order with subsequent contacts. A reference was added in each case to the page number of the former contact with the same patient.

75 Comparable examples include: Johann Anton Grass von Portein (1684–1770), cf. Daems, 1985; on Samuel Thomas Soemmerring, cf. Dumont, 1993; Heinrich Grotjahn, cf. Engel, 1978; Dietrich Tutzke and Regina Engel, 'Tätigkeit und Einkommen eines Allgemeinpraktikers vor der Mitte des 19. Jahrhunderts. Ergebnisse einer historisch-statistischen Analyse', *Zeitschrift für die gesamte Hygiene* 24 (1978), 460–65; Jutta Wolff and Hans-Peter Wolff, 'Das Profil einer ärztlichen Allgemeinpraxis im Jahre 1862', *Das Deutsche Gesundheitswesen. Wochenschrift für die gesamte Medizin* 34 (1979), 568–71; Michel Schüppach (1707–1781), cf. Hartmann Rordorf, 'Ordinationsbücher und das Manual von Michael Schüppach, Chirurgus von Langnau (1707–1781)', *Pharmaceutica Acta Helvetiae* 5 (1934), 72–84; Wehren, 1985.

76 Bönninghausen 1863, 114 (with thanks to Marion Baschin for this information).

Bönninghausen, who explicitly introduced 'exact book-keeping' for his own instruction and in order to 'set the physician's[!] mind at rest'.[77]

Types 1 and 2 favour a particular kind of presentation (for which there is sufficient evidence). While type 3 resembles type 2 in that its recordings are chronological, it is generically different from types 1 and 2 if we focus on its formulated design. The partitioning lines on the paper – whether drawn by hand or pre-printed – have not only a typographic, but also a functional purpose in that the cloze texts organize the writing due to their slot-and-filler structure. Such 'mandatory entries' considerably limit the author's freedom of narrative description (which is the purpose of any form):[78] the tabular set-up of date and event structures the temporal order of the patient history, as is apparent from Ottenthal's patient chart. The tabular order of date and patient history resists narrative anticipation (prolepsis) or revision (analepsis) while it generates repetitive narration patterns. Structuring the diegetic narrative in this way imposes a certain order on the history. Moreover, the arrangement of table elements as used by Bönninghausen for his record sheets ('date', 'prescription', 'success and new signs') predetermines the perspective of the patient history and gives it, through the form, a meaningful structure by connecting the three elements 'in a telling manner'. It is true that such formal restrictions may be broken or overridden, but the form in itself marks alternate ways of recording as violations (non-compliance with the form). We do not suggest that types 1 and 2 are not based on generic formalisation too, but this is much less explicit in the casebook or journal entries. The entering of information or 'filling in' of forms generates a different form of narrative and should therefore constitute a literary form in its own right.[79]

The hybridisation of administrative and scientific techniques as well as the operationalization of record-keeping through schematization, cloze texts and index terms point to the origin of this third type: in typological terms it developed out of the hospital journal, a formularized note-taking technique that

77 Unfortunately we do not know whether the lay healer Clemens von Bönninghausen was considering his own (non-medical) conscience or that of the competing or supervising medical officers.

78 Arndt Brendecke, 'Tabellen und Formulare als Regulative der Wissenserfassung und Wissenspräsentation', in Wulf Oesterreicher, Gerhard Regn and Winfried Schulze (eds), *Autorität der Form – Autorisierung – Institutionelle Autorität* (Münster 2003), 37–53: 37. Cf. Matias Martinez and Michael Scheffel, *Einführung in die Erzähltheorie* (München ^{8}2009).

79 For the example of psychiatry cf. Brigitta Bernet, '"Eintragen und Ausfüllen". Der Fall des psychiatrischen Formulars', in Sibylle Brändli, Barbara Lüthi and Gregor Spuhler (eds), *Zum Fall machen, zum Fall werden. Wissensproduktion und Patientenerfahrung in Medizin und Psychiatrie des 19. und 20. Jahrhunderts* (Frankfurt am Main 2009), 62–91.

emerged in clinical training between 1780 and 1795. Physicians who had learned how to use forms as part of their clinical training were able to buy pre-printed forms from booksellers.[80] Medical practitioners expected them to result in greater descriptive precision, a better overview and – not least – time-saving and efficient documentation from such charts which were advertised as a 'patient diary or medical pocketbook' (*Kranken-Diarium oder Medizinisches Taschenbuch*), a 'patient manual or medical register' (*Kranken-Manual oder Medicinisches Hauptbuch*) or a 'medico-practical business and address calendar' (*medicinisch-praktischer Geschäfts- und Adresskalender*).[81]

Summary

Typologies are by necessity based on idealization, while the 'historical reality' tends to produce combinations of the ideal types described. One type we left out in our survey is that of retrospective journalization. We would like to use this term for a documentation practice that involves the preliminary noting down of case-oriented observations on loose sheets, followed by retrospective chronological assembly and binding. We find examples of this technique in the seventeenth as well as nineteenth century.[82] Journalization combines the advantage of the (scholarly) case history with that of the chronological journal, in that it shows the comprehensive history of an illness and simultaneously documents the medical activities for administrative purposes (billing etc.). This kind of case-related documentation already employs the main features of

80 F.R., 1844; cf. for instance the '*Krankentabellen*' (patient tables): Anzeige, 1799; 'Rezension: Medicinisch-praktisches Geschäfts- und Adress-Buch auf das Jahr 1811 für praktische Ärzte, Chirurgen und Apotheker. Herausgegeben von Dr. Carl Heinrich Ludwig Schulz. Nebst 12 Monatstafeln und einem Bildnisse des Hn. Geh. Hofr. Dr. Wendt. Erlangen, bey Palm', *Jenaische Allgemeine Literaturzeitung* 9 vol. 2, No 69 (1812), coll. 29–31; 'Rezension: Kranken-Manual oder medicinisches Hauptbuch zu angemessener Aufzeichnung der in der medicinischen Praxis vorkommenden bemerkungswerthen Krankheitsfälle nach ihrem Verlaufe, dem Heilverfahren dabey und dem besorglichen wie dem wirklichen Ausgange der Krankheit', *Heidelbergische Jahrbücher der Literatur* 3 (1810), 223; 'Rezension: Kranken-Diarium oder Medicinisches Taschenbuch zu fortgehender Einzeichnung der in der täglichen Praxis vorkommenden Krankheitsfälle und die diesfalls getroffenen medicinischen Anordnungen wie auch des besorglichen und des endlichen Ausgangs der Krankheit', *Heidelbergische Jahrbücher der Literatur* 3 (1810), 223.

81 Küchler, 1844.

82 *Surgical Casebook* (1633–1663) of Joesph Binns (BL MS Sloane 153), *Observations* of Yulfranc Gerdy, (1831–1837), Bibliotheque Interuniversitaire de Médecine (MS 2185).

modern medical recording systems: in the early nineteenth century folders were introduced into hospitals and, in the mid-twentieth century patient record cards arrived in the doctors' offices. It is only the virtual 'binding' of the case-related documentations in archives or as part of filing-systems that distinguishes them from retrospective journalization. Our third type also constitutes such a combination. Its formulaic character combines patient-related and chronological record-keeping: the formalized header individualizes the record while the listed progress report establishes a chronological order. The registration techniques, some of which are very elaborate, also soften the boundaries between the ideal types since they transfer, via the indexation, the journal's chronological order into a case-related one, if only virtually. Our typology, if it is not to be an end in itself, is therefore only justified if the suggested differentiation adds to the historical research or if it offers a starting point for further analysis.

Attempting a Definition

The primary purpose of practice records as a genre of sources is to meet the need for effective access to practice-relevant information. It is the practice – meaning the physicians' activities on and with the patient – that generates the core of the data to be administered and that all the sources presented have in common (recording of the patients' history, diagnosis, choice of therapy as a foundation for recapitulating the treatment process). These notes are complemented by entries/records that depend on how the individual doctor's practice is organized (references to pharmacists, extracts of letters, draft letters, consultation, accounts, post-mortem report, expertise etc.).

Practice records are therefore notes written on paper and bound into a fascicle or book that document and organize the medical practice. *What* (content) was documented for *which* purpose (function) can vary considerably and this needs to be taken into account by historical analysis: with the typological distinction of practice records suggested – into case histories, chronological journals and pre-printed forms – each zooms in on different phenomena, 'data', characteristics or occurrences such as the course of the illness (patient history), the frequency of consultations or the services delivered in the process, prescriptions and fees, or – as with homeopathic forms – the connection between remedies and symptoms. This means that the outer form cannot constitute a criterion since the record-taking followed the prevailing writing traditions. In defining practice records as a type of source, the crucial criterion is the context of their production.

Any definition based on such a limited number of practice records has to be preliminary. Our suggestion needs to be corroborated by further practice records. Our point is that one should do without a rigid definition that rests primarily on formal criteria. Instead, we plead for a definition that, as discussed extensively, focuses on the mediality and the writing technique of the source. This material-based definition has the additional advantage that it does not prematurely categorize content and function of the practice record, but that it operationalizes the definition historically in that the functional character of the records and the contents that arise from everyday medical practice are included in the definition and therefore in the scientific analysis.

The Question of the Genre

The question as to the literary genre is, however, not solved. Our investigation does not supply a historical genealogy. What seems important to us is rather that the most complex form of record-keeping – the classical commonplacing which documents a patient's illness in a pre-determined space – constitutes the oldest form of medical record examined by our Research Group. The journal which is simpler compared to the records in commonplace-style does not appear as a regular means of medical record-keeping until the eighteenth century. We see this on the examples of Götz and Glaser where the humanist tradition of recording a case in its entirety – in the style of Magirus' *observatio* – came to an end. Whether this had historical reasons or is due to the small number of sources available is a question that has to remain unanswered (for now). Synthesis of day-by-day records from the eighteenth and nineteenth centuries is, on the other hand, based on the classical *observatio*, even if this genre was more detailed and chronologically more differentiated (analepses/prolepses). The literary form of the *observatio* means that the physician has more responsibility as the author of his notes than is the case with formalized tables where the writing is governed and regulated by outlines and cloze text.[83]

That we cannot derive a genealogy on this basis is apparent from the example of the Aarhus clinician Christopher Hahn (1744–1822) who continued to keep his records as individual *observationes* as early modern physicians used to do and sorted, indexed and bound them afterwards in alphabetical order according to illness. The policlinic casebooks in our sample also arose from learned practice even though (or because) these sophisticated histories of individual diseases were based on formulaic and tabularized note-taking, because the

83 Cf. Brendecke, 2003.

assistant physicians were urged by the town physician, who was in charge of the policlinic as an institution for the poor, to 'keep a special diary on their patients in the style of the Munich model'.[84] They, moreover, had to submit 'an extract from that diary every month'.[85] The well-formulated academic *observatio* consequently arose from formalized observation which took its cue from the obligation for documentation imposed by the Bavarian poor-relief authorities.

With regard to the journals preferred by eighteenth century physicians, the small number of practice records investigated so far does not show whether examples of this type (other than prescription diaries) also existed in earlier centuries. The example of an eighteenth century British country doctor, edited by Edmund Hobhouse, suggests that the day-by-day note-taking arose from the diary in which private as well as medical events were noted day after day (as we know from many examples).[86] Instances of record-keeping in the style of hospital journals can be established with greater historical precision. Due to their direct association with the development of the institutional recording system, we do not find pre-printed (or pre-drawn) forms until the late eighteenth century. Since then they have governed medical practice documentation, either as card filing systems or as electronic practice programmes. We can therefore hypothesize that the medical documentation practice can be traced back to various literary genres without constituting a literary genre in its own right.

The Significance of the Writing Context

The various types also have different addressees. They are obvious as we can see from the policlinic casebooks: the case histories which were neatly copied out from student notes constitute a selection that was considered interesting and exemplary enough for scientific publication or as material for medical textbooks.[87] The change in documentation especially – from pre-fabricated tables to written out literary case histories – illustrates the importance of the envisaged audience (in this case the poor-relief authorities on the one hand

84 Besuchs-Anstalt, 1820, 30.

85 *Ibid.*, 30. The forms given in the appendix show the tabular structure for listing patients according to 'Cullen's nosological systems' as well as the medicines prescribed and the effect observed.

86 Hobhouse, 1934. The best-known examples are the private diaries of Samuel Pepys (1633–1703).

87 The students in Würzburg presumably submitted – like their peers in Paris and Berlin – the patient histories on loose, folded folio-sheets; cf. Richter's vivid description of the writing of such *observationes* (Richter, 1844).

and the academic world on the other.)[88] We detect a similar trend even with Magirus' notes: in his first practice in Berlin he focused on the exact recording of prescriptions and prices; in Zerbst he noted down prescriptions only roughly according to their ingredients, but devoted more space to the individual phases of illness which he annotated with page references or technical terms from medical literature. This change in his note-taking practice is doubtlessly connected with his position as professor at the Zerbst *Gymnasium* and is apparent also in the third-person entries which were presumably added by students or trainee physicians.

In the journal-type practice records we often find evidence of the – direct or indirect – duty of documentation towards third parties. A typical example of this is the 'practice booklet' (*Praxisbüchli*) of the Swiss lay healer Wachter which he used in order to comply with the regulations of the local medical authorities. The preprinted journals used by the non-licensed homeopathic practitioner Clemens von Bönninghausen to account for his activities are another example. Bloesch's daily note-taking was associated with a joint meteorological project.[89] And even though Götz' bedside observations were intended for scientific use, the pharmacy names added by a third person in the margins suggest at least one other reader (probably Götz' wife) who used the notes as proof and index of prescriptions. The same applies to Ottenthal. It is also noticeable in our sample that some publically appointed physicians – *physici* or town physicians – kept their practice notes in the form of journals. Friedrich Glaser and the Nuremberg town physician Johann Friedrich Götz are examples of this. But there are also cases to the contrary: there are indications that Magirus, who used commonplacing to record his notes, also served as town physician while he practised in Berlin. Especially the changes in his record-keeping when he relocated to Zerbst, which were associated with the use of these records for academic teaching, seem to indicate that the records were by no means merely a continuation of an academic notation practice. It was rather a case of the connection with the medical authorities (in the widest sense) affecting the format and practice of documentation. This 'context of composition' in the wider sense seems to us to be crucial for any future definition of this genre of sources.

Historical Localization

The practice records evaluated by our research group therefore support the idea that even the medical records that seem very private at first glance can be

88 *Audience* is by no means restricted to the historically verified readership. Mah, 2000 illustrates that the public imagined by the writer is as important.

89 Cf. Gafner (in this volume).

seen as a contribution to a *res publica* on two accounts: firstly, as the articulation of a *res publica* in the original sense,[90] when, as part of the emerging reporting culture, representatives of the medical authorities of a state or city wrote down their observations in form of a journal. Medical practice documentation would then appear as an element of the 'good practice' that state-building and administrative regulation relied on.[91] Secondly, the *observationes*, when processed for further application in the academic *res publica litteraria*, were a contribution to the world of scholarship.

From this perspective the third type can also be seen in relation with the change in the academic world: the introduction of tabular recording techniques based on prescribed or pre-printed forms is not only a reflection of the fact that hospitals became more important in medical training; it also reflects the thorough transformation process which, with the growing independence of the medical sciences, led to a reduction of the concept of science to institutionalized disciplines and practices. While the adopting of administrative documentation meant that an essential element of institutional record-taking was taken over too, the techniques of processing, storing and archiving data on paper did not find their way into the general medical practice. These techniques include, as well as statistics and graphic visualisation (as in patient charts), the diagrammatic data assessment that built on the extensive archiving and evaluating of patient files in hospitals. Out-patient practice did not participate in this development. It therefore needs to be considered in what way the merely partial adoption of institutional documentation practices contributed to the scientific depreciation of general medical practice which, starting from the early nineteenth century, has been regarded an extension of the hospital (and therefore, in a sense, as a minor antechamber to the temple of medical science).

Conclusion

We propose that the essential criteria for the categorization of practice records can be grounded in the circumstances of their use, their purpose as well as the observation tradition.[92] Including this material composition context adds analytical depth and precision to the historiography. Because the inconsistencies

90 The notion of the state as public is very modern. For Immanuel Kant the actions of civil servants were still private, while those of scholars were public.

91 Cf. André Holenstein, *"Gute Policey" und lokale Gesellschaft im Staat des Ancien Régime. Das Fallbeispiel der Markgrafschaft Baden(-Durlach)* (Epfendorf 2003).

92 Daston *et al.*, 2011.

of medium and material of each source of this genre are not as much determined by individuality of the recording physician as by the freedom granted him by the prevailing writing tradition (and the underlying record-taking system), as is apparent from the formalisation (and formularisation) of records discussed earlier.[93] While practice records are documents of practical activity, they are not 'minuted records'. They, in fact, fix the 'processing of information'.[94] This conclusion may seem trivial, but practice records are no authentic testimony of practical activity (or of the physician-patient interaction). They provide information that has been filtered in multiple ways and broken up in the process. This fracturing depends on the one hand on the individual physician's perception and interpretation which, in turn, is based on his state of knowledge. On the other hand it is caused by the articulation and fixation of the interpreted observation (material dependence). It manifests in the record-keeping system that underlies our practice records. It is precisely this materially palpable mediality that makes these types of sources so interesting and historically challenging.

---

93 Siegfried Mattl, 'Formalisierung der Welt', *Österreichische Zeitschrift für Sozialwissenschaften* 7 (1996), 173–76; Hess, 'Beobachtung', 2010.

94 Cf. Peter Burke, *A Social History of Knowledge. From Gutenberg to Diderot* (Cambridge 2000), who described this process, in keeping with the concepts of humorism, as a transition from 'raw' to 'concocted' and processed information.

CHAPTER 2

Doctors and Their Patients in the Seventeenth to Nineteenth Centuries

Marion Baschin, Elisabeth Dietrich-Daum and Iris Ritzmann

Introduction

The oil painting shown in Figure 2.1 is titled *In the Doctor's Office*. It was created around 1880, towards the end of the period under investigation, by the Tyrolean painter Franz von Defregger (1835–1921) and shows patients in the waiting room of a country physician in the nineteenth century. The doctor is in the foreground. The patients are not only represented as individuals, but as a group of people waiting to use a service, each with their various biographical and social backgrounds and expectations. With this in mind, and focusing on the patients,[1] we will compare, discuss and categorize the evaluations of patient data gathered from practice records in the context of the DFG-funded research network 'Medical practices (1600–1900).

Over the past thirty years, medical historical research has considered patients primarily in their roles and positions within the changing healthcare system: as the target group of civic medicalization initiatives,[2] as agents in the

* We thank the following members of the individual projects in the DFG-funded research network 'Medical practices (1600–1900),' on whose partly still unpublished results this contribution is largely founded: Sabine Schlegelmilch (J. Magirus), Kay Peter Jankrift / Annemarie Kinzelbach / Susanne Grosser / Marion Ruisinger (Chr. Götz), Ruth Schilling (J.F. Glaser), Stephanie Neuner / Karen Nolte (Ch. Fuchs), Katharina Weikl / Alois Unterkircher / Iris Ritzmann (G. Wachter), Lina Gafner / Philipp Klaas (C.A. Bloesch), Marion Baschin (F.P. von Bönninghausen), Marina Hilber / Eberhard Wolff / Alois Unterkircher (F. von Ottenthal).

1 Eberhard Wolff, 'Perspectives on Patients' History. Methodological Considerations on the Example of Recent German-Speaking Literature', *Canadian Bulletin of Medical History* 15 (1998), 207–28; Karen Nolte, 'Die Erfahrung "zwischen den Zeilen". Eine patientenzentrierte Perspektive als eine andere Geschichte?', in Marguérite Bos, Bettina Vincenz and Tanja Wirz (eds), *Erfahrung. Alles nur Diskurs? Zur Verwendung des Erfahrungsbegriffes in der Geschlechtergeschichte* (Zürich 2004), 273–81.

2 Cf. Ute Frevert, *Krankheit als politisches Problem 1770–1880. Soziale Unterschichten in Preußen zwischen medizinischer Polizei und staatlicher Sozialversicherung* (Göttingen 1984); Claudia Huerkamp, *Der Aufstieg der Ärzte im 19. Jahrhundert. Vom gelehrten Stand zum professionellen Experten. Das Beispiel Preussens* (Göttingen 1985); Francisca Loetz, *Vom Kranken zum Patienten. "Medikalisierung" und medizinische Vergesellschaftung am Beispiel Badens 1750–1850* (Stuttgart 1993);

 | DOI 10.1163/9789004303324_004

FIGURE 2.1 *Franz von Defregger,* In the Doctor's Office. *Oil on canvas, around 1880.* ROSENHEIMER VERLAG.

Francisca Loetz, '"Medikalisierung" in Frankreich, Großbritannien und Deutschland 1750–1850. Ansätze, Ergebnisse und Perspektiven der Forschung', in Wolfgang Eckart and Robert Jütte (eds), *Das europäische Gesundheitswesen. Gemeinsamkeiten und Unterschiede in historischer Perspektive* (Stuttgart 1994), 123–61; Hans-Christoph Seidel, *Eine neue "Kultur des Gebärens." Die Medikalisierung von Geburt im 18. und 19. Jahrhundert in Deutschland* (Stuttgart 1998); Eberhard Wolff, *Einschneidende Maßnahmen. Pockenschutzimpfung und traditionale Gesellschaft im Württemberg des frühen 19. Jahrhunderts* (Stuttgart 1998); Wolfgang Eckart and Robert Jütte, *Medizingeschichte. Eine Einführung* (Köln 2007), 315–16; Martin Dinges, 'Immer schon 60% Frauen in den Arztpraxen? Zur geschlechtsspezifischen Inanspruchnahme des medizinischen Angebotes (1600–2000),' in *idem* (ed.), *Männlichkeit und Gesundheit im historischen Wandel 1850–2000* (Stuttgart 2007), 295–322; Martin Dinges, 'Social History of Medicine in Germany and France in the Late Twentieth Century. From History of Medicine Toward a History of Health,' in Frank Huisman and John Warner (eds), *Locating Medical History. The Stories and Their Meanings* (Baltimore 2004), 209–36: 218–24; for discussions of the concept of medicalization cf. Jean Pierre Goubert, 'Die Medikalisierung der französischen Gesellschaft am Ende des Ancien Régime. Die Bretagne als Beispiel,' *Medizinhistorisches Journal* 17, no. 1–2 (1982), 89–114; Michael Stolberg, 'Heilkundige. Professionalisierung und Medikalisierung', in Norbert Paul and Thomas Schlich (eds), *Medizingeschichte. Aufgaben, Probleme, Perspektiven* (Frankfurt am Main 1998), 69–86; Elisabeth Dietrich-Daum and Rodolfo Taiani, editorial to *Medikalisierung auf dem Lande/Medicalizzazione in aree alpina* [special issue: *Geschichte und Region / Storia e regione* 14, no. 1 (2005)], 5–18; Eckart *et al.*, 2007, 312–15; for a critical view cf. Oliver Stenzel, *Medikale Differenzierung. Der Konflikt zwischen akademischer Medizin und Laienheilkunde im 18. Jahrhundert* (Heidelberg 2005).

process of medical socialization;[3] as sub-groups or differentiated groups, either privileged or disadvantaged by their social affiliation and other factors, in systems of inequality;[4] as clients in the medical marketplace;[5] or as a clientele determined by its 'lifeworld' and first excluded, but later fought over by medicine as a 'functional system' of society.[6] It was not until historiography resolutely shifted its focus away from the physician that the patient moved to the centre of its inquiry.[7] The results of this shift appear in numerous studies on

---

3 Cf. Loetz, 1993.

4 Cf. Reinhard Spree, *Soziale Ungleichheit vor Krankheit und Tod. Zur Sozialgeschichte des Gesundheitsbereichs im Deutschen Kaiserreich* (Göttingen 1981); Jörg Vögele (ed.), *Stadt, Krankheit und Tod. Geschichte der städtischen Gesundheitsverhältnisse während der epidemiologischen Transition (vom 18. bis ins frühe 20. Jahrhundert)* (Berlin 2000); with an intersectional approach cf. Gabriele Winker and Nina Degele, *Intersektionalität. Zur Analyse sozialer Ungleichheiten* (Bielefeld 2009), 49–52; Marlene Ellerkamp, *Industriearbeit, Krankheit und Geschlecht* (Göttingen 1991); Ingrid von Stumm, *Gesundheit, Arbeit und Geschlecht im Kaiserreich am Beispiel der Krankenstatistik der Leipziger Ortskrankenkasse 1887–1905* (Frankfurt am Main 1995); Nicole Schweig, *Gesundheitsverhalten von Männern. Gesundheit und Krankheit in Briefen 1800–1950* (Stuttgart 2009); Susanne Hoffmann, *Gesunder Alltag im 20. Jahrhundert? Geschlechterspezifische Diskurse und gesundheitsrelevante Verhaltensstile in deutschsprachigen Ländern* (Stuttgart 2010); Martin Dinges and Andreas Weigl (eds), *Gesundheit und Geschlecht.* [special issue: *Österreichische Zeitschrift für Geschichtswissenschaften* 22, no. 2 (2011)]; Alois Unterkircher, 'Männer als Patienten. Krankheitsverhalten von Männern im ländlichen Raum in der zweiten Hälfte des 19. Jahrhunderts am Beispiel der Praxisjournale des Südtiroler Arztes Franz von Ottenthal' (PhD diss., University of Innsbruck, 2012).

5 Lina Gafner, Iris Ritzmann and Katharina Weikl (eds), *Penning Patient's Histories – Doctors' Records and the Medical Market in the 18th and 19th Century* [special issue: *Gesnerus* 69, no. 1 (2012)]; Michael Stolberg, *Experiencing Illness and the Sick Body in Early Modern Europe* (Basingstoke 2011), 58–63; Nancy Tomes, 'Patients or Health-Care Consumers? Why the History of Contested Terms Matters', in Rosemary Stevens, Charles Rosenberg and Lawton Burns (eds), *History and Health Policy in the United States* (New Brunswick 2006), 83–110.

6 Stenzel, 2005.

7 For basic methodical considerations on the history of patients cf. Roy Porter, 'The Patient's View. Doing Medical History from Below', *Theory and Society* 14 (1985), 175–98; Anne Digby, *Making a Medical Living. Doctors and Patients in the English Market for Medicine 1720–1911* (Cambridge 1994); Anne Digby, 'The Patient's View,' in Irvine Loudon (ed.), *Western Medicine* (Oxford 1997), 291–305; Wolff, 'Perspectives,' 1998; Katharina Ernst, 'Patientengeschichte. Die kulturhistorische Wende in der Medizinhistoriographie,' in Ralf Bröer (ed.), *Eine Wissenschaft emanzipiert sich. Medizinhistoriographie von der Aufklärung bis zur Postmoderne* (Pfaffenweiler 1999), 97–108; Jonathan Gillis, 'The History of the Patient History since 1850,' *Bulletin of the History of Medicine* 80 (2006), 490–511; Flurin Condrau, 'The Patient's View Meets the Clinical Gaze,' *Social History of Medicine* 20 (2007), 525–40; John Burnham, 'The Death of the Sick Role,' *Social History of Medicine Advance Access* (2012), 1–16 [http://shm

specific patient groups such as children or the elderly, men or women, pregnant and parturient women, religious or cultural minorities or groups with comparable symptoms such as psychiatric patients or tuberculosis sufferers. In addition there are publications that look specifically at particular periods, regions or sources, focusing on the patient's view.[8] This approach dwells more on the aspect of demand, presenting patients not merely as passive objects of externally controlled processes, but as influential agents and clients.[9] The gender, age and social class of patients have been given greater attention and have evolved as central parameters of participation in medical healthcare. The main questions of this contribution will be: who are the patients? Are there noticeable changes in the sample period – seventeenth to nineteenth century – in the consultation of physicians by various patient groups? What afflictions or situations led patients to seek out a physician? What was the doctor-patient-relationship like?

.oxfordjournals.org/content/early/2012/04/18/shm.hks018, doi: 10.1093/shm/hks018]; for the German-speaking world cf. in particular Robert Jütte, *Ärzte, Heiler und Patienten. Medizinischer Alltag in der frühen Neuzeit* (München 1991); Annemarie Kinzelbach, *Gesundbleiben, Krankwerden, Armsein in der frühneuzeitlichen Gesellschaft. Gesunde und Kranke in den Reichsstädten Ueberlingen und Ulm, 1500–1700* (Stuttgart 1995); Jens Lachmund and Gunnar Stollberg, *Patientenwelten. Krankheit und Medizin vom späten 18. Jahrhundert bis zum frühen 20. Jahrhundert im Spiegel von Autobiographien* (Opladen 1995); Michael Stolberg, 'Ärzte und ländliche Patienten. Soziologisch-historische Aspekte einer schwierigen Beziehung', *Die medizinische Welt* 43 (1992), 529–33; Michael Stolberg, 'Patientenschaft und Krankheitsspektrum in ländlichen Arztpraxen des 19. Jahrhunderts', *Medizinhistorisches Journal* 28 (1993), 3–27; Stolberg, *Experiencing Illness*, 2011.

8 For literary overviews see Eckart *et al.*, 2007, 181–90; Philip Rieder, 'L'histoire du "patient". Aléa, moyen ou finalité de l'histoire médicale?', *Gesnerus* 60 (2003), 260–71; Burkhart Brückner, 'Psychiatriegeschichte und Patientengeschichte. Eine Literaturübersicht zum Stand der deutschsprachigen Forschung,' *Sozialpsychiatrische Informationen* 35 (2006), 26–30.

9 For the aspect of demand cf. Michael Stolberg, 'Heilkunde zwischen Staat und Bevölkerung. Angebot und Annahme medizinischer Versorgung in Oberfranken im frühen 19. Jahrhundert' (Med. diss., University of München, 1986); Olivier Faure, *Les Français et leur médecine au XIXe siécle* (Paris 1993); Francisca Loetz, 'Andere Grenzen. Faktoren ärztlicher Inanspruchnahme in Deutschland 1780–1830. Empirische Ergebnisse und methodologische Überlegungen', in Thomas Schnalke and Claudia Wiesemann (eds), *Die Grenzen des Anderen. Medizingeschichte aus postmoderner Perspektive* (Köln 1998), 25–48; Iris Ritzmann, 'Der Faktor Nachfrage bei der Ausformung des modernen Medizinalwesens. Überlegungen am Beispiel der Kinderheilkunde', in Bettina Wahrig und Werner Sohn (eds), *Zwischen Aufklärung, Policey und Verwaltung. Zur Genese des Medizinalwesens 1750–1850* (Wiesbaden 2003), 163–78.

Methods and Sources

Demand for the services of physicians has been the object of various investigations into individual doctors' practices, but these have not usually involved comparison.[10] This contribution will therefore employ universalizing as well as contrasting[11] approaches in investigating the distribution of gender, marital status and age as well as the social structure and regional background of the clientele of medical practices. The individual practices serve to portray change and continuity in the consultation of physicians by various sections of the population, the qualitative aspects of the patients' motivation for choosing a particular doctor, their reasons for seeking treatment and the relationship of physician and patient.

The potential for knowledge acquisition from the subsequent analysis is limited, firstly due to the fact that the number and choice of practices investigated are limited and arbitrary, and secondly due to the divergent structures of the data extracted from the records. The different purposes of these records, their disparity of transmission and the variable ways in which patient data were assessed in the projects render the samples collected rather heterogeneous.[12] Depending on the individual variables, the problems that emerge are also very different.[13] Generally, questions of gender were easiest to evaluate

10 For first attempts see Wolfgang Balster, 'Medizinische Wissenschaft und ärztliche Praxis im Leben des Bochumer Arztes Karl Arnold Kortum (1745–1824). Medizinhistorische Analyse seines Patiententagebuches' (Med. diss., University of Bochum, 1990); Jacalyn Duffin, *Langstaff. A Nineteenth-Century Medical Life* (Toronto 1993, repr. 1999); Andrea Thümmler, 'Rekonstruktion des Alltags eines thüringischen Arztes im 18. Jahrhundert anhand seines Praxistagebuchs 1750–1763' (Med. diss., Free University of Berlin, 2004); Anne Hilde van Baal, 'In Search of a Cure. The Patients of the Ghent Homoeopathic Physician Gustave A. van den Berghe (1837–1902)' (PhD diss., University of Amsterdam, 2004); Dinges, 2007; Martin Dinges, 'Arztpraxen 1500–1900. Zum Stand der Forschung,' in Elisabeth Dietrich-Daum *et al.*, *Arztpraxen*, 2008, 23–61; Marion Maria Ruisinger, *Patientenwege. Die Konsiliarkorrespondenz Lorenz Heisters (1683–1758) in der Trew-Sammlung Erlangen* (Stuttgart 2008); Marion Baschin, *Wer lässt sich von einem Homöopathen behandeln? Die Patienten des Clemens Maria Franz von Bönninghausen, 1785–1864* (Stuttgart 2010) and the publications of the Research Network (cf. the contributions in this volume).

11 Lutz Sauerteig, 'Vergleich. Ein Königsweg auch für die Medizingeschichte? Methodologische Fragen komparativen Forschens', in Norbert Paul and Thomas Schlich (eds), *Medizingeschichte. Aufgaben, Probleme, Perspektiven* (Frankfurt 1998), 266–91: 274.

12 Cf. the explanations in the respective contributions.

13 Physicians commented rarely on religious affiliation, cf. Kay Peter Jankrift and Ruth Schilling: 'Medical Practice in Context. Religion, Family, Politics and Scientific Networks' (in this volume).

from the records even though this parameter could only be, in each of the practice records, indirectly assessed on the basis of names or, in some cases, occupations or, more rarely, from the marital status or other information. Only with babies or infants it sometimes was impossible to come to a definite conclusion. None of the records investigated systematically list the civil status which renders conclusive statements about this parameter impossible.[14] Only the notes of Bloesch and Bönninghausen supply reliable information to that effect. In both practices we notice a predominance of unmarried patients: this can be explained by the large proportion of children and by the large number of single people in Biel and Münster. With regard to the patients' age, the early modern physician Johannes Magirus (1615–1697) and the Swiss lay healer Gottfried Wachter (1776–1861) sporadically used designations such as 'very old' or 'a young girl', but this allows for only rough categorizations. The other physicians – Johann Götz (1688–1733), Conrad Fuchs (1803–1855), Caesar Bloesch (1804–1863), Franz von Ottenthal (1809–1899) and Friedrich von Bönninghausen (1828–1910) – meticulously recorded the ages of their patients, specifying even days and months with babies. While the assessment, description and comparison of gender and age of the various clientele proved relatively satisfactory, the records yielded only limited numerical information on the patients' occupation and socio-economic status. This is due, on the one hand, to the fact that the relevant entries tended to be quite rudimentary; on the other hand it was often not possible to discern – as in Magirus' casebook or Glaser's journal – whether entries referred to the name or occupation of patients.[15] In some cases, information regarding the occupation of the presumed main clientele was missing altogether, as with Wachter and Ottenthal.[16] But even where the physicians, as in the case of Glaser and Bloesch, regularly wrote down the patients' occupation or trade, sometimes in great detail, there were methodical limitations to the evaluation of the socio-economic affiliation of the clientele.[17] From Bloesch's notes we also see that this town physician's patients

14 Balster, 1990, 107; Baal, 2004, 117; Duffin, 1999, 94 or Baschin, 2010, 154 came to the same conclusion. It is possible to derive the civil status from the age and occasionally from the case history. With women, designations like *Fräulein/Jungfer/innupta* (Miss or spinster) or *Frau/uxor/hausfrawe* (Mrs or housewife) or *Witwe/Vidua* (widow) were indicators.

15 The designation *Kellerschenkin* could refer to a woman of that name or to the wife of a publican (= *Kellerschenk*). The designation *Mahler* (miller) could equally refer to the name or occupation of a patient. On Glaser cf. Thümmler, 2004, 72.

16 Unterkircher, 'Männer,' 2012; Thümmler, 2004.

17 On Glaser cf. Thümmler, 2004, 72–5; Volker Hess, 'Ärztlicher Alltag in Thüringen um 1750. Auswertung eines Praxistagebuchs,' in Elisabeth Dietrich-Daum, Martin Dinges, Robert Jütte and Christine Roilo (eds), *Arztpraxen im Vergleich. 18–20. Jahrhundert* (Innsbruck

often had more than one way of earning their livelihood. The same applies to Ottenthal's journals.[18] Allocation of patients to particular occupations would therefore seem largely arbitrary and did not conform to regional and local 'occupation censuses' or the more recent social stratification surveys of the individual catchment areas. It is, furthermore, not possible to infer patients' social circumstances from their job designations.[19] When considering the enormous socio-economic changes that evolved between the seventeenth and nineteenth centuries the data assessed is qualified by the heterogeneity of the specific practices and locations.[20] Research will rely on further investigations into the structure of the physicians' clientele.

The second set of questions presents an entirely different set of challenges: why did patients consult a physician and what was their relationship like? Analysis of the patients' reasons for seeing a doctor necessitates careful inquiry into the historical concepts of illness. Despite major reservations, there has been in-depth discussion in recent decades on ways in which historical descriptions of illnesses could be used in research.[21] This is not the place, however, to

---

2008) 87–107: 98; on Bloesch cf. Nadine Boucherin, 'Die Krankengeschichten von C.A. Bloesch. (1804–1863),' in Dietrich-Daum *et al.*, *Arztpraxen*, 2008, 147–65: 155.

18 On the general problem with establishing the occupation of a selected population in Austria for instance cf. Birgit Bolognese-Leuchtenmüller, *Bevölkerungsentwicklung und Berufsstruktur, Gesundheits- und Fürsorgewesen in Österreich 1750–1918* (Wien 1978), 99–149; for Germany cf. e.g. Kevin Schuerer and Herman Diederiks (eds), *The Use of Occupations in Historical Analysis* (Göttingen 1993); Jürgen Hoffmeyer-Zlotnik, 'Operationalisierung von "Beruf" als zentrale Variable zur Messung von sozio-ökonomischem Status,' ZUMA-*Nachrichten* 32 (1993), 135–41 or Flurin Condrau, *Lungenheilanstalt und Patientenschicksal. Sozialgeschichte der Tuberkulose in Deutschland und England im späten 19. und frühen 20. Jahrhundert* (Göttingen 2000), 283–89.

19 Sylvia Schraut, *Sozialer Wandel im Industrialisierungsprozeß. Esslingen 1800–1870* (Esslingen 1989), 345–51; Michael Vogl, '"Nahe und entfernte Landpraxis". Untersuchungen zu Samuel Hahnemanns Eilenburger Patientenschaft 1801–1803', *Medizin, Gesellschaft und Geschichte* 9 (1990), 165–80: 170–72; Thümmler, 2004, 75.

20 Jürgen Vögele, *Urban Mortality Change in England and Germany 1870–1913* (Liverpool 1998), 113–14.

21 Cf. Charles Rosenberg, 'Framing Disease. Illness, Society and History', introduction to Charles Rosenberg and Janet Golden (eds), *Framing Disease. Studies in Cultural History* (New Brunswick 1992), I–XXVI; Karl-Heinz Leven, *Die Geschichte der Infektionskrankheiten. Von der Antike bis ins 20. Jahrhundert* (Landsberg 1997); Karl-Heinz Leven, 'Krankheiten. Historische Deutung versus retrospektive Diagnose', in Norbert Paul and Thomas Schlich (eds), *Medizingeschichte. Aufgaben, Probleme, Perspektiven* (Frankfurt am Main 1998), 153–85; Iris Ritzmann, 'Leidenserfahrung in der historischen Betrachtung. Ein Seiltanz zwischen sozialem Konstrukt und humanbiologischer Konstanz', *Historische Zeitschrift*

discuss the therapeutic range of the doctors' practices since that is the subject of another contribution.[22] We will focus on the motives that led patients to see a physician by comparing and analysing possible criteria for choosing a doctor, recorded patterns of ailments and information about the actual encounter between physician and patient. We refer to the studies carried out in the 1990s into the doctor-patient relationship that diverged from the assumption that the physicians played the more powerful part in these relationships.[23]

Gender, Age and Social Affiliation

Gender

Gender-specific differences in health and illness behaviour have been the subject of historical research for some decades.[24] The main question has been whether, in the past, more women than men consulted a physician, as is the case today.[25] According to the medical historian Martin Dinges the percentage of men and women in doctors' practices varied widely over time. In the eighteenth century men apparently dominated, as was the case among the patients

Beiheft (2001), 59–72; Jacalyn Duffin, *Lovers and Livers. Disease Concepts in History* (Toronto 2005); Michael Stolberg, 'Möglichkeiten und Grenzen einer retrospektiven Diagnose', in Waltraud Pulz (ed.), *Zwischen Himmel und Erde. Körperliche Zeichen der Heiligkeit* (Stuttgart 2012), 209–27.

22 Cf. Annemarie Kinzelbach, Stephanie Neuner and Karen Nolte: 'Medicine in Practice: Knowledge, Diagnosis and Therapy' (in this volume).

23 Essential reading: Nicolas Jewson, 'Medical Knowledge and the Patronage System in 18th Century England. A Sociological Analysis', *Sociology* 8 (1974), 369–85, introducing the thesis of the 'patronage systems'; Claudia Huerkamp, 'Ärzte und Patienten. Zum strukturellen Wandel der Arzt-Patient-Beziehung vom ausgehenden 18. bis zum frühen 20. Jahrhundert,' in Reinhard Spree and Alfons Labisch (eds), *Medizinische Deutungsmacht im sozialen Wandel des 19. und 20. Jahrhunderts* (Bonn 1989), 57–73; Roy Porter and Dorothy Porter, *Patient's Progress. Doctors and Doctoring in 18th Century England*, (Cambridge 1989) refer to Jewson, while Stolberg, 1993; Lachmund *et al.*, 1995, 123–26; Digby, 1997, 291–305; Loetz, 1998; Philip Rieder, *La figure du patient au XVIIIe siècle* (Genève 2010) qualify and differentiate Jewson.

24 Apart from the studies mentioned in note 4 cf. e. g. Elisabeth Mixa and Elisabeth Malleier (eds), *Körper – Geschlecht – Geschichte. Historische und aktuelle Debatten in der Medizin* (Innsbruck 1996); Jürgen-Helmut Mauthe (ed.), *Krankheit & Geschlecht. Konzepte & Kontroversen* (Sternenfels 1999); Elmar Brähler and Hildegard Felder (eds), *Weiblichkeit, Männlichkeit und Gesundheit* (Opladen [2]1999); Klaus Hurrelmann and Petra Kolip (eds), *Geschlecht, Gesundheit, Krankheit. Männer und Frauen im Vergleich* (Bern 2002).

25 Dinges, 2007.

of barber-surgeons.[26] In the first half of the nineteenth century, the gender distribution varied from one practice to the next with a continually slightly higher representation of males. From 1860/70 onwards, again according to Dinges, women were in the majority.[27] How do the results of the Research Network fit with these findings?

In his Berlin practice the early modern physician Johannes Magirus only treated a few more men than women, while in his practice in Zerbst, a town north of Leipzig, male patients were clearly in the majority.[28] A similar distribution is apparent from the almost 1,600 prescriptions recorded, between 1612 and 1616 in a book of medical prescriptions of the Breslau physician Petrus Kirstenius, around thirty years before Magirus' time. In those four years, 929 prescriptions (58 per cent) were made out to men and 24 per cent to women.[29] The seventeenth-century practices that Dinges investigated also reveal a predominance of male patients.[30] Conversely, the Nuremberg town physician Johann Götz who practised in the early eighteenth century always had more female than male patients, although there were fluctuations.[31] Like Götz, the Bern physician Albrecht von Haller (1708–1777) treated more women[32] while Johannes Glaser, town physician and health officer in Suhl, had a very slightly higher ratio of male patients.[33] These findings, relating to the second half of

---

26 Data are still scarce for the sixteenth and seventeenth centuries.

27 Dinges, 2007, 303; Martin Dinges, 'Medizin- und gesundheitsgeschichtliche Paradigmen zur geschlechterspezifischen Ungleichheit seit ca. 1750,' in *idem* and Andreas Weigl (eds), *Gesundheit und Geschlecht. Österreichische Zeitschrift für Geschichtswissenschaften* Sonderheft 22, no. 2 (2011), 8–49: 31–3.

28 The following proportions emerge 1648 (Berlin, 97 patients) 44.3% (43) men and 43.3% (42) women; 1653 (Zerbst, 100 patients) 46% (46) men and 30% (30) women. These and the following information are reflected in Figure 2. 2.

29 Königliche Bibliothek Stockholm, Ms X 101. We thank Franziska Ofenhitzer for this information. Ms Ofenhitzer (Würzburg) is working on a dissertation on this physician, in connection with the Research Network.

30 Dinges, 2007, 303.

31 1717 (91 patients) 40.7% (37) men, 59.3% (54) women; 1721 (301 Patienten) 45.8% (138) men, 48.2% (145) women; 1726 (254 patients) 43.4% (110) Männer, 56.7% (144) women. In total, of the 646 patients of the three years 44.1% Männer and 53.1% women.

32 Dinges, 2007, 303. These are the only practices that provide data from the first half of the eighteenth century. An attempt at an explanation was made by Hubert Steinke, 'Der junge Arzt und seine Patienten. Albrecht von Hallers Praxis in Bern 1731–1736', in Dietrich-Daum *et al.*, *Arztpraxen*, 2008, 79–86: 83.

33 Of the 2,848 patient contacts assessed for 1753, 1760 and 1763, 46 per cent (1,317) related to men and 44 per cent (1,258) to women. Thümmler, 2004, 78, shows for 556 patients in 1750 that more men (45.9 per cent, 255) than women (44.8 per cent, 249) received treatment.

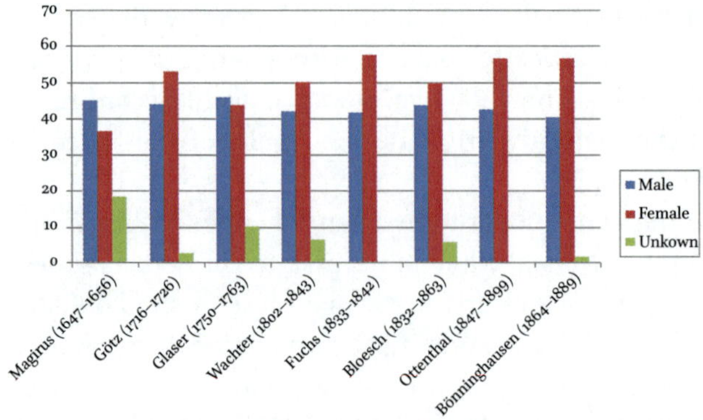

Synopsis	Magirus (1647–1656)	Götz (1716–1726)	Glaser (1750–1763)	Wachter (1802–1843)	Fuchs (1833–1842)	Bloesch (1832–1863)	Ottenthal (1847–1899)	Bönninghausen (1864–1889)
Male	45.2%	44.1%	46%	42.6%	42%	44%	42.9%	40.7%
Female	36.5%	53.1%	44%	50.5%	58.1%	50%	57%	57.1%
Unkown	18.3%	2.8%	10%	6.8%	0%	6%	0.2%	2.2%
Basic number	197	646	2848	279	2275	2071	7827	6832

FIGURE 2.2 *Percentage of women and men in the individual doctors' practices.*

the eighteenth century, are not exceptional either.[34] According to our research results, male patients mostly dominated in the early modern doctors' practices, but no definite trend can be ascertained until around 1800.

Such a trend does not emerge until around the turn of the nineteenth century. The healer Gottfried Wachter, the Swiss physician Caesar Bloesch and the policlinics in Würzburg and Göttingen had more women patients throughout, with the percentage of female patients growing steadily in the course of their years of medical activity.[35] The same was true for the practice of the homeopath

34 Cf. Dinges, 2007, 303; the findings refer to three doctors' practices in the Swiss Canton of Grisons, Bavaria and Lausanne between 1751 and 1799.

35 The calculations for Wachter are based on the years 1805 (163 patients) 42.9% (70) men, 49.7% (81) women and 1843 (27 patients) 40.7% (11) men, 55.5% (15) women; for Bloesch 1834 (515 patients) 44.3% (243) men, 47.2% (228) women and 1852 (801 patients) 41.3% (330) men, 53% (425) women; for the policlinics according to Fuchs' annual reports of 1838/39, 56.7% were women (1,328 of 2,343 patients) and of 1853/54, 59.4% were women

Friedrich von Bönninghausen.[36] Major fluctuations only occurred in the case of Ottenthal, the South Tyrolean rural physician, even though women always predominated in his practice, too.[37] This predominance of female patients is, however, not related to the specific gender distribution in the populations in question, as might be assumed, even though women were in fact overrepresented.[38] The Canadian physician Jacalyn Duffin ascribed the phenomenon to the fact that physicians were increasingly called upon to assist with deliveries. The research of Marina Hilber and Philipp Klaas on Ottenthal and Bloesch shows, on the contrary, that the female overrepresentation is not related to the increasing assistance of physicians during childbirth, but to the growing demand of women for medical assistance with problems of menstruation, pregnancy and childbed. The same applies to the practices of Götz, Glaser, Wachter and Bönninghausen, as we can see from their practice records.[39] These findings prove that, up until the end of the nineteenth century, assistance with childbirth was not among the main tasks of general practitioners.[40] They also support the thesis that the medicalization of the female body, with regard to menstruation and menopause, emerged long before the turn of the twentieth

---

(1,312 of 2,207 patients). Among the 710 edited case histories almost 64% (453) referred to female patients.

36 1864/67 (3,720 patients) 42.8% (1,591) men, 55.3% (2,059) women; 1886/89 (621 patients) 40.9% (254) men, 58.3% (362) women.

37 Highest rate: 1850 (1,229 patients) 59.2% (728) women; lowest rate: 1880 (2,076 patients) 54.8% (1,138) women. The proportion was similarly low in 1860 (1,483 patients) with 55.1% (817) women; cf. Andreas Oberhofer, 'Eine Landarztpraxis im 19. Jahrhundert am Beispiel der Ordination des Dr. Franz von Ottenthal (1818–1899)', in Dietrich-Daum *et al.*, *Arztpraxen*, 2008, 167–91: 181; Unterkircher, 'Männer,' 2012, 11.

38 Gender differentiation of the adult group was only carried out for the practices of Ottenthal and Bönninghausen; the same results for the second half of the nineteenth century in Duffin, 1999, 124; Baal, 2004, 55; Dinges, 2007, 304; Baschin, 2010, 148–52.

39 On Ottenthal cf. Marina Hilber, 'Der Landarzt als Geburtshelfer – Dr. Franz von Ottenthal und der medizinische Markt in Südtirol (1860–1869),' in Gafner *et al.*, 2012, 141–57; on Bloesch cf. Philipp Klaas, 'Cäsar Adolf Bloesch – ein Geburtshelfer? Einblicke in seine ärztliche Praxis zwischen 1832 und 1850', in Gafner *et al.*, 2012, 110–25; on Wachter cf. Katharina Weikl and Iris Ritzmann, 'Ein Auslaufmodell. Spuren einer ländlichen Heilerpraxis im Zeitalter der Akademisierung', in Gafner *et al.*, 2012, 76–94; on Götz cf. Annemarie Kinzelbach, Susanne Grosser, Kay Peter Jankrift and Marion Ruisinger: "*Observationes et Curationes Nurimbergenses*'. The Medical Practice of Johann Christoph Götz (1688–1733)' (in this volume); on Glaser cf. Thümmler, 2004, 134; on Bönninghausen cf. Marion Baschin, *Ärztliche Praxis im letzten Drittel des 19. Jahrhunderts. Der Homöopath Dr. Friedrich Paul von Bönninghausen (1828–1910)* (Stuttgart 2014).

40 Cf. Stolberg, 1993, 23.

century.[41] It is therefore remarkable that in the practices investigated within the Research Network, women predominated among the patients from as early as 1800 and not, as has been assumed so far, from 1860/70, as can be seen in Figure 2.2.[42] However, this research result needs to be further investigated and differentiated, especially with a view to potential differences between rural and urban practices, because in the Bavarian countryside practices investigated by Stolberg men continued to be slightly more predominant even after 1850.[43]

Age

For a long time, historiography regarded babies, infants and the elderly as groups that rarely took advantage of medical assistance compared to other patient groups.[44] Evaluation of the clientele investigated here shows that, in general, all age groups were represented in the practices. Where the age was regularly listed, adults constituted – at between 45 per cent to 70 per cent – the largest patient group.[45] This result is not entirely conclusive since the adult group encompasses most ages. Moreover, it is often not possible to compare the findings with data regarding the age structure of the particular population.

The second largest patient group is that of children and adolescents,[46] whose percentage tended to range between 20 per cent and 30 per cent and

41 For general information on medication for menarche, menstruation and menopause cf. Dinges, 2007; Stolberg 1998; Michael Stolberg, 'The Monthly Malady. A History of Premenstrual Suffering', *Medical History* 44 (2000), 301–22; Michael Stolberg, 'Menstruation and Sexual Difference in Early Modern Medicine', in Andrew Shail and Gillian Howie (eds), *Menstruation. A Cultural History* (New York 2005), 90–101; disorders of these processes of the female body were mentioned by patients of Clemens von Bönninghausen and of the physician Carl Arnold Kortum, cf. Baschin, 2010, 295; Balster, 1990, 159–62; Thümmler, 2004, 134.

42 Dinges, 2007, 303–06.

43 Cf. the examples mentioned by Stolberg, 1993, 20.

44 Stolberg, 1993, 18–9, 26–7; Digby, 1994, 279–81; qualified by Loetz, 1993, 247–49; that this sweeping statement does not apply to children has been pointed out by Iris Ritzmann in her studies of 2003 and 2008. On 'older' patients cf. Daniel Schäfer, *Alter und Krankheit in der Frühen Neuzeit* (Frankfurt am Main 2004); Simone Moses, *Alt und krank. Ältere Patienten in der Medizinischen Klinik der Universität Tübingen zur Zeit der Entstehung der Geriatrie 1880 bis 1914* (Stuttgart 2005).

45 These percentages represent the proportion of 19- to 65-year-old patients among the group where no age is mentioned.

46 The members of the Research Network agreed to divide the parameter 'age' into the following categories: 0–365 days, 1–4 years, 5–14 years, 15–18 years, 19–49 years, 50–65 and over 65 years. This means that the groups were defined according to today's rules and research interests. On the problem of the delineation of sick children and adolescents cf.

was therefore higher than previously estimated.[47] In Götz' medical practice this age group made up around 22 per cent of the clientele in 1717, but their number decreased considerably over the subsequent two years.[48] The development that emerges from Glaser's practice journal is entirely different. From the middle of the eighteenth century, Glaser treated a growing number of children and youngsters.[49] This ratio may largely correspond to the percentage of this age group in the general population. In the Würzburg and Göttingen policlinics the percentage of child patients was also high. 34.2 per cent (243 of 710) of case histories selected by Fuchs described children's cases, a sign of the policlinics' particular interest in this clientele.[50] The growing interest in health politics and health education will also have played a part in this development. Treating children was a means of paving the way towards treating other family members, too.[51] In Bloesch's practice the younger age groups initially constituted more than 40 per cent and still accounted for 30 per cent of patients later on.[52] These figures may well be a reflection of the high number of vaccinations he carried out. In Ottenthal's practice, children and youngsters were initially hardly represented,[53] but between 1850 and 1890, the percentage of patients

Iris Ritzmann, 'Chirurgische Kinderheilkunde im alten Zürcher Spital', *Gesnerus* 58 (2001), 292–99; Iris Ritzmann, *Sorgenkinder. Kranke und behinderte Mädchen und Jungen im 18. Jahrhundert* (Köln 2008), 39–44; the same problem applies to 'older' patients.

47 Dinges, 2007, 303 shows a proportion of children of 7 per cent to 16 per cent for various doctor's practices; Stolberg, 1993, 19–20, 26–7 also found that children were underrepresented in the data available to him from the nineteenth century. In Kirstenius' early seventeenth century pharmacopoeia 17 per cent of prescriptions were for children, cf. Ofenhitzer, see note 29.

48 In 1717, 20 of 91 patients; in 1721, 10% (30 of 301 patients); in 1726, 4.2% (28 of 254 patients).

49 In 1750, 17.3% (131 of 760 contacts); in 1753 22% (113 of 514 contacts); in 1760, 26.8% (310 of 1,155 contacts); in 1763, 25.8% (298 of 1,179 contacts).

50 This proportion of children in the case histories approximately corresponds to the proportion of children in the policlinics.

51 Franz von Rinecker, *Medicinische Statistik der poliklinischen Anstalt an der Julius-Maximilians-Universität zu Würzburg in ihrem vierten Decenium 1837–1847* (Würzburg 1848), 3–4; Ritzmann, 2008.

52 In 1834, 43.7% (225 of 515 patients); in 1842, 30.1% (227 of 755 patients); in 1852, 33.1% (265 of 801 patients).

53 The data also shows a higher representation of male babies and infants or persons over the age of sixty. Unterkircher, 'Männer', 2012, 311 also establishes a slight surplus of male older patients for the 1860s based on absolute numbers; on Friedrich von Bönninghausen cf. Baschin, 'Choice', 2014. Findings are similar for the practice of Clemens von Bönninghausen, cf. Baschin, 2010, 158–60.

under eighteen grew from 16.2 per cent to 20.5 per cent.[54] These findings approximately correspond to the picture Michael Stolberg presented of rural practices in Lower Bavaria.[55] In Bönninghausen's practice, the percentage of under 18-year-olds decreased from around 30 per cent to 20 per cent during the last third of the nineteenth century,[56] which does not necessarily mean that children were underrepresented in this practice. Due to low birth rates the town and district of Münster feature an unusual age structure compared to Prussia as a whole.[57]

Even though, in every century, patients over the age of sixty-five consulted physicians, they – like babies and infants – were seen, for a long time, as patients who very rarely saw a licensed physician.[58] In Götz' practice they constituted 2 per cent at the most (12 of 646 patients) and the proportion was even below 1 per cent in the case of Glaser (21 of 3,608 contacts). In both cases, one can conclude that elderly patients were clearly underrepresented. Although hardly any data is available on the nineteenth century age distribution in the individual catchment areas, meaning that, consequently, no statements are can be made regarding the representativeness of the distribution, the percentages suggest a slight increase in elderly people's utilization of medical services. In the Göttingen policlinic the ratio of over 60-year-old patients was 6 per cent (142 of 2,343 patients). On the other hand, a quarter of Fuchs' compilation of case histories, (179 of 710) refers to that age group, leading us to assume that the policlinic had a special interest in that patient group.[59] A slightly rising trend in the utilization of medical treatment by over 65-year-olds is apparent in the practices of Bloesch, Bönninghausen and, particularly, Ottenthal. Over and above that, the percentages

54 In 1850, 183 of the 1,129 patients; in 1890, 379 of the 1,859 patients. It was even higher in 1880 at 23.7 per cent (491 of 2,076 patients). Generally see Unterkircher, 'Männer,' 2012, 185–87; Alois Unterkircher, 'Ausser Konkurrenz? Strategien und Inanspruchnahme einer Südtiroler Landarztpraxis um 1890 bei Erkrankungen von Kindern', in Gafner *et al.*, 2012, 158–77: 162–64.

55 Stolberg, 1993, 19.

56 In 1864/1867, 1,135 of the 3,720 patients; in 1886/1889, 20.2%, 126 of the 621 Patienten. With Clemens von Bönninghausen the proportion of children and adolescents increased over time, cf. Baschin, 2010, 187.

57 Hans-Jürgen Teuteberg, 'Bevölkerungsentwicklung und Eingemeindungen (1816–1945)', in Franz-Josef Jakobi (ed.), *Geschichte der Stadt Münster*, vol. 2 (Münster 1993), 331–86: 334–35.

58 'Older' patients are those over sixty-five. In the practices where the age group of fifty to sixty-five was also examined, the proportion of 'older' patients was accordingly higher: with Götz 9.9% (64 of 646 patients); with Fuchs 36.5% (259 of 710 patients); with Bloesch 11% (228 of 2,071 patients); with Bönninghausen 13% (891 of 6,832 patients).

59 Karen Nolte, 'Die medizinische Versorgung der Göttinger Stadtarmen durch das Poliklinische Institut der Universität', *Würzburger medizinhistorische Mitteilungen* 29 (2010), 131–57: 142; cf. in general Moses, 2005, 37; for early modern times cf. Schäfer, 2004.

of elderly patients continued to grow to various degrees in these practices.[60] In Ottenthal's practice the percentage of elderly patients largely corresponds to their percentage of the population, with a slight underrepresentation of elderly men.[61] In the practice of the homeopath Bönninghausen elderly people were underrepresented due to the particular age structure in the district of Münster.[62] All in all, these findings support the existing thesis that elderly patients were underrepresented in the doctors' practices.[63] It was nevertheless possible to illustrate that the utilization of medical provision by elderly people slowly increased in the course of the nineteenth century.

Social Class

Consultation of academic physicians by members of the lower and middle classes who did not receive poor relief has been considered to have been low up until the mid-nineteenth century.[64] It was not until the last decades of that century that a definite change in the social composition of the physicians' clientele occurred;[65] from then on, groups who used to be averse to attending academic physicians were able to do so due to their increased purchasing power[66] and the introduction of public health funds.[67] It is not the place here to discuss in more detail whether this process was driven by the authorities, by the growing confidence[68] of patients or by the commodification[69] of medical services on the health market.[70]

---

60 For Bloesch in 1834, 2.5% (13 of 515 patients); in 1852 4.4% (35 of 801 patients); for Ottenthal in 1850, 10.3% (116 of 1,129 patients); in 1890, 14.9% (277 of 1,856 patients); for Bönninghausen in 1864/67, 2.7% (100 of 3,720 patients); in 1886/1889, 3.1% (19 of 621 patients).

61 Unterkircher, 'Männer', 2012, 306–16.

62 Teuteberg, 1993, 334–35.

63 Stolberg, 1993, 26–7.

64 Cf. Frevert, 1984, Huerkamp, 1989.

65 Reinhard Spree, 'Kurpfuscherei-Bekämpfung und ihre sozialen Funktionen während des 19. und zu Beginn des 20. Jahrhunderts', in Alfons Labisch and Reinhard Spree (eds), *Medizinische Deutungsmacht im sozialen Wandel des 19. und frühen 20. Jahrhunderts* (Bonn 1989), 103–21: 112–13; Stolberg, 1998, 69, 80; for France Laurence Brockliss and Colin Jones, *The Medical World of Early Modern France* (Oxford 1997).

66 Huerkamp, 1989, 58, 63; Stolberg, 1993, 26–7.

67 Frevert, 1984, 336.

68 On the category of confidence cf. Niklas Luhmann, *Vertrauen. Ein Mechanismus der Reduktion sozialer Komplexität* (Stuttgart [3]1989); Ute Frevert, 'Vertrauen. Eine historische Spurensuche,' in Ute Frevert (ed.), *Vertrauen. Historische Annäherungen* (Göttingen 2003), 7–66; Stenzel, 2005, 33; Stolberg, 1993, 25.

69 Dinges, 2008, 52; Lachmund *et al.*, 1995, 125.

70 See also Gerd Göckenjan, *Kurieren und Staat machen. Gesundheit und Medizin in der bürgerlichen Welt* (Frankfurt am Main 1985), 193–94.

The examples of the Research Network indicate that the number of consultations rose between the seventeenth and nineteenth centuries and that this increase is mostly due to the rising number of lower- and middle-class patients.[71] Magirus' practice largely corresponds to the image of the early modern physician described in literature, who served the noble and wealthy within a specific social urban network. The clientele in his Berlin practice (in 1648: 184 patient contacts, 97 patients) consisted 'mostly of courtiers and ministers to the Elector's family', lower court officials, military persons, members of the clergy and persons from his own family and religious circle.[72] The percentage of members of Berlin's petty bourgeoisie or poor population seems to have been low and cannot be specified in more detail from the physicians' records. In his later practice in Zerbst most of Magirus' patients (in 1653, 266 consultations, 100 patients) were courtiers, court servants, academics, merchants, administrative clerks and craftspeople. They came from the town itself and from a radius of twenty to twenty-five kilometres around the town.[73]

The Nuremberg town physician Götz also counted aristocrats from the House of Bayreuth-Kulmbach as well as Count Ernst von Metternich among his patients (in 1726, 254 patients).[74] At less than 2 per cent (3 in 130 occupational designations) they were, however, not his main clientele.[75] The bulk of his patients were families (some of them related to him) belonging to the Nuremberg bourgeoisie; in 1726, 32 occupations are named, 25 per cent were artisans or employed in administration, the church or schools, 28 per cent (37) were craftspeople, 9 per cent (12) worked in the transport industry or were innkeepers, nine were male servants, apprentices and day labourers (14 per cent) and 18 were female servants, maids and cooks (27 per cent).[76] Around a

71 Cf. Stolberg, 1993, 9–18, cf. the contributions in this volume.

72 Sabine Schlegelmilch, 'Johannes Magirus. Stadtarzt in Zerbst (1651–1656)', *Mitteilungen des Vereins für Anhaltische Geschichte* 20 (2011), 9–30: 16; Sabine Schlegelmilch, 'Vom Nutzen des Nebensächlichen. Paratexte in den Kalendern des Johannes Magirus', in Klaus-Dieter Herbst (ed.), *Astronomie – Literatur – Volksaufklärung. Der Schreibkalender der Frühen Neuzeit mit seinen Text- und Bildbeigaben* (Bremen 2012), 393–411.

73 Schlegelmilch, 2011, 16–8.

74 Susan Splinter, '"Der Aufrichtige Medicus". Eine Zeitschrift des Nürnberger Arztes Johann Christoph Götz (1688–1733) als Vorläufer des "Commercium Litterarium"', *Jahrbuch für Kommunikationsgeschichte* 13 (2011), 5–15: 6, 8; Kay-Peter Jankrift, Annemarie Kinzelbach and Marion Ruisinger, 'Ernst von Metternich (1656–1727). Ein patientenzentrierter Einblick in den medizinischen Markt um 1720', in Gafner *et al.*, 2012, 12–35.

75 For the year 1726 the social affiliation could be established in 51 per cent of the patient contacts.

76 Cf. Kinzelbach *et al.*, '*Observationes*' (in this volume).

quarter of the fees he omitted to charge were for the treatment of impoverished patients, such as farm girls, farm hands, orphans and day labourers.[77]

In the mid-eighteenth century, the town physician Glaser treated quite a representative cross-section of the population, including members of the regional political, clerical and academic elite and their families.[78] Nevertheless Glaser, like Götz, cannot be regarded as a physician of the upper classes.[79] Glaser's journal boasts a broad variety of occupations and a clientele (in 1750, 556 patients) that is typical of highly complex rural and proto-industrial regions.[80] It includes poor as well as rich craftsmen's families (in 1750, 199 occupations were named; 57 (29 per cent) of these patients came from 'rich' and 50 (25 per cent) from 'poor' craftsmen's families), village teachers or sheriffs and patients from the town's bourgeoisie, such as teachers, cantors or court stenographers (30 designations, 15 per cent) and – at 30 per cent (60) the largest group – servants on farms or in noble households, including shepherds, gamekeepers, cooks, farmhands and maids. The catchment area extended beyond the town to include villages within a radius of twenty to thirty kilometres.[81] The social diversity of Glaser's clientele is certainly comparable to that of the physician Karl Kortum (1745–1824), who practised circa fifty years later in the German mining town of Bochum. While the percentage of aristocratic patients (20 per cent, 61 of 303 occupational designations) and of clerics (15 per cent, 46) was still very high in his practice, the majority of his patients were miners (48), 'citizens' (51), farmers and farmhands (43), craftsmen (23) and servants (23).[82] We see a similar distribution in Samuel Hahnemann's (1755–1843) Eilenburg practice between 1801 and 1803, where we have evidence that, besides the considerable percentage of aristocrats (circa 6 per cent) and courtiers, lawyers and

77 In 1721 Götz treated 45 per cent of his patient free of charge, cf. Kinzelbach *et al.*, '*Observationes*' (in this volume).

78 For Glaser's range of patients cf. Ruth Schilling, Sabine Schlegelmilch and Susan Splinter, 'Stadtarzt oder Arzt in der Stadt? Drei Ärzte der Frühen Neuzeit und ihr Verständnis des städtischen Amtes', *Medizinhistorisches Journal* 46 (2011), 99–133: 122; Ruth Schilling: 'Social Mobility and Medical practice: Johann Friedrich Glaser (1707–1789)' (in this volume).

79 Thümmler, 2004, 74–5 pointed this out, criticizing the image of the town physician who treated primarily noble and rich patients. The young physician Albrecht von Haller in Bern treated mostly 'citizens' and only a few prominent families and noble patients, cf. Steinke, 2008.

80 Glaser designated no occupation in over 80 per cent of cases in the sample years 1750, 1753, 1760 and 1763; cf. Schilling (in this volume).

81 Hess, 2008, 98–9.

82 Balster, 1990, 109–15.

administrators (circa 25 per cent), 45 per cent of patients were merchants or tradespeople and around 15 per cent belonged to the urban and rural lower classes.[83]

The social composition of patients consequently began to change as early as the early eighteenth century. At the beginning of the nineteenth century members of low-wage social groups, from trade and commerce, as well as domestic servants and farm labourers already made up a considerable, even if not yet representative, percentage of the clientele. This observation applies only to the urban medical practices, however, since patients in rural areas continued to seek out traditional healers or, if need be, barber surgeons.[84] Licensed physicians would remain in the minority among the providers of medical services in rural areas for another fifty years. Small villages and hamlets held little attraction for them, a fact that is also reflected in the remoteness of the rural population from academic physicians.[85] Medicalization initiatives also concentrated on towns and cities. The Göttingen and Würzburg policlinics are an instructive example of this trend: they treated the sick among the cities' lower classes in their own homes, using funds provided by the university, the Apprentices' Institute and the municipal poor relief.[86]

While only a few individual physicians settled in the villages and hamlets before publicly funded physicians' practices[87] were instituted in or around the year 1800, such practices had already become accessible – at least in a geographical sense – to patients in the towns. The percentage of patients from a relatively narrow catchment area was accordingly high. In the sample year 1842, 67 per cent (503 of 643 mentioned) of Bloesch's patients came from the town of Biel itself and another 11 per cent (85) from nearby Nidau. The regular clientele of his well-attended practice consisted mostly of bourgeois families in Biel, including their servants and the workforce of the town's craft

83 Vogl, 1990, 33. No absolute numbers are given. Further results on Hahnemann's clientele are found in the evaluated journals, cf. Martin Dinges, 'Patients in the History of Homoeopathy,' introduction to *idem* (ed.), *Patients in the History of Homoeopathy* (Sheffield 2002), 1–32: 7–8; Dinges, 2008, 44–5.

84 Sabine Sander, *Handwerkschirurgen. Sozialgeschichte einer verdrängten Berufsgruppe* (Göttingen 1989).

85 In Prussia in the 1820s almost six times more non-academic and apprenticed persons offered their services than learned physicians, cf. Huerkamp, 1989, 58; for Bavaria see Stolberg, 1993, 6.

86 On the Hamburg policlinic cf. Frevert, 1984, 111.

87 On Bavaria cf. Stolberg, 1993, 8–11; on Austria cf. Joseph Knolz, *Darstellung der Medicinal-Verfassung in den k.k. Staaten Oesterreichs, in Beziehung auf den Wirkungskreis der Kreiswundärzte, der Civil-, Stadt- und Landwundärzte und der Landesthierärzte* (Wien 1829).

and industrial businesses. More than 60 per cent (1,760 consultations) of the physician-patient contacts (2,874) fell to 15 per cent of his patients (74 persons), with each of them seeing Bloesch more than ten times. Apart from this remarkable result, the structure of his clientele reflects the social make-up in his catchment area: in 1842, almost 60 per cent of the patients were merchants and tradespeople (339 of 571 occupations mentioned), another 12 per cent (68) were servants and labourers in their service.[88] That Bloesch himself was a member of the town bourgeoisie might well have contributed to the fact that more than 23 per cent (134) of his clientele belonged to the educated middle classes and included physicians, teachers, apothecaries, vicars, lawyers, engineers or employees of the municipal and cantonal administration. As one would expect, not many farm labourers (14) attended this practice (2.4 per cent).

The healer Wachter, who practised in the Swiss Canton of Thurgau at the same time, mostly called on his patients in their homes, as did Bloesch. In Wachter's case notes we find, if any, occupations primarily in the small craft businesses of the area. Hardly any mention of agricultural workers appears even though analysis of the villages suggests that they must have constituted the majority of his clients. In 1818, Wachter's notes mention a 'Baron' who sought treatment for a venereal disease, and in 1843 we read about several patients from the families of a town councillor and of a teacher – a clientele that one would rather expect to see in the practice of a licensed practitioner.

As with Bloesch, Ottenthal's clientele of the years between 1848 and 1899 probably also reflects the composition of the regional population. Even if Ottenthal mentioned his patients' occupation in less than 5 per cent of cases, in-depth study of St Peter, a village of around 500 inhabitants in his catchment area, reveals that where no occupation is mentioned, the patients mostly came from farming or other agricultural backgrounds.[89] Within this group the social

---

88 This applies also to the practice of the medical practitioner Heinrich Grotjahn (1794–1872) in the Harz region of Germany. According to the social statistics in the administrative office in Schladen around 45 per cent of his patients were farm workers and around 25 per cent were craftsmen between 1833 and 1841, cf. Regina Engel, 'Das Patientengut eines praktischen Arztes um die Mitte des 19. Jahrhunderts, gezeigt am Beispiel des Patientenjournals von Dr. Heinrich Grotjahn in Schladen am Harz' (Med. diss., University of East-Berlin, 1978), 64. In the practice of the Belgian homeopath van den Berghe the structure of the clientele reflected 'Ghent's population and society at large', cf. Baal, 2004, 120.

89 Cf. the pilot study by Elisabeth Lobenwein, 'Höfe- und Häusergeschichte von St. Peter in Ahrn. Die Patientinnen und Patienten (aus St. Peter in Ahrn) des Franz von Ottenthal' (preliminary analysis presented in an unpublished typescript, University of Innsbruck, 2012).

position was, however, a decisive factor for the frequency with which the academic physician was consulted. The number of consultations by the owners of larger farms and their families was, in this remote South Tyrolese valley, significantly higher than that by small farmers or the inhabitants of the modest holdings (*Soldhäusler*) that were too small to even support a single family. Compared to the patients from a farming or forestry background, the percentage of craftspeople, miners, innkeepers, as well as officials employed in administration and schools, declined sharply. Patients with health insurance, with the exception of the miners, also only constituted a minority due to the occupational structure in the valley. A relatively small percentage of people became Ottenthal's patients because of his office as publicly appointed doctor to the poor,[90] as health police officer (vaccinations, support with local epidemics) and as official court physician (post-mortems, statements of mental illness, investigation of accidents). These findings suggest that this extensive area with almost 10,000 inhabitants could not always be covered by the physician and the surgeon appointed by the community and that people were relying on the services of the private physician. In 1850 Ottenthal recorded 1,229 and in 1890, 1,856 patient contacts, showing that in the second half of the nineteenth century the demand for medical services in the country had risen sharply and that it was also on the rise in the places that already had medical provision. A quarter to a third of patients came from Sand, the town where the community doctor practised, 175 (122) from the Ahrn valley, the catchment area of the surgeon, and circa 1.5 per cent (10) from the district of Bruneck, a district that had a hospital in addition to resident physicians.[91] We see the reverse trend in the social composition of the clientele of the homeopath Bönninghausen in the city of Münster.[92] Even though he was consulted – as it says in an obituary – by 'all

90 The local poor relief funds paid 1 per cent of the fees.

91 Christine Roilo, 'Medizin auf dem Lande. Die Historiae morborum des Franz von Ottenthal', in Elisabeth Dietrich-Daum, Werner Matt und Hanno Platzgummer (eds), *Geschichte und Medizin. Forschungsberichte – Fachgespräche. Dokumentation zur internationalen Tagung "Geschichte der Medizin", 5. Dornbirner Geschichtstage, 9. Bis 12. Juni 1999* (Dornbirn 2001), 151–72: 160; a sample of 696 patients with family names starting with 'R' were selected from the years 1870–1879.

92 The majority of his patients from the district and the city of Münster travelled less than fifty kilometres to get to his practice which means that the main catchment area of Bönninghausen's homeopathic practice was comparable to that of an allopathic physician of the same period, cf. Balster, 1990, 117–21, 142–7; Robert Jütte, 'Samuel Hahnemanns Patientenschaft', in Martin Dinges (ed.), *Homöopathie. Patienten, Heilkundige, Institutionen. Von den Anfängen bis heute* (Heidelberg 1996), 23–45: 39 considers a radius of around forty kilometres to be the norm for the average 'allopathic' practice.

social strata',[93] of the patients who can be categorized according to occupation almost half belonged to the lower classes (420 of 910 occupational designations), 37 per cent (337) to the middle classes and 16 per cent (153) to the upper classes.[94] In his early years, patients from the lower social strata (299 of 541 designations, 59 per cent) even constituted the largest group. The same applies to the Belgian homeopath Gustave Adolphe van den Berghe (1837–1902).[95] The lower classes also comprised the main part of society in the provincial capital of Münster.[96] After ten years in practice, the occupations dominating in Bönninghausen's practice belonged, however, to the middle classes and included members of the trade and transport industries, of public administration as well as education, science, the church and the arts. Farmhands and maids, servants and publicans were represented at a rate of 6 per cent (10 of 166 occupational designations) and those working in farming at 5.4 per cent (9). The lower strata are increasingly underrepresented in the course of Bönninghausen's activity as a practitioner.[97] The development is similar with Hahnemann, in whose practice the members of the lower classes only made up 5 per cent between 1833 and 1835 (as opposed to 15 per cent from 1801 and 1803), while the percentage of aristocrats remained more or less stable at 6 per cent, as it had been at the beginning of his professional career.[98] Again, we see the opposite trend with the homeopath Clemens von Bönninghausen, the father of Friedrich. In his first years of practice, he treated mostly upper class patients, primarily aristocrats, with an increase in lower class patients occurring around the middle of the nineteenth century.[99] The trend in the homeopathic practices is clearly not linear and yields rather an ambivalent result.

93 Hermann Schnütgen, 'Sanitätsrath Dr. Friedrich von Bönninghausen†', *Allgemeine Homöopathische Zeitung* 58 (1910), 351–52.

94 On this and on the problem with assigning individual occupations to particular social strata cf. Wolfgang von Hippel, Ute Mocker and Sylvia Schraut, 'Wohnen im Zeitalter der Industrialisierung. Esslingen am Neckar 1800–1914,' *Esslinger Studien* 26 (1987), 47–180: 61–2.

95 Van den Berghe even offered special hours for the poor and treated many poor people free of charge, cf. Baal, 2004, 120.

96 Wolfgang Köllmann, 'Die Bevölkerung des Regierungsbezirks Münster im 19. Jahrhundert', *Westfälische Forschungen* 40 (1990), 195–222: 208.

97 Baschin, 'Choice', 2014.

98 No absolute figures are mentioned. Information on occupations could be established in this example for 67 per cent of entries, cf. Monika Papsch, 'Sozialstatistische Auswertung von Samuel Hahnemanns (1755–1843) homöopathischer Praxis von Dezember 1833 bis Mai 1835 anhand des Krankentagebuches "D38"', in Dietrich-Daum *et al.*, *Arztpraxen*, 2008, 129–45: 134–36.

99 Baschin, 2010, 162–69; for further reading see also Dinges, 2002, 8–18.

For the nineteenth century these findings, with the exception of Friedrich von Bönninghausen's homeopathic practice, show a clearly growing use of medical services by members of the lower social classes.[100] Even though the social groups of aristocracy, administration, clergy and educated bourgeoisie were still represented, the composition of the individual clienteles was progressively adjusting to the social structure of the catchment areas, becoming more richly textured. With the growing availability of physicians the geographical catchment areas of the practices also declined slightly. The bulk of the clientele now came from places which were not further than a day's walk from the practices.

But, as we can see from the entries of Hahnemann, Bloesch and Ottenthal, the frequency of contacts with physicians was still considerably higher for aristocrats and clients from the educated and property-owning bourgeoisie than for other occupational groups. Right up to the nineteenth century the social standing therefore remains a more important factor for differentiation than the gender of patients. Wherever the sources permit this, the conclusions regarding the representativeness of the individual patient groups therefore need to be compared with the corresponding population data.

The Therapeutic Relationship

While the parameters outlined above offer themselves to quantitative evaluation and the assessment of trends, analysis of the physician-patient-relationship is considerably more challenging. We will look into patients' motives for consulting a physician, their reasons for seeking treatment and the balance of power between physicians and patients as factors determining the therapeutic relationship. The problem is that all the patient statements we find in the practice records were written down by the physicians.[101] Not all record the patients'

100 Cf. Stolberg, 1993, 17, 25.

101 Unlike studies based on ego-documents, which include the patients' perspective, cf. Lachmund *et al.*, 1995; Otto Ulbricht, 'Supplikationen als Ego-Dokumente. Bittschriften von Leibeigenen aus der ersten Hälfte des 17. Jahrhunderts als Beispiel', in Winfried Schulze (ed.), *Ego-Dokumente. Annäherung an den Menschen in der Geschichte* (Berlin 1996), 149–74; Michael Stolberg, '"Mein äskulapisches Orakel!" Patientenbriefe als Quelle einer Kulturgeschichte der Krankheitserfahrung im 18. Jahrhundert', *Österreichische Zeitschrift für Geschichtswissenschaften* 7 (1996), 385–404; Stolberg, *Experiencing Illness*, 2011; Iris Ritzmann, '"Die der Welt und sich selbst zur Last sind". Behinderte Kinder und Jugendliche in der Frühen Neuzeit', *Traverse. Zeitschrift für Geschichte / Revue d'histoire* 3

descriptions as comprehensively as those of Götz and Ottenthal, who often wrote down literally what was said to them, as can be seen from the colloquial or German passages among the otherwise Latin entries. Götz would add the word *narrat* to denote what a patient said during the medical examination. *Credit* or *timet* are comments Ottenthal used to refer to the concerns expressed by a patient during history taking, as in the case of the miner T. from Prettau who feared he might have caught syphilis from his neighbour (*timet syphilim ex No 1785/huj*[*us*] *anni*).[102] Bloesch, for whom patient reports were highly significant, included comments such as 'she believes' or 'complains about' to indicate authorship.[103] While we do not find such indications in all practice records, most physicians noted at least how patients rated their condition or the therapy,[104] whether they followed doctor's orders, rejected them or asked for alternative treatments. These annotations ultimately reveal that the conversation between physician and patient remained a key part of the treatment.

(2006), 73–86; Martin Dinges and Vincent Barras (eds), *Krankheit in Briefen im deutschen und französischen Sprachraum 17.–21. Jahrhundert* (Stuttgart 2007); Ruisinger, *Patientenwege*, 2008; Christina Vanja, 'Supplikationen als Quelle der Patientengeschichte', in Arnd Friedrich, Irmtraut Sahmland and Christina Vanja (eds), *An der Wende zur Moderne. Die hessischen Hohen Hospitäler im 18. und 19. Jahrhundert* (Petersberg 2008), 163–72; Alexandra-Kathrin Stanislaw-Kemenah, 'Von der Hand Gottes berührt?! Krankheit, Alter und Armut im Spiegel von Bittgesuchen zur Aufnahme in Dresdner Hospitäler', in Cordula Nolte (ed.), *Homo debilis. Behinderte – Kranke – Versehrte in der Gesellschaft des Mittelalters* (Korb 2009), 225–44; Philipp Osten (ed.), *Patientendokumente. Krankheit in Selbstzeugnissen* (Stuttgart 2010); Susanne Hoffmann, *Gesundheit und Krankheit bei Ulrich Bräker (1735–1798)* (Dietikon 2005); Hoffmann, 2010.

102 Elisabeth Dietrich-Daum and Andreas Oberhofer, 'Medizingeschichte als Sozialgeschichte. Die Krankengeschichten des Südtiroler Landarztes Franz von Ottenthal (1818–1899), ein Interreg III A-Projekt. Österreich–Italien (2002–2007),' in Christoph Haidacher and Richard Schober (eds), *Bericht über den 24. Österreichischen Historikertag in Innsbruck* (Innsbruck 2006), 646–55: 653.

103 Cf. Lina Gafner: 'Administrative and Epistemic Aspects of Medical Practice: Caesar Adolf Bloesch (1804–1863)' (in this volume).

104 Cf. the case of Georg Grosch in the contribution on Fuchs: Stephanie Neuner and Karen Nolte: 'Bedside Training and Healthcare for the Poor in the Würzburg and Göttingen Policlinics in the First Half of the Nineteenth Century' (in this volume); on the special significance of patients' statements in Bönninghausen's practice cf. Marion Baschin: 'A Special Kind of Practice? The Homeopath Friedrich von Bönninghausen (1828–1910)' (in this volume).

Looking for the Right Doctor

The use of a physician's services is, as Martin Dinges pointed out, a 'valuable, but debatable indicator for medical behaviour.'[105] For consulting a doctor was (and is) only one way of dealing with illness. Self-medication is another. There has been hardly any research into the extent and practice of self-medication, and indeed the subject is only marginally addressed in this contribution.[106]

The search for the right doctor might lead patients to try healer after healer. All the records contain references to rival physicians or healers who had been consulted before the physician in question was seen. The respective notes show once again that patients, wherever possible, chose their doctor autonomously in the diversified medical market.

In asking about decisive factors for choosing a particular therapist it must be considered whether the patients had a genuine choice. When patients decided to see a physician, their choice often depended on a predetermined premise such as a physician's geographic accessibility or his position within a regional healthcare system, as in the case of welfare office referrals, advisory examinations or vaccinations. Apart from these factors, which were independent of the patient's own views, confidence in a doctor and his therapeutic approach is certainly a central determinant. Within the families and house communities or even the patient group that is referred to as 'bystanders',[107] previous confidence-inspiring experiences will often have informed the decision in favour of a particular physician. In all the practices[108] we find relatives of families who saw the physician in question as their 'family doctor.' The successful treatment of one family member often led to further requests for Magirus' services.[109] Götz could well be seen as a physician for the 'whole household.'[110] With both of them, however, this kind of 'all-round' treatment

105 Dinges, 2007, 303; *idem*, 2011, 31.

106 Gerhard Wilke, 'Die Sünden der Väter. Bedeutung und Wandel von Gesundheit und Krankheit im Dorfalltag', in Alfons Labisch and Reinhard Spree (eds), *Medizinische Deutungsmacht im sozialen Wandel des 19. und frühen 20. Jahrhunderts* (Bonn 1989), 123–40: 132 and, on self-help, also David Green, *Working-Class Patients and the Medical Establishment. Self-Help in Britain from the Mid-Nineteenth Century to 1948* (Aldershot 1985); Charles Rosenberg (ed.), *Right Living. An Anglo-American Tradition of Self-Help Medicine and Hygiene* (Baltimore 2003) and Baschin 2012, including bibliography.

107 Lachmund *et al.*, 1995, 195.

108 Due to the disparate recording systems it is rarely possible to determine how many families entrusted themselves to one particular physician.

109 Similarly Baschin, 2010, 177–86 and Steinke, 2008, 82–3.

110 Cf. Kinzelbach *et al.*, '*Observationes*' (in this volume).

applied chiefly to the wealthier patients.[111] In the nineteenth century practice records we find numerous references of this kind, also in relation to patients from other social strata, with physicians starting to become relatively firmly established as family practitioners.[112] The policlinics again constitute a special case because family members 'in need of treatment' often unintentionally caught the attention of a physician. Generally speaking, treating one family member seems to have provided a good basis for extending the medical services to the wider social environment, primarily the patient's relatives.[113]

As we will show in the last section of this contribution, the physician's willingness to listen to the ideas and wishes of his patients was an important factor in a patient's search for the right doctor.

Attempt at Analysing the Rationale for Treatment

What were the complaints, illnesses or afflictions that led patients to see a physician? Were they always severe and did patients really come 'too late' as physicians liked to claim? How can the complaints described be evaluated whilst taking the problem of the historical relativity of body concepts and syndromes into account? Research has so far dealt with categorizing complaints according to organs, parts of the body, disease entities or therapies.[114] The records investigated in this

111 Schilling *et al.*, 2011, 111; Hess, 2008, 98.

112 Lachmund *et al.*, 1995, 201.

113 This is confirmed by the case histories of all the doctors' practices investigated. For the policlinic cf. the arguments of the physicians according to Nolte, 'Versorgung,' 2010, 137–38. Wachter's services were also recommended among the members of some families, cf. Katharina Weikl *et al.*, in Gafner *et al.*, 2012, 76–94.

114 Cf. Johanna Bleker, Eva Brinkschulte and Pascal Grosse (eds), *Kranke und Krankheiten im Juliusspital zu Würzburg 1819–1829. Zur frühen Geschichte des Allgemeinen Krankenhauses in Deutschland* (Husum 1995); Johanna Bleker, 'Die medizinische Funktion des frühmodernen Krankenhauses. Zur Analyse des Diagnosespektrums der Inneren Abteilung des Würzburger Juliusspitals 1819–1829,' in Alfons Labisch and Reinhard Spree (eds), *"Einem jeden Kranken in einem Hospitale sein eigenes Bett." Zur Sozialgeschichte des Allgemeinen Krankenhauses in Deutschland im 19. Jahrhundert* (Frankfurt am Main 1996), 123–44; Iris Ritzmann, 'Schweizerische Medizingeschichte in Zahlen. Zum Kapitel "Krankheiten und Todesursachen in der Historischen Statistik der Schweiz"', in Christoph Döbeli, Christian Pfister, Hannes Schüle and Remigius Wagner (eds), *Landesgeschichte und Informatik* (Muttenz 1996), 106–12; Thümmler, 2004; Marion Baschin, 'Sozial- und medizingeschichtliche Untersuchung einer württembergischen Oberamtsstadt im 19. Jahrhundert,' in Marion Baschin and Andreas Kozlik (eds), *Studien zur südwestdeutschen Demographie. Die Sterblichkeit in Württemberg im 18./19. Jahrhundert und in Esslingen im 19. Jahrhundert*

context show, however, that the disparate concepts of illness in early modern practices do not really lend themselves to in-depth comparison. When Magirus described his patients as suffering from 'scorbutic' illnesses or *humores acres*, their ailments are not comparable with the stomach pain or colics recorded by Götz or with Glaser's reference to 'much bile and mucus'.[115]

What comparative criteria do we have then for these early modern practices? The differentiation between acute and chronic diseases depending on the duration of an illness allows us at least to state that the majority of consultations in Magirus' practice were due to acute conditions. Götz' notes, on the other hand, do not allow for any such differentiation. Although Glaser wrote down events that had occurred years earlier as possible causes of an ailment, this cannot be interpreted as chronicity in today's sense of the word. Categorization according to the causes of illness is equally problematic for early modern times.[116] The reasons put forward by patients usually referred to eating behaviours and generally did not differentiate between body and soul. In the course of the nineteenth century, the nomenclature became somewhat more unified due to medical standardization, despite the fact that a variety of schools and approaches continued to exist.[117] While Bloesch refrained from designating diseases in his journals, he used precise specialist terminology such as *febris nervosa mucosa rheumatica* when establishing causes of death. The academic physicians Fuchs and Ottenthal regularly used Latin or German specialist terms such as *carcinoma uteri, pneumonia* or *Lungenentzündung* (inflammation of the lungs). The records of these two doctors display a relatively standardized medical jargon, which makes it possible to compare them.[118] Not even Bönninghausen restricted himself to the description of

(Remshalden 2008), 11–139; Elisabeth Dietrich-Daum, 'Reporting Death. The Case of Austrian Tuberculosis Mortality Registration. Critique and Consequences for Historical Epidemiology,' *Prague Medical Report* 110 (2009), 146–58; Unterkircher, 'Männer,' 2012.

115 Cf. Kinzelbach *et al.*, 'Medicine' (in this volume); Sabine Schlegelmilch: '"What a Magnificent Work a Good Physician is." The Medical Practice of Johannes Magirus (1615–1697)' (in this volume).

116 Cf. Stolberg, *Experiencing Illness*, 2011, 82–85.

117 This process can be traced in the uniform nomenclature developed for the statistical assessment of causes of death, cf. Ritzmann, 1996; Robert Lee, Michael Schneider and Jörg Vögele, 'Zur Entwicklung der Todesursachenstatistik in Preußen im 19. und frühen 20. Jahrhundert', in Josef Ehmer, Ursula Ferdinand and Jürgen Reulecke (eds), *Herausforderung Bevölkerung. Zu Entwicklungen des modernen Denkens über die Bevölkerung vor, im und nach dem "Dritten Reich"* (Wiesbaden 2007), 7–16; Dietrich-Daum, 2009.

118 For analysis of medical diagnoses from the physicians' point of view see the chapter Kinzelbach *et al.*, 'Medicine' (in this volume); on the transition from eighteenth to nineteenth

symptoms in homeopathic terms, but added, in a third of his cases, the commonly used names for diseases. The non-academic healer Wachter tended to use descriptions such as 'thick blood', 'weak stomach' or 'full of bile', which did not correspond to any standard terminology and were probably phrases used by the patients themselves.

The dietetic interpretation of illness, which followed the tradition of the *sex res non naturales*, informed the entries even as late as the nineteenth century, particularly the frequent references to emotional upheavals.[119] Physicians often also noted down moral judgements on everyday behaviours, and there is no way of telling whether the assessment is that of the physician or that of the patients themselves. Fuchs wrote of a young woman that her illness was probably 'due to an excessive lifestyle, dancing.' Bönninghausen's records also list intemperance in eating and drinking as causes of disease. Fuchs mentioned 'strict work regimes', Bönninghausen emotions such as fear, fright, irritation or joy, or – like Bloesch – the weather. The causes of disease mentioned in Ottenthal's journal can be divided into the four categories of emotional upset, bad eating habits, accidents and infection.

While the information about the cause of diseases could have been provided by either the physician or the patient, descriptions of individual symptoms can be attributed more definitely to the patients. Fuchs' severely ill patients spoke of fear, pain, fever and loss of appetite, occasionally also of sleeping problems, fatigue and moodiness as well as tightness in the chest. Bloesch's patients referred most frequently to gastro-intestinal or breathing problems, pain and febrile conditions. Evaluation so far also shows increased incidence of respiratory and digestive disorders in Ottenthal's practice. The medication used by Ottenthal, on the other hand, reveals without doubt that patients came primarily because of pain. Evaluation of the data taken from Bönninghausen's practice records suggests the most frequent symptoms to have been menstrual problems, 'feverish conditions', loss of appetite, coughs, headaches and digestive problems. While a class-specific evaluation of Bönninghausen's practice divulges no notable differences in the range of complaints, analysis based on

century cf. Volker Hess, *Von der semiotischen zur diagnostischen Medizin. Die Entstehung der klinischen Methode zwischen 1750 und 1850* (Husum 1993).

119 Cf. Michael Stolberg, *Experiencing Illness*, 59–64; Elisabeth Dietrich-Daum and Elena Taddei, 'Curare – segregare – amministrare. L'assistenza e la gestione dei "mentecatti" in un contado del Tirolo. L'esempio del medico generico Franz von Ottenthal (1818–1899) di Campo Tures,' in Elisabeth Dietrich-Daum and Rodolfo Taiani (eds), *Psychiatrielandschaft / Oltre il manicomio* [special issue: *Geschichte und Region / Storio e Regione* 17, no. 2 (2008)], 83–102; Baschin, 2010, 210–21.

age groups produces more conspicuous results. Young children who were unable to verbalize their complaints suffered mostly, according to the statements of adults, from coughs, nausea and indigestion, rashes and sleep disorders. With older children 'fever' was often added. The problems are similar in the case of the homeopath's over 65-year-old patients, where one also sees an increased incidence of headaches, stomachaches and cardiac pain rather than rashes.[120] Women often sought out a doctor when they suffered from menstrual problems or sleep disorders.[121] The pattern of complaints in the homeopathic practice is therefore not extraordinary, but reads like a documentation of disease-related restrictions of everyday life. This pattern also does not deviate much from that apparent in the practices of the physicians Bloesch or Ottenthal[122] and of the lay healer Wachter. Furthermore, the doctors' notes show evidence that patients, contrary to what was usually said against them, tended not to wait to see a physician until the situation seemed hopeless, and rarely only saw him when their conditions were life-threatening, but mostly hoped for a cure or at least mitigation of their various troubling or weakening – and often chronic – symptoms.

Therapy as the Centre of the Physician-Patient-Relationship[123]

What was the physician-patient encounter like? Were patients generally submissive or did they rather comport themselves as clients demanding service? Can we discern any particular developments?[124]

Physicians, as we showed earlier, wrote down what their patients described in more or less detail. In the seventeenth century, the patients' feedback was seen as important, certainly by Magirus. Praise, as well as criticism, of his prescriptions had a bearing on his behaviour. If a patient claimed that a remedy worked well, Magirus would continue to prescribe it while he would rate unpleasant side-effects or ineffectiveness as failures. If patients rejected prescriptions – such as pills which had an unpleasant taste or bitter elixirs – Magirus would change the medication or replace the offensive ingredient.

120 Unterkircher, 'Männer,' 2012, 244 also states for male patients that they mostly saw a physician for diarrhea and cough as well as pain and other chronic ailments. On child patients cf. Unterkircher, 'Konkurrenz,' 2012.

121 Similar findings in Balster, 1990; Thümmler, 2004; Baal, 2004; Baschin, 2010.

122 Unterkircher, 'Männer,' 2012.

123 Stolberg, *Experiencing Illness*, 2011, 64–76.

124 Cf. Lachmund *et al.*, 1995, 123–6; Loetz, 1993; *eadem*, 1998.

Some of the children in his practice categorically refused any medication – a problem that other physicians and apothecaries also frequently experienced.[125] Götz even mentioned patients who discontinued treatment because they found the medicines repulsive or unfit for consumption. Fuchs described a severely ill woman who, even after repeated exhortations, refused to take creosote oil because of its smoky flavour.[126] He failed similarly with his instructions to resort to cooling remedies because the patients or their relatives could clearly not be convinced of the wisdom of this approach. This relatively strong position of policlinic patients may be explained by the young medical students' dependency on them for their instruction. The 'Regulations for the Policlinic Hospital' even specified explicitly that the patients' prejudice must not be met with violence. Treatment was to be gentle and humane.[127]

The nineteenth-century practice records also hint at patients refusing to take medication. Bloesch wrote: 'did not take the powder' and a number of Ottenthal's entries also suggest that patients rejected medicines because they did not like their taste. Ottenthal often referred to disagreements with his patients regarding their eating and drinking habits, choice of clothes or lifestyles in general. According to Bönninghausen's notes, some of his patients took traditional as well as homeopathic medicines, treated topical complaints themselves or even continued their treatment in a hospital or spa. These findings support previous research results which claim that the views of the physician met with scepticism on the part of the patients.[128] We find allusions to self-medication mostly in the doctors' notes about patients who asked for specific medicines, not because the physician had advised them, but because they were themselves convinced of their efficacy. The records also reveal evidence of patients and physicians discussing particular procedures. From Magirus, patients were able to purchase '4 flasks of *aquae antiscorbuticae*', a 'gargling water' and even snuff without consulting him first as a doctor. From Götz'

125 Iris Ritzmann, 'Medikamentöse Kinderbehandlung im 18. Jahrhundert,' *Arzneimittel-, Therapie-Kritik & Medizin und Umwelt* 41 (2009), 55–62 and 191–8.

126 Karen Nolte, 'Ärztliche Praxis am Sterbebett in der ersten Hälfte des 19. Jahrhunderts', in Walter Bruchhausen and Hans-Georg Hofer (eds), *Ärztliches Ethos im Kontext. Historische, phänomenologische und didaktische Analysen* (Göttingen 2010), 39–58: 52–3.

127 Staatsarchiv Würzburg, Regierungsbezirk von Unterfranken: Physikate, Würzburg: 6244: Physikat an der ambulanten Klinik Würzburg Stadt 1809–1821: 'Normativ für die Klinik der ärztlichen Besuchs-Anstalt,' §. 15; cf. Neuner *et al.* (in this volume).

128 E.g. Porter *et al.*, 1989; Loetz, 1993; Baal, 2004, 192–8; Robert Jütte, 'Die Arzt-Patient-Beziehung im Spiegel der Krankenjournale Samuel Hahnemanns', in Dietrich-Daum *et al.*, *Arztpraxen*, 2008, 109–27; Ruisinger, *Patientenwege*, 2008; Ritzmann, 2008, 115–20; Baschin, 2010, 354–68.

practice journal we gather that he was largely guided by the wishes of his clients. He sold proprietary medicines such as *Carlsche Pillen* or 'Stahlian Pills', powders and mixtures that were inspired by Pietist medicine. Glaser even tolerated patients who wore charms as long as they 'were not harmful or superstitious remedies.' The lay healer Wachter presumably ensured that some of his clients continued to return by selling them purgatives, in particular a laxative herbal wine. The patients of the homeopath Bönninghausen also tried to interfere with their treatment: they would buy powders to alleviate childbirth or relieve sore throats without necessarily consulting Bönninghausen beforehand. Some of the adult patients who consulted Ottenthal came with their own ideas and expectations. They would present the doctor with their view of how their bodies were to be cleansed from the *materia peccans* by humoral means and which laxative they preferred (*laxans desiderat/cupit*).[129] Ottenthal gave a special goitre salve to a man who kept coming back for it for years, as well as large amounts of opium, often when pressed by patients, to alleviate pain, for relaxation and as a sleeping draught.[130]

All in all, analysis of the practice records shows that the physician-patient-contact up until the nineteenth century consisted mostly of communication about complaints and the negotiation of possible therapies. Despite the application of the most varied diagnostic procedures,[131] the conversation remained the central part of the examination. Patients rarely came to their doctor in cases of great emergency, but often because they suffered from persistent symptoms. Not only did they describe their own interpretations, they also expressed expectations, which the physicians largely tried to satisfy. This accommodating attitude might well be the key to the success of their practices. Up until at least the late nineteenth century patients remained influential agents in a system of relatively balanced relationships.

129 In the decade between 1860 and 1869 laxatives were asked for in 11.5 per cent of consultations by men and in 7.1 per cent by women; cf. Alois Unterkircher, 'Die Praxis des Südtiroler Landarztes Franz von Ottenthal. Krankheitsspektrum und Gesundheitsverhalten seiner Patientenschaft um 1860', in Rainer Alsheimer and Roland Weibezahn (eds), *Körperlichkeit und Kultur 2004. Interdisziplinäre Medikalkulturforschung. Dokumentation des 7. Arbeitstreffens des "Netzwerk Gesundheit und Kultur in der volkskundlichen Forschung", Würzburg, 31. März–2. April 2004* (Bremen 2005), 215–37.

130 Lachmund *et al.*, 1995, 125; cf. Kinzelbach *et al.*, 'Medicine' (in this volume).

131 Cf. Kinzelbach *et al.*, 'Medicine' (in this volume).

Summary

How can these findings be interpreted in conclusion? Analysis has revealed firstly that, depending on the chosen period, the socio-geographical situation and the profile of the individual doctor's practice, the clientele varied widely in terms of gender, age and social rank. The consultation behaviour of men and women changed noticeably. Findings overall suggest that up until 1800 the gender distribution varied in the individual practices. There was a trend for women to be overrepresented in urban practices during the earlier period. But in general, from the mid-nineteenth century, they predominated – in towns as well as in the country, in allopathic as well as homeopathic practices. The absence of children, which was bemoaned by many physicians, did not apply to the practices under investigation. On the contrary: the percentage is consistently high while older patients remained underrepresented right up until the end of the period under investigation, even though their proportion increased in the individual practices during the course of the nineteenth century. In each of the nineteenth century practices investigated – and increasingly among the lower and middle classes – the physicians' services were used by several members of the same family.

We have found no evidence to support the thesis that up until the nineteenth century academic physicians were mainly consulted by aristocratic or wealthy bourgeois patients. The theory probably applies only to early modern urban doctors. In the practices examined here, from the middle of the eighteenth century, patients from all social strata went to consult physicians. The participation of members of the lower classes or from an artisanal, (proto) industrial or agricultural background clearly increased over time 'despite ubiquitous economic and cultural barriers.'[132] That the annual numbers of consultations per physician increased – despite the growing number of physicians available – suggests that for economically disadvantaged social groups also, the consultation of learned physicians became more common: in towns from the first half of the nineteenth century and in the country from the middle of the century. In addition, the individual findings reveal that, prior to the introduction of statutory health insurance for salaried persons, patients of more secure social standing consulted a physician considerably more frequently in the course of the year than lower class patients.

132 Thomas Schnalke, *Medizin im Brief. Der städtische Arzt des 18. Jahrhunderts im Spiegel seiner Korrespondenz* (Stuttgart 1997), 212; Loetz, 1998 qualifies these barriers.

While the patient structure clearly changed around 1800, the relationship between physician and patient continued without major changes from the seventeenth to the nineteenth century. The therapeutic encounter up until the end of the investigated period can be summarized as a negotiation process. Patients were discerning in their choice of healer and did not refrain from using rival services. They sought help for unpleasant symptoms such as indigestion, pain or fever, and only rarely in cases of emergency. Therapy was decided on after an exchange between a critical and autonomous client and the medical specialist who was generally willing to compromise.

CHAPTER 3

Daily Business: The Organization and Finances of Doctors' Practices

Philipp Klaas, Hubert Steinke and Alois Unterkircher

During the night of 6 December 1826, Franziska Gross, a midwife in the Austrian town of Innsbruck, was called to a heavily pregnant woman who had just gone into labour. When the woman, after suffering severe convulsions, lost consciousness the midwife sent two messengers to the nearby physician. But because 'there was no bell',[1] the attempt to find immediate help in the middle of the night failed. Driven by despair, the two messengers hurried on to the nearby military hospital where they were given tea and ointment for the dangerously ill woman. She died in the evening of the following day.

This example of a failure to establish contact between a physician and a patient or her relatives, tragic as it was, holds important clues as to how physicians organized their availability outside the usual practice hours. Ringing the night-bell should have woken the doctor. The search for medical help often ended outside the physician's house, however, as in the case above, when there was no bell, when no bell could be found or when the residents could not be aroused from their deep sleep.

The physician, on the other hand, could not – when packing his instrument bag and saddling his horse – be sure that he would still find an acute case of emergency at the house he was led to by the messenger that the distraught relatives had sent. In many cases medical intervention had become obsolete after the many hours taken to travel to some remote village. On 26 October 1889 the South Tyrolean country physician Franz von Ottenthal arrived in a farmhouse in Luttach in the Ahrn Valley, a village, which had 400 inhabitants at the time. He went there every day to see a 73-year-old man who had, for

* We thank the following authors of the individual projects in the DFG Research Network 'Medical practices (1600–1900)', on whose partly still unpublished results this contribution is largely founded: *Sabine Schlegelmilch* (*J. Magirus*), *Kay Peter Jankrift/Annemarie Kinzelbach/Susanne Grosser/Marion Ruisinger* (*Chr. Götz*), *Ruth Schilling* (*J.F. Glaser*), *Stephanie Neuner/Karen Nolte* (*Ch. Fuchs*), *Katharina Weikl/Alois Unterkircher/Iris Ritzmann* (*G. Wachter*), *Lina Gafner/Philipp Klaas* (*C.A. Bloesch*), *Marion Baschin* (*F.P. von Bönninghausen*), *Marina Hilber/Eberhard Wolff/Alois Unterkircher* (*F. von Ottenthal*).

1 TLA, Jüngeres Gubernium 1826, Sanität, Faszikel 2448, Zl. 1990: 'Schreiben des Kais. Königl. Guberniums von Tirol und Vorarlberg vom 21.1.1826 an das Magistrat Innsbruck'.

 | DOI 10.1163/9789004303324_005

some time, needed catherisation to void his bladder. But on this day the bladder had emptied itself – *proprio motu* – and the physician had made the long journey for nothing. The case history in the patient journal begins with the comment: *iter incassum* (journey in vain).[2]

The experience – documented in the sources – of not being accessible or of arriving too late, was certainly part of every physician's daily business. Supply and demand not only had to coincide in the socio-economic, but also in the spatial sense: in a defined place for treatment and examination. These locations of diagnostic and therapeutic encounters were constantly subject to renegotiation and communication between physician and patient. One such specific location of medical encounter – the sickbed – has been extensively investigated in recent decades with regard to the stronger position it placed the patients in from their own point of view.[3] More historical-medical research is needed on how physicians organized and scheduled their visits to sickbeds, which often lay far apart, within a normal working day.

Our comparative analysis will therefore not focus primarily on the biographies of physicians or on medical interventions, but ask instead direct practical questions regarding the everyday running of the doctor's practice: where did the contact between physician and patient mostly take place? How many patients did the practitioner or healer see on average in a day? How did physicians' practices change over time? How did physicians organize their daily routine between receiving patients in his own house, house calls and, in some cases, teaching activities? What were their earnings from all these activities? One might think that these questions are easy to answer, but they are not. Research has so far – and even that to a limited extent only – focused on financial questions. There is little information on other aspects of the physician's daily business. This contribution is concerned with the everyday routine of physicians and other healers, a perspective on which the recent research into doctors' practices has increasingly focused.[4]

---

2 '*iter incassum proprio motu ur. abiit*'. Südtiroler Landesarchiv (SLA), Nachlass Franz v. Ottenthal, Historiae Morborum 1889, HM 820/1889, entry of 26 May.

3 Cf. Nicolas Jewson, 'Medical Knowledge and the Patronage System in 18th Century England. A Sociological Analysis', *Sociology* 8 (1974), 369–85; Michael Stolberg, *Experiencing Illness and the Sick Body in Early Modern Europe* (Basingstoke 2011), 53–5, 64–76.

4 Cf. Martin Dinges, 'Arztpraxen 1500–1900. Zum Stand der Forschung', in Elisabeth Dietrich-Daum, Martin Dinges, Robert Jütte and Christine Roilo (eds), *Arztpraxen im Vergleich. 18.–20. Jahrhundert* (Innsbruck 2008), 23–61: 23–6.

The Doctor's Practice as the Space Where Treatment Takes Place

The *doctores medicinae* of earlier times did not need to see their clients in person in order to establish what pathological processes were going on in the body or to prescribe the necessary medication. Most physicians and healers were prepared to determine the cause of physical complaints by examining the patient's urine[5] and sometimes just from descriptions they received by letter.[6] In one out of ten cases the Nuremberg town physician Johann Christoph Götz learnt of a patient's symptoms through a third person and the extent of consultations by letter was also considerable. The frequent references, made by Franz von Ottenthal, to messengers who explained the course of illness or came to fetch medicines, show how wide-spread this kind of contact was, certainly in rural regions, up until the late nineteenth century. Our notion of the encounter between physician and patient must therefore, in spatial terms, be more flexible than it is today.

Comparison between the doctors' practices investigated within the Research Network has revealed a number of ways and locations for this contact to take place. What all practices had in common was that the patient's home and the doctor's house were the most frequent 'meeting places'. It was, however, only possible in very few cases to establish the actual percentages of house calls and consultations in the physician's own house. We can assume that these figures varied depending on the type of practice: town and communal physicians provided some of their services by calling on their patients, since this was part of their job description. Country practitioners who had a more extended catchment area might be delayed by a difficult, drawn-out birth and have to keep patients waiting for them to return home. It has been almost impossible to reconstruct such processes or journeys for any of the Research Network projects and the sources evaluated by them. The development from the practice of calling on patients to that of offering specific consultation hours, on the other hand, was described early on in medical historiography by Ingrid Vieler and others.[7] Based on their synopsis, it has long been assumed that early modern

5 Cf. Michael Stolberg, *Die Harnschau. Eine Kultur- und Alltagsgeschichte* (Köln 2009), 19–21.

6 Cf. Sévérine Pilloud, Stefan Hächler and Vincent Barras, 'Consulter par lettre au XVIIIe siècle', *Gesnerus* 61 (2004), 232–53; Martin Dinges and Vincent Barras (eds), *Krankheit in Briefen im deutschen und französischen Sprachraum 17–21. Jahrhundert* (Stuttgart 2007); Thomas Schnalke, *Medizin im Brief. Der städtische Arzt des 18. Jahrhunderts im Spiegel seiner Korrespondenz* (Stuttgart 1997).

7 Cf. Ingrid Vieler, 'Die deutsche Arztpraxis im 19. Jahrhundert' (Med. diss., University of Mainz, 1958).

physicians mostly visited their patients at home and that fixing times during which the physician received patients in his own house did not become customary until later in the nineteenth century. But the Basel physician Felix Platter (1536–1614) already had a room designated especially for seeing patients.[8] Brian Nance was able to prove that the London physician Théodore Turquet de Mayerne (1573–1654/55) was regularly available to his patients at his own house.[9] There are also entries in the practice journal of Magirus and Götz, which reveal that these early modern physicians had examination and treatment rooms in the libraries or studies of their houses. The example of the town physician Friedrich Glaser (1707–1789) even suggests that fixed times during which the physician hours could be consulted in his own house were a 'common phenomenon in pre-modern medicine'.[10]

The nineteenth century homeopathic physician Friedrich von Bönninghausen received almost all his patients in a specially adapted room in his house, the country doctor Ottenthal the greatest part of his clientele. At a slightly earlier date, on the other hand, the town physician Caesar Adolf Bloesch mostly made house calls and hardly ever examined patients in his house. Gottfried Wachter, the lay healer from Thurgau in Switzerland, visited almost all his patients in their homes, but we find the occasional entry in his journal that mentions messengers coming to collect medicines for their sick employers – as in the case of a publican from Bohlingen, a village near the German border, who sent someone to fetch a preparation to treat a dangerous concentration of bile in his body.[11]

The examples investigated do not contradict Vieler's conclusion that the transition from house calls as the primary kind of patient contact to consultations in the physicians house occurred mostly in the second half of the nineteenth century. Neither are they in opposition to the assumption that this process generally happened later in the country than in towns. But they warn us against interpreting this trend as a fixed and defined situation. Early modern physicians might have been interested in receiving – as Turquet de Mayerne

---

8 Felix Platter, *Tagebuch (Lebensbeschreibung) 1536–1567*, ed. by Valentin Lötscher (Basel 1976), 330.

9 Brian K. Nance, *Turquet de Mayerne as Baroque Physician. The Art of Medical Portraiture* (Amsterdam 2001), 24.

10 Ruth Schilling, 'Raum und medizinischer Markt. Die Praxis des Stadt- und Amtsarztes Johann Friedrich Glaser (1707–1789) in Suhl', in Lina Gafner, Iris Ritzmann and Katharina Weikl (eds), *Penning Patient's Histories – Doctors' Records and the Medical Market in the 18th and 19th Century* [special issue: *Gesnerus* 69, no. 1 (2012)], 36–53: 39.

11 Archives of the Institute and Museum of Medical History at the University of Zurich (MHIZ) MsH 17.4, folio 55.

did – their patients in an imposing study packed with books and scientific objects that would enhance the impression of their superior knowledge. Maybe Bloesch consciously gave preference to house calls because he found it important to have this intimate contact with his patients so he would get to know them better and be better able to respond to their personal circumstances.[12] In the case of the Canadian country physician James Langstaff (1849–1889) the same wish for close supervision of his patients' state might have been the reason for his consistently high rate of house visits (77 per cent).[13] In the case of Franz Peter Siffert, a physician from Freiburg in Switzerland, the house calls constituted only a quarter of all treatments in 1866 and were further reduced in subsequent years. He was probably keen to increase patient numbers by making sure that he was available at home.[14] Ottenthal also tried to avoid long absences on house calls so as to be more permanently available.[15]

At the policlinics in Würzburg and Göttingen the physician-patient contact – if the patients were poor members of the cities' lower classes – played a particular role because it was part of the medical training. It consequently took place in the houses of the patients, despite the fact that Göttingen boasted special treatment rooms for the indigent who were mobile, and even some hospital beds. The fact that students discussed and reflected on interesting cases with the physicians in the houses of the professors – and in absence of the patients – added an extra dimension to the treatment situation in this case.

Growing Pains

We know from numerous self-testimonies and ego-documents how physicians experienced the difficult and slow establishment of their practices. Against the social background of the seventeenth century it seemed natural for Johannes Magirus that 'young practici have little to do and need to practise for some time before they fully arrive in their profession'.[16] Magirus therefore, after having relocated to Berlin, sought close contact with the cathedral

12 Cf. Michael Stolberg, 'Patientenschaft und Krankheitsspektrum in ländlichen Arztpraxen des 19. Jahrhunderts', *Medizinhistorisches Journal* 28 (1993), 3–27: 5.

13 Jacalyn Duffin, *Langstaff. A Nineteenth-Century Medical Life* (Toronto 1993, repr. 1999).

14 Urs Boschung, 'Franz Peter Siffert (1837–1881), Landarzt in Kleingurmels und Wallenbuch', in, *Deutschfreiburg im Aufbruch* (Freiburg im Üechtland 1999), 36–60: 54; Dinges, 2008, 40.

15 Elena Taddei, *Franz von Ottenthal. Arzt und Tiroler Landtagsabgeordneter (1818–1899)* (Wien 2010), 90.

16 Johannes Magirus: *Alter und Neuer Schreibcalender für das Jahr 1648*. Nürnberg 1648, Bl. D4v.

parish and Court in order to accelerate the tiresome process of building up a new clientele.[17] Albrecht von Haller struggled with this in the eighteenth century, finding access to the important families difficult. Often he had to be content with being employed as subordinate to an established physician.[18] These problems continued to be wide-spread in the nineteenth century.[19] It took James Langstaff a decade to arrive at an average number of ten patient visits a day.[20] Of the eight physicians in Lower Bavaria, who at the request of a colleague communicated their annual patient numbers for the years 1861 and 1863, the one with the most recent qualification had the lowest number of patients.[21] Moving to a town that was closer to the more frequented trade routes occasionally improved the situation. The workload of the Belgian homeopath Gustave Van den Berghe, for instance, was hardly heavy during his first practice years in tranquil Zwevegem (1865–1869) where he saw 173 new patients per year on average, but once he had moved to Ghent in 1869, 'the quiet days were over'.[22] In this flourishing trade town the number of new patients climbed to an impressive 1,400 per year in a very short time. But even though most physicians, from early modern times to the nineteenth century, could only build up their practices very slowly and with great difficulty, there were also cases that demonstrate the opposite pattern: Ottenthal had as many as six consultations a day after only a month.[23] The homeopath Bönninghausen was able to take over a very successful practice from his father. Often – and this is the case with the extensive journals of Götz and Bloesch, the very early stages are not documented. Even though we can assume that growth was mostly based on word-of-mouth and family networking, research so far has little to say to the question as to how physicians actively promoted those processes.[24]

---

17 Cf. Sabine Schlegelmilch, 'Vom Nutzen des Nebensächlichen. Paratexte in den Kalendern des Johannes Magirus', in Klaus-Dieter Herbst (ed.), *Astronomie – Literatur – Volksaufklärung. Der Schreibkalender der Frühen Neuzeit mit seinen Text- und Bildbeigaben* (Bremen 2012), 393–411: 395–8.

18 Hubert Steinke, 'Der junge Arzt und seine Patienten. Albrecht von Hallers Praxis in Bern 1731–1736', in Elisabeth Dietrich-Daum *et al.*, *Arztpraxen*, 79–86.

19 Cf. Claudia Huerkamp, *Der Aufstieg der Ärzte im 19. Jahrhundert. Vom gelehrten Stand zum professionellen Experten. Das Beispiel Preussens* (Göttingen 1985), 119–25.

20 Duffin, 1993, 36, Table 2.2.

21 Stolberg, 1993, 15.

22 Anne Hilde van Baal, 'In Search of a Cure. The Patients of the Ghent Homoeopathic Physician Gustave A. van den Berghe (1837–1902)' (Rotterdam 2008), 48.

23 Taddei, 2010, 73.

24 On family networks cf. Schilling, 2012, 43–5.

Frequency of Daily Visits

Because of the difference in how the various records were kept, the doctors' practices featuring in our research only offer themselves to a limited extent to comparative quantitative analysis of visits per day. It is unlikely that Johannes Magirus wrote down consultations that did not bring much income or failed to promise a long-term 'patient attachment'. Gottfried Wachter never noted the day or month of treatments and, because of numerous insertions and cross-references, it is difficult to establish the overall number of individual consultations for each patient. The average figures also conceal the – sometimes considerable – fluctuations between the various days of the week, on some days there was as good as no demand for medical assistance.

Despite the methodological problems that are concealed behind the average figures, we can discern a developmental pattern, especially with those practices that were documented over a long period of time. A considerable number of physicians were able to increase their patient numbers in the course of their professional lives, often needing to adjust their growing workload to other activities such as teaching (Magirus), writing (Glaser) or political commitments (Bloesch, Ottenthal). With the Southern Tyrolese country physician Franz von Ottenthal, daily patient fluctuations are easy to quantify and can be seen as almost typical of practices that remained in the same location. Based on four samples taken in the early years of his practice, the time when it was becoming established, the flourishing years and the final years, respectively, we see first a steadily growing patient frequency which then sharply subsides after a peak in the 1880s (see Fig. 3.1).[25] Glaser's eighteenth-century practice developed similarly, with a growing number of patients in the beginning that then settled in at a relatively constant level. Bloesch's clientele took another climb towards the end of his working life due to the fact that the economic boom in his home town of Biel attracted a great number of new residents.

Jacalyn Duffin[26] and Marion Baschin[27] examined the long-term development of two practices – that of a Canadian general practitioner and that of a

25 On the development of Ottenthal's patient numbers between 1847 and 1899 cf. Andreas Oberhofer, 'Eine Landarztpraxis im 19. Jahrhundert am Beispiel der Ordination des Dr. Franz von Ottenthal (1818–1899)', in Dietrich-Daum *et al.*, *Arztpraxen*, 2008, 167–91: 181, diagram 1.

26 Cf. Duffin, 1993.

27 Marion Baschin, 'A Special Kind of Practice? The Homeopath Friedrich von Bönninghausen (1828–1910)' (in this volume).

physician/healer/ institution	patient contacts per day (average and approximate figures)	patient contacts per period of time (absolute figures)	sample years
Johann Magirus	0.5	184	1648 (Berlin)
(1651–1656)	0.7	266	1653 (Zerbst)
Johann Christoph	1	404	1717
Götz (1688–1733)	3	1,216	1721
	3	1,170	1726
Johann Friedrich	1	514	1753
Glaser	3	1,155	1760
(1707–1789)	3	1,179	1763
Caesar Adolph	8	2,869	1834
Bloesch	18	6,389	1842
(1804–1863)	13	4,801	1852
	15	5,548	1862
Franz von	7	2,763	1857
Ottenthal	8	3,045	1867
(1818–1899)	13	4,813	1887
	7	2,307	1897
Friedrich von	4–10	11,242	1864–1867
Bönninghausen	3–6	7,022	1872–1875
(1828–1910)	1–2	2,207	1879–1882
	1–2	2,063	1886–1889
Gottfried Wachter (1776–1861)	1.5	52 entries	27/11/1831–31/12/1831
Physicians for the	0.2–3	3,526	1837–1841
poor in two	2–3	4,907	1842–1846 (Würzburg)
policlinics (around	4–10	14,778	1838/39–1842/43
1840)	10–11	18,497	1843/44–1847/48 (Göttingen)

FIGURE 3.1 *Average number of doctor-patient contacts per day in the various doctors' practices of the seventeenth to nineteenth centuries.*

German homeopath – based on four samples taken at various stages in their career. Both authors concluded from these two very different practices that, after an initial period of slow growth, they had strongly increasing patient numbers, which began to decrease slightly following two decades of medical activity.[28] Both Langstaff and Bönninghausen, when they were older, noticeably reduced the numbers of patients they saw per day. While the decision in the case of the homeopath is likely to have been influenced by growing competition,[29] Langstaff felt that his strength declined with advancing years. The question of age must have been more relevant for country physicians, since they had a wider catchment area compared to town physicians. Around 1910, an elderly Tyrolese physician applied to be relocated to a nearby healthcare district to allow him to reduce patient numbers.[30]

The gradual process of establishing a doctor's practice can also be derived from the number of work-free days at specific stages in a physician's life. In the case of Ottenthal, we only find periods of rest – a few days several times a year – in the first half of his professional life, while later on there is not a single day when he did not see at least one patient, with the exception of special holidays such as Christmas Day. Bloesch also devoted much time to his patients, building up very close medical relationships. While the 500 to 800 patients treated by the Swiss physician in one year might not have been very different from the number of patients of other physicians in small towns at that time,[31] his number of consultations per year was, at 5000–6000, extraordinarily high. Most of the figures referring to the doctors' practices examined within the Research Network, including those for individual sample years, therefore support the hypothesis that, with a physician's growing expertise and continued practising in the same place, the trust of the population in his medical skills grew and patient numbers rose measurably.

This was apparently not the same with physicians who moved several times in their lives: they had to prove their skilfulness again and again in their new places of residence. The study by Stolberg mentioned earlier showed that out of eight physicians, the one who relocated to another town had patient numbers that were below average.[32] Not even a luminary like Samuel Hahnemann was safe from such fluctuations in the various places where he practised.[33]

---

28 Duffin, 1993, 35–6; Marion Baschin, *Wer lässt sich von einem Homöopathen behandeln? Die Patienten des Clemens Maria Franz von Bönninghausen, 1785–1864* (Stuttgart 2010), 321–3.

29 Cf. Marion Baschin, 'Warum besucht man einen Homöopathen? Eine Praxis in Münster im 19. Jahrhundert', in Lina Gafner *et al.*, 2012, 129.

30 Wolfgang Raffeiner, *Der alte Doktor. Eine wahre Erzählung aus dem alten Tirol* (Innsbruck 2002), 81.

31 Stolberg, 1993, 14.

32 *Ibid.*, 15.

33 Robert Jütte, 'Samuel Hahnemanns Patientenschaft', in Martin Dinges (ed.), *Homöopathie. Patienten, Heilkundige, Institutionen. Von den Anfängen bis heute* (Heidelberg 1996), 23–45: 27–9; Baschin, 2010, 322; Dinges, 2008, 45.

We generally observe, with almost all practices documented over a longer period of time, considerable fluctuations in terms of patient numbers and consultations between the beginning and final years of practice. The workload also varied substantially from one physician to the other, a phenomenon that cannot be fully interpreted without taking into account variable factors such as the population-to-physician ratio, population development or the regional healthcare provision by hospitals and almshouses. The striking dissimilarities between the average patient numbers for the two policlinics investigated, for instance, are due to the fact that Würzburg had a more tight-knit network of care institutions for the poor.[34] Poor patients in Göttingen had no alternative to the policlinic if they were looking for inexpensive healthcare. The drop in patient numbers in Bönninghausen's case may be explained with the appearance in Münster of three other homeopathic physicians who even promoted the scientific-critical approach to homeopathy that was seen as 'more modern'.[35] On the other hand, the country doctor Franz von Ottenthal faced problems of an entirely different nature: his catchment area included Tyrolese districts with a comparatively low population-to-physician ratio. In 1885, for instance, one physician was responsible for 2,940 inhabitants.[36] At that time Ottenthal was obliged to provisionally step in as community and forensic physician, since both posts were vacant at the time, and this also led to an increase in patient numbers. Country physicians, therefore, found it more difficult to understand the complaints about the 'oversupply' of doctors that physicians in towns and cities expressed in the late nineteenth century.[37] Setting up practice in the long Alpine valleys, which were far removed from any cultural centres, was not a particularly attractive option for young physicians.

These few examples illustrate the strong correlation between patient numbers and the location of a practice (whether it was in a town or in the country),

34 Cf. Stephanie Neuner and Karen Nolte, 'Bedside Training and Healthcare for the Poor in the Würzburg and Göttingen Policlinics in the First Half of the Nineteenth Century' (in this volume).

35 Marion Baschin, 'A Special Kind of Practice? The Homeopath Friedrich von Bönninghausen (1828–1910)' (in this volume).

36 Cf. Alois Unterkircher, 'Männer als Patienten. Krankheitsverhalten von Männern im ländlichen Raum in der zweiten Hälfte des 19. Jahrhunderts am Beispiel der Praxisjournale des Südtiroler Arztes Franz von Ottenthal' (PhD diss., University of Innsbruck, 2012), 80. In 1885 there was one physician for 1,836 inhabitants in the entire Tyrol.

37 For Austria cf. Elisabeth Dietrich-Daum, 'Die Klage der Ärzte. Marktprobleme und Professionalisierung in der zweiten Hälfte des 19. Jahrhunderts in Österreich', in Alois Unterkircher, with Reinhard Bodner and Kathrin Sohm (eds), *Medikale Kulturen* [special issue: *Bricolage* 5 (2008)], 202–21: 210–21.

the time a physician remained in one particular place, and his acquiring a 'good reputation'. Once physicians in rural areas had gained the trust of the population, their practices might soon be overcrowded, especially if the patients had no alternatives to choose from. Despite the greater physician-to-population ratio in towns, the physicians there benefited from the multi-layered social structure, with wealthy citizens who could afford to make use of a variety of medical services. Homeopaths especially filled therapeutic market niches in the towns, and the health officers and poor-physicians 'specialized' in patient groups that tended to be underrepresented in the practices of private physicians.[38]

Week Days and Market Days

The nineteenth century doctors' journals that included the dates of consultations reveal that weekends were a preferred time for seeing a physician. In the case of Friedrich von Bönninghausen, the four selected four-year periods suggest that Saturday was the favourite day for consultations in Münster. In the case of Caesar Bloesch in Biel, another urban practitioner, analysis according to the days of the week revealed that for April 1852 Sunday was, with an average of 15 patient contacts, among the most active days, even though he conducted over 20 consultations on some workdays in the same month. Receiving patients or calling on patients on Sundays and Holidays was not only habitual in the practices investigated here. Where the data available allows for interpretation, we see similar arrangements with physicians in the eighteenth century. The figures for the physician Friedrich Glaser, who practised in the German town of Suhl, show that he was busiest on Sundays (see Fig. 3.2), and for the Nuremberg physician Johann Christoph Götz, consultations on Sundays and holidays were also the rule rather than the exception.

In the few doctors' practices investigated so far the lack of work-free Sundays, as in the case of the Canadian practitioner James Langstaff, is sufficiently described.[39] Gustave Van den Berghe, on opening his practice, even advertised with the fact that he was available every day of the week.[40] Of the three nineteenth-century practices (Bloesch, Ottenthal, Bönninghausen) that of Franz

38 Karen Nolte, 'Die medizinische Versorgung der Göttinger Stadtarmen durch das Poliklinische Institut der Universität', *Würzburger medizinhistorische Mitteilungen* 29 (2010), 131–57: 133–8.

39 Cf. Duffin, 1993, 43–4.

40 Cf. Baal, 2008, 39.

Weekday of consultation in the years	Monday	Tuesday	Wednesday	Thursday	Friday	Saturday	Sunday
1753, 1760, 1763 (Glaser)	12,7 (365)	14,8 (426)	13,9 (400)	14,3 (410)	12,2 (351)	10,5 (302)	21,3 (611)
1857, 1867, 1877 (Ottenthal)	13,8 (1334)	13,4 (1304)	12,2 (1182)	13,3 (1287)	13,6 (1316)	13,9 (1347)	19,8 (1923)

FIGURE 3.2 *Percentage of patient contacts of the physicians Glaser and Ottenthal on the various days of the week, for three sample years (absolute total in brackets).*

von Ottenthal was visited most frequently on Sundays. It is possible that in his mostly agricultural catchment area in the Alpine Ahrn Valley, Sundays constituted one of few opportunities when rural employers allowed their employees to see a doctor without being acutely ill, usually after attending mass. Patients from the more remote areas had to walk for several hours to reach the physician's house. On four out of five Sundays in September 1867, Ottenthal saw 10 to 18 patients, two or even three times as many as on an ordinary weekday.

The practices of Ottenthal and Bloesch were particularly busy during the annual fair, those of the two homeopaths Bönninghausen senior and junior on Wednesdays and Saturdays, both market days in Münster.[41] People clearly made use of market days to obtain, besides objects of everyday use, medical preparations or to consult a physician about an affliction they might have endured for some time. We do not observe these Sunday peaks for all months or with any regularity, either in Ottenthal's practice or any of the others. Epidemics, absences due to other obligations and unquantifiable cases of emergency resulted in considerable fluctuations in patient numbers on the various days of the week.

We nevertheless note a general trend towards the increased use of medical services on Sundays and market days. For private practitioners of the eighteenth and nineteenth century it was certainly propitious to be available on such non-working days. It was not a coincidence that itinerant tradespeople and surgeons, who were not organized in guilds, offered – as an 'amorphous

41 Baschin, 2010, 331–2.

group of the most diverse provenience, competence and respectability'[42] their services, remedies and essences on just those days, a fact which often enraged the licensed healers in the area.[43] As part of the healthcare reform introduced from the eighteenth century onwards, it became increasingly difficult for the doctors and medicine sellers who travelled from fair to fair to obtain the necessary permits.[44] Considering that, up to the nineteenth century, the medical market was hardly monopolized, it would not have been wise for any private practitioner, who did not have the comfort of a number of lucrative private contracts, to keep his practice closed on days that might provide him with a large number of clients.

The Doctor's Practice in the Course of the Year

The later practices especially, with their serial journals and exact documentation of each individual physician-patient contact, provide an excellent basis for the comparative investigation into the distribution of contact frequency during particular months of the year. Quantitative evaluation of sample years in the journals of Glaser, Bloesch, Ottenthal and Bönninghausen showed that the four physicians were most often consulted in the spring (March, April, May), while consultations were very low in the summer (July and August) and reached their lowest levels towards the autumn. During the coldest season of the year consultation numbers gradually rose again (cf. Fig. 3.3). The town physician Götz could also reckon with more patients during the winter months and with extended 'free periods' in July and August. We see a similar picture with the Würzburg policlinic between 1838 and 1848. Even for the homeopath Clemens von Bönninghausen, there is statistical evidence of consultation peaks in the spring during his later years as a practitioner, between 1830 and 1860.[45]

Generally speaking, the methodical and differentiated synthesis of complaints and seasonal peaks should show whether the increase in patient numbers during the spring months coincided with a greater susceptibility to illness

42 Sabine Sander, *Handwerkschirurgen. Sozialgeschichte einer verdrängten Berufsgruppe* (Göttingen 1989), 51.

43 Cf. Christian Probst, *Fahrende Heiler und Heilmittelhändler. Medizin von Marktplatz und Landstraße* (Rosenheim 1992); Robert Jütte, *Ärzte, Heiler und Patienten. Medizinischer Alltag in der frühen Neuzeit* (München 1991), 29; Sander, 1989, 66–7.

44 Cf. Probst, 1992, 16–7, 128–9.

45 Cf. Baschin, 2010, 326–9.

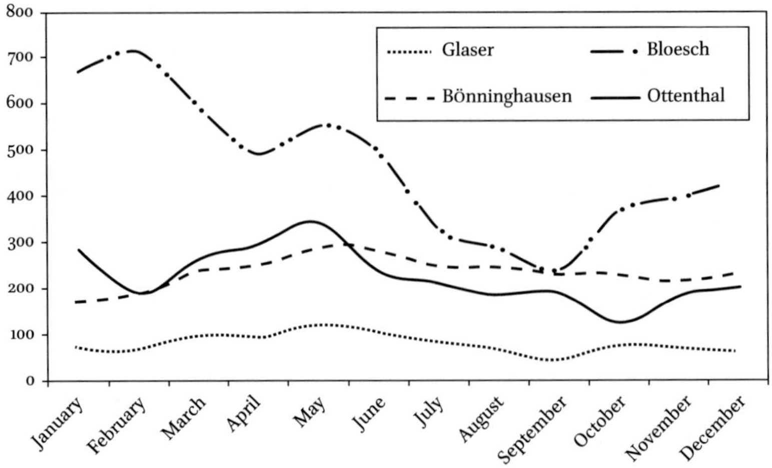

FIGURE 3.3 *Seasonal fluctuations in the number of treatments with Glaser (1753, 1760, 1763), Bloesch (1842, 1852, 1862), Bönninghausen (1864–1867) and Ottenthal (1857, 1867, 1897) (Median values of the sample years stated).*

due to people's constitutions being weakened at the end of the winter, especially elderly or very young patients. Around 1860, the medical statistician Friedrich Oesterlen concluded after studying numerous death registers 'that in the temperate zones of Europe the maximum number of deaths occurred on average towards the end of winter, especially in March, January and February, while the mortality was lowest in summer or early autumn'.[46] The atypical peak in Bloesch's practice in January and February is due to the severe typhoid epidemic of 1842 that led to a sudden increase in patient numbers. With seasonal fluctuations we must include basic conditions such as the weather or road conditions. It was not unusual for Ottenthal not to see some of his patients all winter, until the snow had melted and they were able to take upon themselves again the long journey to see their physician. In Langstaff's case patient numbers also rose suddenly in late spring, when the long and hard Canadian winters came to an end and the roads and paths were again free of snow and mud.[47]

But seasonal proneness to illness can only partly explain the similar annual distribution of patient numbers for town and country practices in regions with very different climates. What seems equally important is the especially high demand for medical treatment in the spring ('spring cures'). Barbara

46 Friedrich Oesterlen, *Handbuch der medicinischen Statistik* (Tübingen [2]1874), 304.

47 Cf. Duffin, 1993, 40.

Hansch-Mock showed in her investigation into the German-Swiss calendars of the nineteenth century that spring and autumn were considered favourable for prophylaxis against certain afflictions by means of bloodletting and purging.[48] These findings confirm the observations Robert Jütte made in the case of a family called Weinsberg in early modern Cologne. Twenty-five venesections were carried out in this family between 1550 and 1594, nineteen of them in May and three in September.[49] Analysis of the reasons for adult men between the ages of twenty-five and forty-two to consult Ottenthal revealed that around 11 per cent of all physician contacts in the 1860s alone were due to patients asking for purges or having 'laxative treatment'.[50] Study of the calendar designed by Johannes Magirus to show the dates favourable for venesection, cupping or the ingestion of laxatives might reveal a surprising continuity in the demand of patients for these treatments.[51]

In contrast to the increased demand for physicians in the spring, there was a drop in patient numbers during the summer months, which was certainly related to the rising workload of the rural population during harvest time. Urban physicians might have been less busy in the hottest time of the year because the wealthy citizens – including the physicians themselves – tended to withdraw to the country for their 'summer retreat'. In the cases of Bönninghausen and Bloesch we have evidence of private journeys during the summer months. Bönninghausen tended to take a break from his practice regularly in September.[52] Ottenthal as delegate for the rural districts of the Puster Valley was obliged to remain in Innsbruck during the parliamentary session (mostly in February and March) and was unable to see patients during that time.[53]

Daily Routine and Organization

The patient records only allow for indirect and imprecise conclusions regarding the daily routine and organization of practices. For Bönninghausen we have relatively reliable findings. We can assume that he – like his father – received

48 Cf. Barbara Hansch-Mock, *Deutschschweizerische Kalender des 19. Jahrhunderts als Vermittler schul- und volksmedizinischer Vorstellungen* (Aarau 1976), 19–40.

49 Cf. Jütte, *Ärzte*, 1991, 236, Table 3.

50 Cf. Unterkircher, 'Männer', 2012, 242, Table 3.3.3.a.

51 On the calendars cf. Schlegelmilch, 2012.

52 Cf. Baschin, 2010, 328; Marion Baschin, *Ärztliche Praxis im letzten Drittel des 19. Jahrhunderts. Der Homöopath Dr. Friedrich Paul von Bönninghausen (1828–1910)* Stuttgart 2014.

53 Cf. Taddei, 2010, 173–209.

patients for five hours every day, receiving 5–10 patients in that time.[54] The average length of his consultations was therefore – as Hahnemann also suggested – between 30 and 45 minutes.[55] It is possible that he – like his homeopathic colleague Gustave Van den Berghe in Ghent – saw poor patients between the hours of seven and nine and patients for 'special consultations' between 11 a.m. and 1 p.m.[56] Since he only rarely made house calls these must have been the main times for his practical medical activity.

With all other practices we must base our conclusions on vague statements and assumptions. Like all physicians who made house calls, Bloesch and Ottenthal must have made arrangements for follow-up visits directly with their patients. Since many of Wachter's treatments required intensive follow-ups he must have negotiated exact times for his next call. In serious cases, Wachter probably stayed with the patient until the end of the critical phase was reached. When in 1835 a woman from the small hamlet of Schlatt had a serious fall in her chamber, suffering 'an injury to her head and concussion' and 'lying unconscious for more than 8 hours with her eyes closed', the healer administered salves, inunctions and compresses.[57] In the patient journal we read: 'After ten hours had passed she recovered and gained consciousness again with open eyes'. The relevant journey entry is composed in a way that suggests that Wachter stayed with the injured woman for the entire period of time. Although we find occasional entries of this kind in the records we know little about how these three healers planned their visits, how much time they allocated to each patient or how long their visits lasted on average. We can only assume that the physicians with high patient numbers must have had precise time schedules. If the urban physician Bloesch called on fifteen to twenty patients a day in an average year, on twenty to thirty patients per day in the spring, on thirty to fourty during epidemics and on up to fifty patients on some days, he surely planned the duration as well as sequence of his house calls. He must have had a travel schedule like the one suggested by his fellow physician Heinrich Küchler (1811–1873) in the German city of Darmstadt:

> The physician who sets off early in the morning to call on thirty or more patients one after the other, must make sure he orders the names of his

---

54 These figures apply to the first decade of his practice; the numbers were considerably lower in later years.

55 Richard Haehl, *Samuel Hahnemann. Sein Leben und Schaffen*, 2 vols (Leipzig 1922), 412–3; Baschin, 2010, 332–3.

56 Baal, 2008, 39.

57 MHIZ MsH 17.13, folio 59.

> patients in a way that allows him to visit as many sick people as possible on one journey, without having to walk to and fro too much or retrace his steps. Only older physicians can know their routine so well that they can manage without such a written plan, and some will never be able to manage it. I personally use such a schedule for my own purpose.[58]

As an essential part of his medical activity consisted in his regular house calls, Bloesch was able to foresee on the previous day or even earlier, approximately what each day would look like. In Bloesch's case we find no indications to regular employment of an assistant as we do with Johann Friedrich Dieffenbach (1792–1847) who met his assistant in the houses of his patients. The assistant would prepare the operations which Dieffenbach, the director of the Berlin University Hospital, would carry out in the mornings.[59] For the policlinics in Göttingen and Würzburg we have plenty of reliable information because of the structures and standardization of routine processes that were characteristic of these institutions. The assistant doctors would meet with the 'clinicists' (students in practical training) – in the summer at 7 a.m., in the winter at 10 a.m. – in the clinic and go from there to the various city districts to visit the sick in their homes.[60] House calls were generally carried out in the mornings.

It is particularly difficult to draw any conclusions regarding the physicians' ordinary schedule in the early modern period, since physicians at that time – more so than in the nineteenth century – tended to hold official, political or academic positions as well. We do not know how much time Götz devoted to his editing tasks or Glaser to his duties as health officer. We do know that Glaser complained about having to draw up so many statistics that he did not have enough time to practise medicine.[61] But such laments might have been pure rhetoric.

The question as to the extent to which physicians were supported by family members or relatives can only be answered for individual cases. We have definite evidence that Götz' wife dispensed medicines in his absence. Bloesch referred in writing to a *famulus*. With Ottenthal we have reason to assume that

58 Heinrich Küchler, *Bemerkungen über die Erkenntniss der Krankheiten nebst einer Anleitung zur Führung einer geordneten Krankencontrole* (Darmstadt 1844), 100.

59 Cf. Wolfgang Genschorek, *Wegbereiter der Chirurgie. Johann Friedrich Dieffenbach. Theodor Billroth* (Leipzig 1983), 88.

60 Cf. Staatsarchiv Würzburg, Reg.v. Ufr.: Physikate, Würzburg: 6244: Physikat an der ambulanten Klinik Würzburg Stadt 1809–21: 'Normativ für die Klinik der ärztlichen Besuchs-Anstalt'.

61 Leopoldina, Archiv, Matrikelmappe Sign. 630, Johann Friedrich Glaser's letter of 9 July 1770 to Ferdinand Jakob Baier, who was president of the Leopoldina at the time.

he had some kind of assistance because of the different handwritings in some of his journals. So far, however, this assumption has not been confirmed by archival findings.

Medicine as the Main Occupation? An Outline of Medical Earnings

The management of the practices under investigation also included the financial aspects of the healthcare business. While from the 1980s onwards, scholars, especially from the English-speaking world have investigated the financial situation and business activities of physicians and their patients in the past,[62] the topic has so far found little interest with German-speaking authors.[63] The projects presented here can fill this gap to an extent. The picture arising from the information gathered nevertheless remains largely fragmentary. It can generally be said that the chance of earning a living with medicine varied depending on region and period, a fact that makes it difficult to draw up meaningful comparisons between individual practices.[64] The main reason why many questions regarding the cost-effectiveness of the practices – whether or not the practitioners were academically trained – could not be answered lies in the fact that in some practice records or account books we find only fragmentary or no records at all of earnings. In addition, it was difficult to reconstruct the personal circumstances of most physicians and this rendered it almost impossible to

62 For Great Britain cf. Irvine Loudon, *Medical Care and the General Practitioner 1750–1850* (Oxford 1986); Roy Porter, *Health for Sale. Quackery in England, 1660–1850* (Manchester 1989); Anne Digby, *Making a Medical Living. Doctors and Patients in the English Market for Medicine 1720–1911* (Cambridge 1994); Anne Digby, *The Evolution of British General Practice, 1850–1948* (Oxford 1999); for France cf. Jacques Léonard, *Les médecins de l'Ouest au XIXe siècle*. 3 vols (Lille 1978); Matthew Ramsey, *Professional and Popular Medicine in France, 1770–1830. The Social World of Medical Practice* (Cambridge 1988).

63 According to Robert Jütte, '"Und es sammelte sich ohne Verdruß von Seiten des Kranken in des Arztes Beutel". Samuel Hahnemann und die Honorarfrage', *Medizin, Gesellschaft und Geschichte* 18 (1999), 149–67: 150; cf. nevertheless Anette Drees, *Die Ärzte auf dem Weg zu Prestige und Wohlstand. Sozialgeschichte der württembergischen Ärzte im 19. Jahrhundert* (Köln 1988); Tilmann Walter, 'Ärztehaushalte im 16. Jahrhundert. Einkünfte, Status und Praktiken der Repräsentation', *Medizin, Gesellschaft und Geschichte* 27 (2008), 31–74.

64 Cf. Irvine Loudon, 'A Doctor's Cash Book. The Economy of General Practice in the 1830s', *Medical history* 27 (1983), 249–68; Loudon, 1986, 100–13, 249–66; Ramsey, 1988, 56–7, 124; Digby, 1994, 135–96; Francisca Loetz, 'Andere Grenzen. Faktoren ärztlicher Inanspruchnahme in Deutschland 1780–1830. Empirische Ergebnisse und methodologische Überlegungen', in Thomas Schnalke and Claudia Wiesemann (eds), *Die Grenzen des Anderen. Medizingeschichte aus postmoderner Perspektive* (Köln 1998), 25–48.

establish the role of their medical work as a source of income. Our knowledge is even patchier when it comes to the financial situation of the patients or to the question of how much their health was 'worth' to them. It is clear that patients in the past often had a clear idea of what it would cost them to consult a healer. Their choice of therapist was therefore generally co-determined by financial calculations and their economic behaviour tended to have a strong influence on the thriving of a doctor's practice.[65]

One way in which physicians made sure they were not solely dependent on their income as medical practitioners was to find other economic opportunities.[66] This proved particularly advantageous if the business in question could be combined with their activities as healers. Johannes Magirus, while he practised in Zerbst, also held the chair for physics and mathematics at the local *Gymnasium Illustre*, which does not mean that he taught subjects outside his own sphere of knowledge, since physics was a fundamental part of medical training.[67] Magirus' private practice continued to play an important, if varying, part in his professional career, as we can see from his earnings in 1652. As health officer in Zerbst and professor at the *Gymnasium* Magirus earned 140 thalers. Private lectures and the accommodation of students in his own house gave him a further 31 thalers, and, on top of that he earned 50 thalers by editing a medical 'Old and New Writing-Calendar'.[68] All in all, Magirus earned in that year 221 thalers from his non-medical activities alone. His income from medical practice during that period amounted to 149 thalers and 18 groschen,[69] which means that only 40 per cent of his annual income was from medical practice. Magirus' social ascent and the subsequent improvement of his financial situation after having left Berlin made him increasingly independent of practical medicine as a means of earning a living.[70]

The eighteenth- and nineteenth-century sources also illustrate that physicians were happy to try other income opportunities as long as they seemed promising or if they offered themselves as a result of various social affiliations. Johann Friedrich Glaser[71] combined his medical activities, and added to his income, by taking on an official post in the town of Suhl in Electoral Saxony.

65 Cf. Loetz, 1998, 32.

66 Cf. the examples in Walter, 2008, 41–3.

67 Cf. Sabine Schlegelmilch, 'Johannes Magirus. Stadtarzt in Zerbst (1651–1656)', *Mitteilungen des Vereins für Anhaltische Geschichte* 20 (2011), 9–30: 22.

68 Cf. Schlegelmilch, 2012.

69 Cf. Universitätsbibliothek Marburg, Ms. 96, S. 231–3 (1652).

70 Cf. Schlegelmilch, 2011, 14–6 and Lötscher, 1976, 519–36 on Felix Platter (see note 8).

71 On Glaser's biography cf. Ruth Schilling, Sabine Schlegelmilch and Susan Splinter, 'Stadtarzt oder Arzt in der Stadt? Drei Ärzte der Frühen Neuzeit und ihr Verständnis des städtischen Amtes', *Medizinhistorisches Journal* 46 (2011), 99–133: 118–24.

While he treated his patients and provided them with medicines, he had the right, as health officer of Suhl, to sell beer and charge a fee for supervising the examination of lay-healers. Being well acquainted with the utilitarian thinking of the Enlightenment, Glaser published smaller scientific writings and even sold a fire-proof coating for wood that he manufactured himself. As he was the son of a wealthy executioner's family, it was, however, probably not financial necessity that induced him to pursue all these commercial activities, but an attempt to raise his social status and make up for his family background through such official and scientific aspirations.[72] The management of the medical practice therefore implied considerations other than the need for earning a living. The 'advantage' or 'gain' the individual historical agents had from their non-medical work can therefore not be quantified – even if it existed.

While the financial situation of Franz von Ottenthal was probably more than satisfactory, not least due to his income as a medical practitioner, it was not this income that was mainly responsible for his life-long prosperity. The fees he charged his clientele may have added up to a sizeable sum at the end of the year, but the notes do not reveal how much of this was actually paid to him.[73] What is certain is that Ottenthal's family owned several properties as well as land and that they would have been able to live on the income from those estates. Ottenthal, moreover, acted for many years as a money lender, making money available not only to private persons, but also considerable sums to his local commune.[74]

We can say in summary that medical activities in a wider sense encompassed, in the cases investigated, more than the treatment and cure of the sick. Physicians were often not only medical practitioners, even if their medical work was what they mainly identified themselves with. In parallel they worked as writers, sold medicines, officiated as health officers or university lecturers, lent money or were land-owners, drawing an income from a variety of sources. Considering the fluctuations – short- or long-term – that the individual doctor's income or the economic situation of the profession at large was subject to, it

72 Cf. *ibid.*, 124–5.

73 Cf. Taddei, 2010, 126.

74 Cf. *ibid.*, 65–6. For further examples on land ownership of physicians cf. Duffin, 1993, 53–6; Charles G. Roland and Bohodar Rubashewsky, 'The Economic Status of the Practice of Dr. Harmaunus Smith in Wentworth County, Ontario', *Canadian Bulletin of Medical History* 5 (1988), 29–49: 38 and of surgeons cf. Willem F. Daems, *Johann Anton Grass von Portein, 1684–1770. Arzt, Chirurg, Zahnarzt, Harndiagnostiker, Pharmazeut, Viehdoktor und Dorfpolitiker. Ein Beitrag zur Kultur- und Medizingeschichte des Domleschgs und Heinzenbergs im 18. Jahrhundert* (Chur 1985), 237–42; Sander, 1989, 88–91.

certainly made sense to exploit other sources of income.[75] But even if the physicians had a relatively secure existence or even achieved a degree of prosperity, the 'business' of medicine never became a mere side-line for them. It is therefore worthwhile to look more closely at the business practices of individual physicians and patients and at medical practice as a 'means of subsistence'.

Business Transactions between Physicians and Patients

It is interesting that we have so far not been able to find evidence of private physician's contracts in any of the practices investigated. According to sources surviving for other physicians, such contracts frequently referred to the treatment of all illnesses of a patient or a whole family during the year against a payment of a fixed salary. There are a number of conclusions we can derive from this fact that seems so inconspicuous at first glance, with regard to the managing of a practice as well as to the doctor-patient-relationship. At the practical level it shows that the physicians examined in this research project tended to have their own modes of bookkeeping and billing. Even if one or other of them felt little inclination or skill for that kind of work,[76] they all knew that it was a necessary job. Their administrative work must have been greater than it would have been with private contracts. If a physician, instead of drawing a fixed amount, had to charge individual fees to each patient, he had to record his services in greater detail as a means of legitimisation. In addition to that, the lack of private contracts in the practices under investigation might corroborate Claudia Huerkamp's hypothesis that research had for a long time overemphasized the social exclusivity of the medical clientele.[77] While such a permanent appointment, almost like a patronage, of physicians by members of the bourgeoisie or aristocracy was regarded in earlier research as the most common form of doctor-patient relationship,[78] our research projects suggest a wider picture: in almost all practices the clientele included not only the social

75 Cf. Duffin, 1993, 49. Changes in the eigteenth and nineteenth century are indicated by Loudon, 1986, 258 and Drees, 1998, 229–36.

76 For Langstaff's often sketchy bookkeeping and the help given by his son cf. Duffin, 1993, 46–51.

77 Cf. Claudia Huerkamp, 'Das unterschiedliche Verhältnis von Arzt und Patient in der Krankenhauspraxis und der privaten ärztlichen Praxis im 19. Jahrhundert', in Peter Schneck and Hans-Uwe Lammel (eds), *Die Medizin an der Berliner Universität und an der Charité zwischen 1810 und 1850* (Husum 1995), 254–68: 255–6.

78 Cf. Huerkamp, 1985, 27–8, 137; Claudia Huerkamp, 'Ärzte und Patienten. Zum strukturellen Wandel der Arzt-Patient-Beziehung vom ausgehenden 18. bis zum frühen 20.

elite, but many members of the middle and even lower classes, that is to say people who could not afford to employ a physician all year.[79] And, as we can see from the examples of Magirus and Glaser, early modern physicians were well able to survive without tying themselves to a small and exclusive group of clients, as long as they had other sources of income.

The sources evaluated here suggest that the bookkeeping in medical practices from the seventeenth to the nineteenth century was entirely left to the individual physician's approach and preference. Questions concerning the actual financial administration of physicians in the past can therefore only be answered if these sources, which are very individual in some cases, are amenable to evaluation. Most of the projects are also limited because of a lack of sources: in only two of the practices investigated within the Research Network (Wachter, Bönninghausen) account books could be found, and they are still in the process of being evaluated. The projects in this publication consequently had to content themselves with examining partial aspects.

As the examples of Johannes Magirus and Franz von Ottenthal show, it is especially the mode of recording that makes interpretation difficult: Magirus' *Diarium*, for instance, lists the names of patients and the treatments given, but the respective dates are often missing so that an in-depth investigation of the services used when and by which patients, is impossible.[80] In addition, the *Diarium* changed its purpose over time: while it initially combined the assessment of patients and treatments with the function of an account book, the financial notes disappeared in later years – probably due mostly to the decline of the formerly lucrative medicine sales. While still in Berlin, Magirus tended to write down prices next to the prescriptions he issued or he recorded when patients paid directly for the medicines. Such notes are missing altogether for his time in Zerbst. The impossibility of clearly allocating medical services and treatment times to individual patients is a problem also in the case of Ottenthal. The Tyrolese physician added a kind of 'account column' to his patient journals where he entered the fees he had added up. While this way of recording can easily be decoded for individual treatments, it becomes more difficult if other treatments are noted down under the same name: in such cases Ottenthal continued to add amounts to the 'account column', crossed out others, included subtotals and rough calculations, without showing which fee related to which

Jahrhundert', in Reinhard Spree and Alfons Labisch (eds), *Medizinische Deutungsmacht im sozialen Wandel des 19. und 20. Jahrhunderts* (Bonn 1989), 57–73; Drees, 1988, 229.

79 Cf. Huerkamp, 1995, 256.

80 Cf. Schilling *et al.*, 2011, 107.

service. But despite this 'cumulative' bookkeeping, Ottenthal did not seem to lose sight of outstanding or settled accounts.

In principle not even the well-off physicians could afford to neglect the accounting side of their medical businesses, if only to avoid making losses in their treatment of patients. The Biel physician Caesar Adolf Bloesch is a good case in point for this tension between conscientious monitoring and personal wealth. Coming from a respected bourgeois family, Bloesch not only established a flourishing practice in his home town, he also held, throughout his professional life, a number of administrative and honourable posts. A tax report from 1859, four years before his death, sets the value of Bloesch's property and capital at 179,000 Swiss Francs – an impressive sum which was nonetheless exceeded by many other citizens of Biel.[81]

The consultation fees he added up in his journal in the same year were not exactly low at 8,000 Swiss Francs. Professors at Bern University were paid 2000 to 3000 Francs per annum at the time.[82] We can nevertheless assume that Bloesch had a number of patients who took their time with paying their doctor's bills. A list of open accounts in the 1832 journal mentions 19 debtors and unpaid fees to the amount of 182 Swiss Francs, the equivalent of two months' salaries for the young physician.[83] The oldest bills were, at the time, from four years previously. Bloesch obviously did what many physicians did and carried open accounts through the years rather than write off the debt.[84] The monthly accounts in the patient journal reveal that each year only a small percentage of doctor's bills were paid immediately. The majority of patients would have received their bills for the calendar year, as was habitual in other practices too.[85] We can assume that the physician was paid the greatest part of the money due to him. In the tax accounts which were based on self-declaration Bloesch stated an income of 5,600 Swiss Francs for the year 1859. Assuming that the conscientious physician had stated his actual income and that he had earned this income from medical practice, he must have collected 70 per cent of his bills. It was also typical for physicians of his standing at the time to excuse some of the poorer patients from payment. The fees Bloesch charged

81 Cf. Stadtarchiv (StA) Biel, Tellrodel 1859, S. 303–4.

82 Cf. StA Biel, 4 A 1, Tagebuch, vol. 50 (1859), 271 and the references in Pietro Scandola (red.), *Hochschulgeschichte Berns 1528–1984* (Bern 1984), 370.

83 Cf. *ibid.*, vol. 3, 1832, 1.

84 Cf. Roland *et al.*, 1988, 39; Duffin, 1993, 48–9.

85 Cf. Paul Weindling, 'Medical Practice in Imperial Germany. The Casebook of Alfred Grotjahn', *Bulletin of the History of Medicine* 61 (1987), 391–410; Jutta Wolff and Hans-Peter Wolff, 'Das Profil einer ärztlichen Allgemeinpraxis im Jahre 1862', *Das Deutsche Gesundheitswesen. Wochenschrift für die gesamte Medizin* 34 (1979), 568–71.

were based on price lists for medical treatment: in 1841 a house call within the precincts of Biel cost 75 rappen; consultations in the physician's house were cheaper at 50 rappen. Writing out a prescription cost 40 rappen.[86] For poorer members of the community it was quite expensive to see a doctor: industrial cotton workers in Zurich earned between 66 rappen to 1.80 francs a day depending on their qualification.[87] The prescription alone cost as much as two pounds of bread.[88] This was not dissimilar from earlier centuries. The fees charged for individual house calls by Bloesch's late eighteenth century compatriot Louis Odier in Geneva, corresponded to the amount an ordinary worker would take home after a day's work or to the cost of six pounds of bread.[89] Even earlier physicians expected a fair amount of money for their services. The clients of Johannes Magirus had to pay 12 *groschen* or half a *thaler* for a simple uroscopy. Regular patients might end up with annual treatment and medicine bills of up to 8 *thalers*, an amount that would buy them six live rams or almost an entire ox.[90]

These figures illustrate how, right into the nineteenth century, patients, especially if they suffered from long-term illness, could easily find themselves in financial straits if not serious hardship or decline into poverty.[91] It is, nevertheless, not altogether correct to conclude, as has often been done, that academic physicians of earlier times were unaffordable for most people and that they were therefore rarely consulted.[92] There were plenty of options, some of them firmly institutionalized, to help the sick overcome their financial plight, even in the early modern period.[93] The support funds for craftspeople and

---

86 See for instance the accounts for April 1841 in: StA Biel, 4 A 1, Tagebuch, vol. 19 (1841), 198–202.

87 Cf. Albert Hauser, *Schweizerische Wirtschafts- und Sozialgeschichte. Von den Anfängen bis zur Gegenwart* (Zürich 1961), 318.

88 Cf. *ibid.*, 320.

89 Cf. Philip Rieder, 'The Physician Louis Odier and the Medical Market in Geneva (1774–1817)', *Gesnerus* 69 (2012), 54–75: 64.

90 Cf. Sabine Schlegelmilch, *Die ärztliche Praxis des 17. Jahrhunderts – Johannes Magirus (1615–1697)* [scheduled for 2015].

91 On the historical investigation of the complex 'Sickness and poverty' cf. the detailed survey in Martin Krieger, *Arme und Ärzte, Kranke und Kassen. Ländliche Gesundheitsversorgung und kranke Arme in der südlichen Rheinprovinz (1869 bis 1930)* (Stuttgart 2008), 40–51.

92 Cf. Ramsey, 1988, 118.

93 Cf. Jütte, *Ärzte*, 1991, 195–204 and also Annemarie Kinzelbach, *Gesundbleiben, Krankwerden, Armsein in der frühneuzeitlichen Gesellschaft. Gesunde und Kranke in den Reichsstädten Ueberlingen und Ulm, 1500–1700* (Stuttgart 1995).

servants gave employers, especially in the country, relatively concrete instructions regarding their duty of support in the case of illness. And many of the poorer patients benefited from support within their own families.[94]

We know from past research that financial transactions between healers and patients are very instructive, not only with regard to the financial aspects of healthcare.[95] This is exactly what makes the in-depth evaluation of the account books researched within the Project Network particularly interesting. As Francisca Loetz was able to show, on the basis of the debtors' journal of the surgeon Gramp in Baden, almost a fifth of the debtors did not have to pay for the services they had received from the therapist because they were paid for by relatives or employers.[96] The willingness of some family members to pay for part of their relatives' medical costs by offering services, underlines how important a person's integration into a social network was.[97] Even though detailed results, as they exist for Langstaff and Gramp,[98] on the business behaviour of physician and patients are not yet available for the practices investigated here, the patient journals do reveal a broad spread. But it is difficult to discern trends or 'normal cases', as is apparent from the physicians' willingness to accept services or labour instead of money, an approach that was especially widespread in rural areas. It was a way for physicians to adapt to the lack of solvency in their clients. In the case of Ottenthal, alternatives such as payment in kind or in instalments or pawnbroking seem to have been quite common.[99] Johann Friedrich Glaser was not in favour of such methods and accepted them less and less: while in 1750 still two of 819 patients managed to negotiate alternative ways of paying, there were no instances of this in 1760 or 1763. We consequently find strictness as well as generosity in response to patients' individual payment wishes. Ottenthal's willingness to be accommodating must not hide the fact that he did not consider giving his services away for free. Some journal entries included the Latin word 'nil' or comments such as 'lost' to show that no payment had been forthcoming. His displeasure about

94 Cf. Ute Frevert, *Krankheit als politisches Problem 1770–1880. Soziale Unterschichten in Preußen zwischen medizinischer Polizei und staatlicher Sozialversicherung* (Göttingen 1984), 154–5; Gerhard Siegl and Guenther Steiner, *Ja, jetzt geht es mir gut…Entwicklung der bäuerlichen Sozialversicherung in Österreich* (Wien 2010), 46.

95 Cf. also Dinges, 2008, 50–51.

96 Cf. Francisca Loetz, *Vom Kranken zum Patienten. "Medikalisierung" und medizinische Vergesellschaftung am Beispiel Badens 1750–1850* (Stuttgart 1993), 212.

97 Cf. Loetz, 1993, 212 and Jütte, *Ärzte*, 1991, 195–204.

98 Cf. Duffin, 1993, 46–56 and Loetz, 1993, 209–12.

99 Cf. Taddei, 2010, 124.

tardy debtors is noticeable from terse remarks such as 'does not pay'.[100] This attitude is understandable if one bears in mind that many physicians had to register high losses and that even wealthy patients were sometimes reluctant to pay their bills.[101] But physicians could also express their loyalty to patients and secure their goodwill if they did not appear too petty about payment. A certain financial bolster was of course prerequisite to such generosity. Johann Christoph Götz seems to have been able to rely on such a financial cushion from an early stage, since he often proved generous in financial matters. He would sometimes enter outstanding totals in distinctive red ink into his practice record, specifying if and how the patients had paid their debts. Of the 242 such entries we find in 1721, 110 – or just below 45 per cent – bear the symbol '//', which signifies that Götz had waived the debt. A small proportion of these clients were day labourers, maids, farmhands or orphans, people, in other words, who were always in danger of falling into poverty. The majority of them were people who were close to Goetz, such as relatives, friends, neighbours or some members of the aristocracy. Especially with the prestigious aristocratic patients, the physician might have hoped for other advantages in return for his generosity.[102]

The question as to whether and to what extent physicians adhered to official medical tariffs has not yet been sufficiently investigated for most of the projects.[103] For even if such recommendations existed and could serve as guidelines for financial transactions between therapist and patient, we still need to investigate the economic and social effect this had on patients who sought medical help. It was possible for the practice of Alfred Grotjahn to demonstrate that he referred to the nomenclature of the official lists in 61 per cent of his services, but the fees he charged only coincided with those tariffs in 36 per cent. The normative power of the medicinal tariffs was obviously more binding

100 SLA, Nachlass Franz v. Ottenthal, Historiae Morborum 1880, HM 1555/1880, entry of 1 August.

101 Cf. Duffin, 1993, 49–50; Roland *et al.*, 1988, 45–6.

102 Cf. Kay-Peter Jankrift, Annemarie Kinzelbach and Marion Ruisinger, 'Ernst von Metternich (1656–1727). Ein patientenzentrierter Einblick in den medizinischen Markt um 1720', in Lina Gafner *et al.*, 2012, 12–35: 29–32.

103 Cf. Jütte, 1999, 154. Regarding the context cf. Martin Dinges, '"Medicinische Policey" zwischen Heilkundigen und "Patienten" (1750–1830)', in Karl Härter (ed.), *Policey und frühneuzeitliche Gesellschaft* (Frankfurt am Main 2000), 263–95; Gerd Göckenjan, *Kurieren und Staat machen. Gesundheit und Medizin in der bürgerlichen Welt* (Frankfurt am Main 1985), 169–78; Uwe Bannert, *Diskussionen um ärztliche Gebührenordnungen im 19. Jahrhundert. Ein Beitrag zur Geschichte des ärztlichen Honorars* (Neumünster 1986).

in some parts of his practice than in others.[104] That the situation did not essentially change up to the early twentieth century is apparent from the journal of the homeopath Friedrich von Bönninghausen in Münster, whose practice was certainly profitable even though that is not obvious from the journal entries or his 'account book'. That his journals mention charges for treatments and their settlement in only a very few cases may be due to the fact that Bönninghausen's patients paid him immediately before or after their treatment, as that would have rendered any note-taking redundant.[105] But not all patients were the same in this respect: since most entries in the account book refer to aristocratic clients it can be assumed that these nobler members of his clientele did not pay immediately for each treatment.

All personal and stratum-specific payment preferences aside, the 1857 scale of charges and fees for medical practitioners (*Taxe für practische Aerzte*) as well as the 1896 scale of charges for licensed physicians (*Gebührenordnung für approbierte Ärzte*) called on orthodox physicians as well as homeopaths to consider the local circumstances and especially the personal and financial situation of their patients when charging them for medical services.[106] Bönninghausen followed this advice and even reduced the charges for individual services in the course of his medical career: while his fee for a simple consultation cost 1 *thaler* before the money reform, and was therefore at the upper end of the recommended scale, he afterwards asked 5 *marks* – a fee which was more in the middle range – for the 'first consultation of a patient in the doctor's house. But the homeopathic therapy remained expensive for many patients despite such demonstrations of goodwill. The outcome of the negotiations between physician and patients generally varied in the case of Bönninghausen's practice and some patients would walk away without consultation or medicine.[107] In 1864, on the other hand, he waived the fees for a maid because of her obvious impecuniousness.[108] In other cases the parties were able to negotiate

---

104 Cf. Dietrich Tutzke and Regina Engel, 'Tätigkeit und Einkommen eines Allgemeinpraktikers vor der Mitte des 19. Jahrhunderts. Ergebnisse einer historisch-statistischen Analyse', *Zeitschrift für die gesamte Hygiene* 24 (1978), 460–65: 464. There are also cases in the late nineteenth century where physicians and patients strictly adhered to the official fee system; cf. August Föppl, *Lebenserinnerungen. Rückblick auf meine Lehr- und Aufstiegsjahre* (München 1925), 7.

105 Cf. Baschin (in this volume).

106 Cf. Baschin, 2010.

107 Cf. Institute for the History of Medicine of the Robert Bosch Foundation (IGM) P 117 Fol. 78.

108 Cf. IGM P 109 Fol. 44.

solutions such as advance payment, payment in instalments or a combination of payment in kind and in cash.

Conclusion

Inquiry into the daily medical business reveals, above all, two things: the first is that our knowledge regarding the organization, daily routine and bookkeeping in physicians' practices remains quite fragmentary despite the fact that the more recent research has shown an enhanced interest in the matter. For many of the questions, which interest us, we cannot find exhaustive answers in the practice records or in the other sources available. The most reliable points of reference are the patient numbers, a factor on which most practice records yield information. These figures allow us to estimate the – temporal and financial – significance medical practice had in the life of a physician or healer. This leads to the second central – and not unexpected – conclusion of our presentation: the diversity of the practices and the breadth of the physicians' scope for action. Even though there were external conditions and, from early modern times up to the nineteenth century, increasingly also laws that governed the medical market and restricted this medical freedom to an extent, within these restrictions the scope for action remained relatively broad. As serial sources, the practice records show that physicians had the opportunity, day by day, to either continue or modify their habitual medical routine. Physicians could decide whether to accommodate a patient's request to call on him at home or whether to suggest the patient come to him and, in doing so, he could gradually deviate from an established regimen. He could also advise patients that they needed more intensive treatment or that no further treatment was necessary, thereby steering the number of patients or the regularity of his visits. He could choose between admonishing patients who were tardy with their payment or too poor to pay and waiving part or even all the fee they owed. Depending on what they decided on in the individual case, they would change the character of their practice. What the sources do not tell us is whether a physician did all this, whether he steered his activities or whether he used any definite strategies at all. At this point in time we are restricted to the description of external features such as fluctuating patient numbers, while we can only draw fragmentary conclusions with regard to motivations or internal mechanisms. We can assume, however, that future investigations will add further parts to the puzzle.

CHAPTER 4

Medicine in Practice: Knowledge, Diagnosis and Therapy

Annemarie Kinzelbach, Stephanie Neuner and Karen Nolte

Introduction

> Not just in the writings of physicians but in the real world also, pathologies always conformed to the foremost and generally accepted system… At the time of the Stahlians haemorrhoids were everywhere; when the theory of gastric and bilious conditions prevailed, gastric and bilious illnesses were ubiquitous; as soon as neuropathology gained the upper hand and people had reached a greater degree of luxury there was nothing but nervous disease and ailments arising from weakness; and thus parasites, concealed venereal diseases and many others came to play their part, depending on the prevailing theory of any given time.[1]

In this historical outline of the late eighteenth century the Erfurt medical professor August Friedrich Hecker (1763–1811) claims that new medical theories directly affect medical practice. The extent to which theoretical innovations found immediate expression in diagnosis and therapy is one of the main questions of this chapter, which aims at reconstructing the diagnostic and therapeutic practices and their relative significance. These practices will therefore be juxtaposed with the understanding of body and illnesses held by the academically trained physicians, non-academic practitioners and patients at particular periods of time. A diachronic comparison of the long-term development, changes and continuity in history across eight medical practices will form the centre of this investigation. The practices are those of Johannes Magirus (seventeenth century), Johann Christoph Götz (early eighteenth century), Johann Friedrich Glaser

* We thank the following authors of the individual projects in the DFG Research Network 'Medical practices (1600–1900)', on whose partly still unpublished results this contribution is largely founded: *Sabine Schlegelmilch* (*J. Magirus*), *Kay Peter Jankrift / Annemarie Kinzelbach / Susanne Grosser / Marion Ruisinger* (*Chr. Götz*), *Ruth Schilling* (*J.F. Glaser*), *Stephanie Neuner / Karen Nolte* (*Ch. Fuchs*), *Katharina Weikl / Alois Unterkircher / Iris Ritzmann* (*G. Wachter*), *Lina Gafner / Philipp Klaas* (*C.A. Bloesch*), *Marion Baschin* (*F.P. von Bönninghausen*), *Marina Hilber / Eberhard Wolff / Alois Unterkircher* (*F. von Ottenthal*).

1 August F. Hecker, *Des achtzehenten Jahrhunderts Geschichte der Erfindungen, Theorien und Systeme in der Natur- und Arzneywissenschaft* (Gotha 1799), xxxv.

 | DOI 10.1163/9789004303324_006

(eighteenth century), César Adolph Bloesch (first half of the nineteenth century), Gottfried Wachter (non-academic healer, first half of the nineteenth century), Franz Ottenthal (nineteenth century), Friedrich von Bönninghausen (homeopath, late nineteenth century) and the policlinics at the universities of Würzburg and Göttingen, which, in the first half of the nineteenth century, were under the direction of Conrad Heinrich Fuchs.

We will first show what kind of medical knowledge formed the foundation of the medical approaches implemented in the eight practices under investigation, what kind of knowledge can be directly identified from the practice records and whether these bodies of knowledge were in any way class-specific. It is not only the physicians' own knowledge we were interested in, however. We also investigated the extent to which the physicians' descriptions divulge the knowledge and ideas of medical 'lay-persons'[2] such as the patients and the people around them. This question contributes to the research into the histories of medicine, gender and the human body all of which describe and examine 'knowledge' from a theoretical point of view. The last decade has seen increased discussion regarding the convergence or divergence,[3] the coexistence,[4] competition, conflict and marginalization or appropriation of various forms of knowledge[5] or of knowledge transfer.[6] As the early modern philosophical discourse illustrates, 'knowledge' can be defined in a variety of

2 The term 'lay-persons' will be used to distinguish patients from professional practitioners. Use of the term will imply no judgement on the validity of the medical knowledge of such persons.

3 Cf. Michael Stolberg, *Experiencing Illness and the Sick Body in Early Modern Europe* (Basingstoke 2011), 79–81, 284–85; Marion Maria Ruisinger, *Patientenwege. Die Konsiliarkorrespondenz Lorenz Heisters (1683–1758) in der Trew-Sammlung Erlangen* (Stuttgart 2008), 142–43, 208; Michael Stolberg, *Die Harnschau. Eine Kultur- und Alltagsgeschichte* (Köln 2009), 26–38.

4 Cf. Johanna Geyer-Kordesch, *Pietismus, Medizin und Aufklärung in Preußen im 18. Jahrhundert. Das Leben und Werk Georg Ernst Stahls* (Tübingen 2000); Londa Schiebinger, 'Women of Natural Knowledge', in Katharine Park and Lorraine Daston (eds), *The Cambridge History of Science, vol. 3. Early Modern Science* (Cambridge 2006), 192–205.

5 Cf. Barbara Duden, *Geschichte des Ungeborenen. Zur Erfahrungs- und Wissenschaftsgeschichte der Schwangerschaft, 17.–20. Jahrhundert* (Göttingen 2002); Wulf Oesterreicher, Gerhard Regn and Winfried Schulze (eds), *Autorität der Form, Autorisierung, institutionelle Autorität* (Münster 2003); Monika Mommertz, 'Das Wissen "auslocken". Eine Skizze zur Geschichte der epistemologischen Produktivität von Grenzüberschreitung, Transfer und Grenzziehung zwischen Universität und Gesellschaft', in Yuka Nakamura, Christine Böckelmann and Daniel Tröhler (eds), *Theorie versus Praxis? Perspektiven auf ein Missverständnis* (Zürich 2006), 19–51; Matthias Blanarsch, 'Die Arzt-Patienten-Beziehung zu Beginn des 18. Jahrhunderts, untersucht anhand Johann Storchs Kasuistik zu Molenschwangerschaften', *Medizin, Gesellschaft und Geschichte* 28 (2009), 121–52.

6 Cf. Johann A. Steiger, Sandra Richter and Marc Föcking (eds), *Innovation durch Wissenstransfer in der Frühen Neuzeit. Kultur- und geistesgeschichtliche Studien zu Austauschprozessen in Mitteleuropa* (Amsterdam 2010).

ways.[7] Scholars today tend to be wary of resorting to the notion of knowledge as a clearly defined entity.[8] Contemporary encyclopaedias describe how, in the late eighteenth century, a fundamental differentiation between 'knowledge' and 'science' established itself, while prior to that practical knowledge and skills were also seen as science.[9]

More recent studies based on ego-documents of patients in the eighteenth century question a strict separation and hierarchy of medical knowledge and 'lay knowledge' and even provide evidence of a broad congruence of the knowledge about the body and diseases held by 'lay-persons' and physicians.[10] The ego-documents that were analysed consist exclusively, however, of letters or diaries written by patients sufficiently educated to absorb the medical literature of their time. The question as to how the knowledge of less educated people can be derived from medical practice and how physicians dealt with knowledge that was not academically generated has remained largely unexplored. Inquiry into how the knowledge held by those not academically educated can be methodically gleaned from the practice notes of an academically trained physician is another aspect addressed within the Project Network.[11] Other sections inquire into the connection between routine actions and knowledge from a praxeological point of view.[12] This theoretical approach

---

7 Cf. Stephan Meier-Oeser, 'Wissen. (Frühe Neuzeit)', in Joachim Ritter (ed.), *Historisches Wörterbuch der Philosophie, vol. 12* (Basel 2004), 881–84.

8 Gerd Antos and Tilo Weber, introduction to Tilo Weber (ed.), *Typen von Wissen. Begriffliche Unterscheidung und Ausprägungen in der Praxis des Wissenstransfers* (Frankfurt am Main 2009), 1–10: 2, refer to more than a dozen different forms.

9 Cf. under 'Wissen' in Jacob Grimm and Wilhelm Grimm, *Das Deutsche Wörterbuch*, 16 vols (Leipzig 1854–1961) with register of sources (Leipzig 1971). [http://woerterbuchnetz.de/DWB]; Johann Heinrich Zedler, *Grosses vollständiges Universal-Lexicon aller Wissenschafften und Künste, Welche bißhero durch menschlichen Verstand und Witz erfunden und verbessert worden* (Halle/ Leipzig 1732–1754) [online: http://www.zedler-lexikon.de, 11 August 2014]; Johann C. Adelung, *Versuch eines vollständigen grammatisch-kritischen Wörterbuches der hochdeutschen Mundart* (Leipzig 1786), vol. 5,1 (W–Z); Johann C. Adelung, *Grammatisch-kritisches Wörterbuch der hochdeutschen Mundart. Zweyte vermehrte und verbesserte Ausgabe* (Wien 1811), vol. 4 (Seb–Z).

10 Cf. Stolberg, *Experiencing Illness*, 2011, esp. 79–81; Marion Maria Ruisinger, *Patientenwege. Die Konsiliarkorrespondenz Lorenz Heisters (1683–1758) in der Trew-Sammlung Erlangen* (Stuttgart 2008).

11 Cf. Schiebinger, 2006, who, in her analysis, sees the knowledge of various social groups as equivalent.

12 Cf. Andreas Reckwitz, 'Grundelemente einer Theorie sozialer Praktiken. Eine sozialtheoretische Perspektive', *Zeitschrift für Soziologie* 32, no. 4 (2003), 282–301; Karl H. Hörning and Julia Reuter (eds), *Doing Culture. Neue Positionen zum Verhältnis von Kultur und sozialer Praxis* (Bielefeld 2004).

explores[13] the specific knowledge of the historical agents that is concealed in the doctors' practices and is known as 'implicit knowledge'.[14]

Our investigation follows the encounter between physician and patient, starting with examination techniques and exploring in particular how practices, findings and diagnoses related to one another. The following section asks whether a diagnosis in the modern sense of the word was established or what alternative disease descriptions can be found in the sources. Despite the difficulties associated with the term 'diagnosis' in this analysis – considering the term's specific historical connotations – we use it in this chapter to focus on the descriptions, explanations and classifications of disease in the history of medicine.

The thesis that, in the middle of the nineteenth century, a profound change took place in the medical understanding of illness as a result of which illness was for the first time seen as a process of cause and effect is discussed controversially in the history of medicine.[15] Volker Hess, in particular, emphasizes that the semiotic understanding that had prevailed in medicine previously was replaced around 1850 by the diagnostic view.[16] More recent investigations question the assumption that pre-modern physicians focused primarily on symptoms rather than thinking of diseases in terms of distinct entities.[17] Going on from this research controversy, we will look at the development of the doctor's practices across three centuries in order to investigate the views that informed the medical actions of the physicians and how symptoms of a disease were described and classified. In this context we critically evaluate the thesis that diagnosis based on theoretical concepts was a novelty in nineteenth-century medicine. This leads us to another problem regarding theory and method: what were the concepts applied by physicians at various times to the symptoms of an illness or ailment and what kind of understanding were these concepts

13 Cf. Reckwitz, 2003, 289, 294.

14 On the concept of 'agent' cf. Alf Lüdtke, 'Alltagsgeschichte. Aneignung und Akteure. Oder – es hat doch kaum begonnen!', *WerkstattGeschichte* 17 (1997), 65–80.

15 Cf. Volker Hess, *Von der semiotischen zur diagnostischen Medizin. Die Entstehung der klinischen Methode zwischen 1750 und 1850* (Husum 1993), 296.

16 Cf. Hess, *Semiotische Medizin*, 1993. On this thesis cf. Jacalyn Duffin, *Langstaff. A Nineteenth-Century Medical Life* (Toronto 1993, repr. 1999), 64–65.

17 Cf. Hubert Steinke and Urs Boschung, 'Nützliche Medizin. Theorie und Praxis bei Albrecht von Haller', in André Holenstein, Martin Stuber and Gerrendina Gerber-Visser (eds), *Nützliche Wissenschaft und Ökonomie im Ancien Régime. Akteure, Themen, Kommunikationsformen* (Heidelberg 2007), 133–47: 140–41: Steinke/Boschung point out that Haller spoke of disease entities as early as the eighteenth century, demanding that physicians ought to be able to form a diagnosis beyond any doubt, on the basis of a series of symptoms. Cf. also Michael Stolberg, 'Medizin und Krankheit in der Frühen Neuzeit. Eine Einführung', *Geschichte in Wissenschaft und Unterricht* 2 (2008), 85–95: 89–91.

based on. The concepts prevalent in the period investigated either go back to antiquity, such as *symptoma* or *accidens*, or were literal translations such as *Zufall* for *accidens*[18] or descriptive, such as *krankhafte Erscheinungen* (pathological manifestations).[19] The term *symptoma* was used for any deviation from the healthy state ('against nature').[20] This understanding of 'symptom' in itself implied a causal relationship with the pathological process. In Zedler's *Universallexikon* the term *symptoma* refers to the effect of illness, while the German term *Zufall* denotes phenomena 'arising from a multiplicity of circumstances surrounding the patient'[21] that happen to coincide with the illness.[22] We will use the term 'symptom' in the latter, contemporary, sense and explain, in each case, the pathological concept underlying the *symptomata*.

We will analyse the actual therapeutic practices, again in relation to the pertaining explicit and implicit knowledge, and ask what the doctor-patient relationship was like during the period of treatment. The intention is to demonstrate also to what extent the varying therapeutic practices were related to changing social parameters among the clientele. Starting from the premise that the success of the treatment was crucial for the reputation of the treating physician we will ask in conclusion what understanding of healing the medical sources reveal, how important a cure was for physicians or patients and with what kind of expectations patients came to the physician. The importance attached to 'healing' can also be derived from the way physicians commented on the death of a patient.

Knowledge as Prerequisite and Objective

While knowledge of the body and its diseases was a necessary prerequisite for any medical practice and therapy, the recorded *observationes et curationes*,[23] the diaries of 'medical attendance and consultations'[24] etc. were, certainly for

18 Zedler used both forms: *symptoma* (singular) as well as *symptomata* (plural), cf. Zedler, 1732–1754, vol. 41, 763–67.

19 Cf. Franz Willibald Nushard, *Theoretische Medicin für Wundärzte als Leitfaden zu Vorlesungen entworfen* (Prag 1824), vol. 2 (*Die Grundzüge der allgemeinen Therapie, Arzneimittellehre, Krankendiätetis und Receptirkunde*), 16.

20 Cf. Zedler, 1732–1754, vol. 41, 763.

21 Cf. *ibid.*, 767.

22 On the 'semiotic view' of accidents and signs, cf. Hess, *Semiotische Medizin*, 1993, 54–60.

23 Götz, cf. Universitätsbibliothek Erlangen (UBE), Ms. 1201/1, 1. For *observationes* cf. Lorraine Daston and Elizabeth Lunbeck (eds), *Histories of Scientific Observation* (Chicago 2011).

24 Cf. the contribution of Lina Gafner, 'Administrative and Epistemic Aspects of Medical Practice: Caesar Adolf Bloesch (1804–1863)' (in this volume).

some physicians, a means of generating and documenting explicit forms of knowledge.

The historiographical discussions on the lack of uniformity,[25] on changes,[26] even revolutions[27] and 'paradigm shifts'[28] in scientific medicine point to differences in the theoretical training of the individual writers. It seems therefore a logical conclusion that they had also acquired and used their knowledge in different ways since they lived in different centuries.[29] The time available to each of the physicians for testing, widening and revising their knowledge varied from less than two decades in the cases of Götz and Fuchs to more than half a century in the cases of Wachter and Ottenthal. The social spectrum of these German-speaking authors of practice records and of the Würzburg patient histories was also very broad. It ranged from the 'lay healer'[30] – who most likely trained as a 'craftsman', most certainly followed a family tradition and practised in a rural area to urban physicians with scholarly aspirations[31] to aristocratic physicians who practiced either in cities or in the countryside.[32]

---

25 Cf. Hubert Steinke, 'Anatomie und Physiologie', in Hubert Steinke, Urs Boschung and Wolfgang Proß (eds), *Albrecht von Haller. Leben – Werk – Epoche* (Göttingen 2008), 226–54: 227.

26 Cf. Thomas H. Broman, 'The Medical Sciences', in Roy Porter (ed.), *Eighteenth-Century Science* (Cambridge 2003), 463–84; Katharine Park and Lorraine Daston, 'The Age of the New', introduction to Park, 2006, 1–17.

27 Cf. Harold J. Cook, 'Medicine', in Katharine Park and Lorraine Daston (eds), *The Cambridge History of Science. 3. Early Modern Science* (Cambridge 2006), 407–34: 432–34.

28 Cf. Claudia Wiesemann and Nelly Tsouyopoulos (eds), *Asklepios und die Philosophen. Paradigmawechsel in der Medizin im 19. Jahrhundert* (Stuttgart 2008).

29 Cf. Kay P. Jankrift and Ruth Schilling, 'Medical Practice in Context. Religion, Family, Politics and Scientific Networks' (in this volume); Philipp Klaas, Hubert Steinke and Alois Unterkircher, 'Daily Business: the Organization and Finances of Doctors' Practices' (in this volume).

30 Cf. Katharina Weikl and Iris Ritzmann, 'Ein Auslaufmodell. Spuren einer ländlichen Heilerpraxis im Zeitalter der Akademisierung', in Lina Gafner, Iris Ritzmann and Katharina Weikl (eds), *Penning Patient's Histories – Doctors' Records and the Medical Market in the 18th and 19th Century* [special issue: *Gesnerus* 69, no. 1 (2012)], 76–94.

31 Cf. Nadine Boucherin, 'Die Krankengeschichten von C.A. Bloesch. (1804–1863)', in Elisabeth Dietrich-Daum, Martin Dinges and Robert Jütte (eds), *Arztpraxen im Vergleich. 18.–20. Jahrhundert* (Innsbruck 2008), 147–65; Susan Splinter, '"Der Aufrichtige Medicus". Eine Zeitschrift des Nürnberger Arztes Johann Christoph Götz (1688–1733) als Vorläufer des "Commercium Litterarium"', *Jahrbuch für Kommunikationsgeschichte* 13 (2011), 5–15; Ruth Schilling, 'Stadt und Arzt im 18. Jahrhundert. Johann Friedrich Glaser, Stadtphysicus in Suhl', *Würzburger medizinhistorische Mitteilungen* 30 (2011), 310–33; Ruth Schilling, Sabine Schlegelmilch and Susan Splinter, 'Stadtarzt oder Arzt in der Stadt? Drei Ärzte der Frühen Neuzeit und ihr Verständnis des städtischen Amtes', *Medizinhistorisches Journal* 46 (2011), 99–133.

32 Cf. Marion Baschin, 'Warum besucht man einen Homöopathen? Eine Praxis in Münster im 19. Jahrhundert', in Lina Gafner *et al.*, 2012, 126–40; Elena Taddei, *Franz von Ottenthal. Arzt und Tiroler Landtagsabgeordneter (1818–1899)* (Wien 2010).

The physicians Bloesch, Fuchs, Götz and Magirus, who came from established town families with some of them having prominent academic fathers, were able to use the positions of their families,[33] their academic education and subsequent study-tour to establish a professional network that offered or facilitated access to the most up-to-date scientific insights. Marriage was also an important means of extending and securing the necessary communicative networks,[34] providing additional familial contacts, as in the case of Magirus and Götz. Magirus' marriage with the daughter of a court preacher paved his way into the religiously and socially exclusive community of court physicians where he was able to make lucrative use of his specific knowledge.[35] Götz, who descended from artisans and merchants, married into a family with firm academic roots as well as intimate connections with the social and cultural elite in the area.[36] That he knew how to make good use of these connections is apparent from the fact that he co-founded a scientific society in Nuremberg and from publications he released with a view to making the Stahlian theory and practice accessible to a wider public.[37] His journal and the communicative activity it

33 Cf. Theodor Husemann, 'Fuchs, Konrad Heinrich', *Allgemeine Deutsche Biographie* 8 (1878), 168–69; Johanna Bleker, Eva Brinkschulte and Pascal Grosse (eds), *Kranke und Krankheiten im Juliusspital zu Würzburg 1819–1829. Zur frühen Geschichte des Allgemeinen Krankenhauses in Deutschland* (Husum 1995); Boucherin, 2008; Schilling *et al.*, 2011.

34 On communication networks cf. Regina Dauser, Stefan Hächler, Michael Kempe, Franz Mauelshagen and Martin Stuber (eds), *Wissen im Netz. Botanik und Pflanzentransfer in europäischen Korrespondenznetzen des 18. Jahrhunderts* (Berlin 2008); Marco Bresadola, 'A Physician and a Man of Science. Patients, Physicians, and Diseases in Marcello Malpighi's Medical Practice', *Bulletin of the History of Medicine* 85 (2011), 193–221.

35 Cf. Sabine Schlegelmilch, 'Vom Nutzen des Nebensächlichen. Paratexte in den Kalendern des Johannes Magirus', in Klaus-Dieter Herbst (ed.), *Astronomie – Literatur – Volksaufklärung. Der Schreibkalender der Frühen Neuzeit mit seinen Text- und Bildbeigaben* (Bremen 2012), 393–411; on the significance of such associations at court cf. Silvia de Renzi, 'Medical Competence, Anatomy and the Polity in Seventeenth-Century Rome', in Sandra Cavallo and Davis Gentilcore (eds), *Spaces, Objects and Identities in Early Modern Italian Medicine* (Oxford 2007), 551–86.

36 His father-in-law was Dean of the Law School at the University of Altdorf; his sister-in-law, Dorothea, was a member of the German literary society *Pegnesischer Blumenorden* (Pegnitz Flower Society), cf. *Die Nürnbergischen Adress- und Schreibkalender: Das jetzt lebende Nürnberg: bey einem Calender auf das Jahr 1705/06. Verzeichnus der Republic Nürnberg Regenten, Beamten und Bedienten* (Nürnberg 1705) [microfiche edition, Erlangen 2000], 80; Renate Jürgensen, *Utile cum dulci. Die Blütezeit des Pegnesischen Blumenordens in Nürnberg 1644–1744* (Wiesbaden 1994), 110.

37 Cf. Georg E. Stahl and Johann C. Götz, *Scripta D. Georg. Ern. Stahlii Sacr. Reg. Maiest. Boruss. Consil. Aulici & Archiatri Primarii Aliorumque Ad Eius Mentem Disserentium Serie Chronologica Recenset D. Joh. Christoph Goetzivs* (Nürnberg 1729) [http://reader.digitale-sammlungen.de/de/fs1/object/display/bsb10054543_00005.html]; Tilman T.R. Rau, *Das*

reveals are proof that he cultivated these connections and established new ones as means of accumulating 'knowledge'.[38]

That a physician's social background had an effect on his participation in scientific discourse is also apparent from the difficulties and triumphs of Glaser, an executioner's son. His uncles, who were both established university-trained physicians, supported him during his academic and doctoral studies, thereby helping him to acquire what he needed to take part in scientific discourse. It was, however, not until the end of his career, in 1760, that Glaser managed to overcome social prejudice and become a registered member of a renowned scientific society, the *Leopoldina* (which had accepted Götz in 1726). But Glaser was not able to publish his *observationes*, the store of knowledge he had gathered and systematically catalogued, in the society's organ, the *Ephemerides*.[39]

With all authors record-keeping was a way of documenting information relevant to their practice.[40] Other purposes, which applied in individual cases only, included the possibility of, and interest in, recording knowledge or the wish to play a part in changing the existing body of knowledge. Magirus, a seventeenth-century physician, held on to the knowledge of mathematical medicine or medical mathematics he had acquired and was keen to pass it on to his pupils,[41] although many of his contemporaries had turned to other theoretical foundations.[42] His dedication to medical mathematics in no way meant that he did not value practical experience. Götz, who was active in the

Commercium Litterarium. Die erste medizinische Wochenschrift in Deutschland und die Anfänge des medizinischen Journalismus (Bremen 2009), 44–47.

38 Cf. Annemarie Kinzelbach, Susanne Grosser, Kay P. Jankrift and Marion Ruisinger, '*Observationes et Curationes Nurimbergenses*'. The Medical Practice of Johann Christoph Götz (1688–1733)' (in this volume).

39 Cf. Ruth Schilling, 'Amtsträger und Wissenschaftler – die Repräsentationsstrategien eines Scharfrichtersohns in der Mitte des 18. Jahrhunderts', *Kunstgeschichte. Open Peer Reviewed Journal* (2011), [http://www.kunstgeschichte-ejournal.net/129/]; *Nova Acta Physico-Medica Academiae Caesareae Leopoldino-Carolinae Naturae Curiosorum Exhibentia Ephemerides sive Observationes Historiae et Experimenta a celeberrimis Germaniae et Exterrarum Regionum Viris Habita et Commmunicata singulari studio collecta* 1 (1727), Catalogus No. 391.

40 Cf. Volker Hess and Sabine Schlegelmilch, '*Cornucopia officinae medicae*: Medical Practice Records and their Origins' (in this volume).

41 Cf. Schilling *et al.*, 2011, 103–11.

42 Cf. Wolf-Dieter Müller-Jahncke, *Astrologisch-magische Theorie und Praxis in der Heilkunde der frühen Neuzeit* (Stuttgart 1985), 135–258.

first decades of the eighteenth century, collected information from observations he made among his own patients as well as from reports he received from his own personal environment, without social discrimination.[43] Over and above that, Götz showed interest in the teachings of his contemporary Stahl as well as in other theoretical concepts or movements, for example, Pietism. In the attempt to transform his knowledge into 'science' he presented a selection of his insights in a German-language journal[44] of which he was editor, and in other scientific journals. Among the 28 *observationes* on a wide range of topics – from post-mortems to *vomitus autocraticus* – which were published in the *Ephemerides* we also find information that had come to him 'second hand'.[45] With Bloesch, who lived in the first half of the nineteenth century, we find such emphatic 'empiricism' only in relation to his own person, in his retrospective statement that he used to have little use, initially, for medical 'theory'.[46] Bloesch's writings reveal, however, that he kept a record of his experiences in particular fields, such as gastro-intestinal disease or psychiatric disorders, with a view to offering them for scientific publication. In the field of epidemics he even made an attempt at theory construction, an approach he tended to be critical of.[47] His contemporary C.H. Fuchs went a step further when he included the 'patient histories' he had asked to be compiled from

43 He wrote down information from members of the same class, but also from servants and craftsmen who lived nearby, UBE Ms. 1201/2, pp. 24, 59, 101, 110, 127, 144, 339, 480.

44 Cf. Splinter, 'Medicus', 2011.

45 Cf. Johann C. Götz: 'Observationes Physico-Medicae. Observatio CLXXXI-Observatio CCVII', in *Nova Acta Physico-Medica Academiae Caesareae Leopoldino-Carolinae Naturae Curiosorum Exhibentia Ephemerides sive Observationes Historiae et Experimenta a Celeberrimis Germaniae et Exterrarum Regionum Viris Habita et Commmunicata Singulari Studio Collecta* 2 (1730), 423–63.

46 Cf. Boucherin, 2008, 152. Many physicians who saw themselves as scientists considered the empiricism of medical practitioners as unscientific because it appeared to be unsystematic and hostile to theory. Cf. Georg August Richter (ed.), *Die specielle Therapie nach den hinterlassenen Papieren des verstorbenen August Gottlieb Richter* (Berlin 1813), vol. 1 (*Die acuten Krankheiten*), preface; cf. Wiesemann *et al.*, 143.

47 Cf. Caesar A. Bloesch, *Beytrag zur Pathologie und Therapie der gastrischen Krankheiten* (Bern 1832); *idem*, 'Gutachten über einen gerichtlich-psychologischen Fall', *Schweizerische Zeitschrift für Medizin, Chirurgie und Geburtshülfe* (1842), 369–80; *idem*, 'Allgemeine Grundsätze der theoretischen und der praktischen Medizin, ausgezogen aus meinem Tagebuch ärztlicher Besuche und Konsultationen', *Mitteilungen der medizinisch-chirurgischen Gesellschaft des Kantons Bern* 1 (1866), 1–150.

observations made in the policlinics into a systematic textbook on nosology for the use of his students.[48]

There are other, less ambitious, physicians of whom there are no known scientific publications: as an aristocrat and member of the local parliament, the Tyrolese medical practitioner Ottenthal was busy with other, political and social, activities in the second half of the nineteenth century.[49] Of Bönninghausen we know that he made use of the empirical knowledge his father had published in his 'Therapeutic Pocketbook'. Even though the son occasionally added to his father's medicinal knowledge by describing newer drugs, he did not pass on the experiences he recorded.[50] The same is presumably true for the 'lay healer' Gottfried Wachter who equally named his father as his teacher. While Wachter held various important offices in his home town[51] he appeared to have neither the intention nor the opportunity to join the 'republic of scholars' (*Gelehrtenrepublik*).[52] The bodies of knowledge acquired by Bönninghausen, Ottenthal and Wachter in particular can, to a large extent, be derived indirectly from the descriptions of examinations and treatments in their records.[53]

It is not only today that the communication[54] between physicians and their clientele – as well as the related recourse to established knowledge – plays an

48 Cf. Conrad Heinrich Fuchs: *Lehrbuch der speciellen Nosologie und Therapie* (Göttingen 1845), vol. 1 (*Klassen und Familien*); *idem*, Lehrbuch der speciellen Nosologie und Therapie (Göttingen 1847), vol. 2,2 (*Gattungen und Arten. Zweite Abtheilung: Hämatonosen. Vierte bis sechste Ordnung*); *idem*, *Lehrbuch der speciellen Nosologie und Therapie* (Göttingen 1848), vol. 2,3 (*Gattungen und Arten. Dritte Abtheilung: Neurosen, Morphonosen und Register*).

49 Cf. Taddei, 2010, 173–209; Elisabeth Dietrich-Daum, Marina Hilber and Eberhard Wolff, 'An Alpine Doctor's Private Practice: Franz von Ottenthal (1847–1899)' (in this volume).

50 Cf. Andreas Oberhofer, 'Eine Landarztpraxis im 19. Jahrhundert am Beispiel der Ordination des Dr. Franz von Ottenthal (1818–1899)', in Dietrich-Daum *et al.*, *Arztpraxen*, 2008, 167–91; Andreas Wegener, 'Mittelfindung nach dem Therapeutischen Taschenbuch von Bönninghausen', in Thomas Genneper and Andreas Wegener (eds), *Lehrbuch Homöopathie. Grundlagen und Praxis der klassischen Homöopathie* (Heidelberg 2001), 139–51 (= Stuttgart ^{3}2011, 118–30); Baschin, 'Homöopath', 2012.

51 Cf. Weikl *et al.*, 2012.

52 On the 'republic of scholars' cf. Hubert Steinke and Martin Stuber, 'Haller und die Gelehrtenrepublik', in Steinke *et al.*, 2008, 381–414.

53 For an example of the analysis of implied knowledge cf. Duffin, 1999, 59–73.

54 On the definition of communication in historiography cf. Rudolf Schlögl, 'Vergesellschaftung unter Anwesenden. Zur kommunikativen Form des Politischen in der vormodernen Stadt', in Rudolf Schlögl (ed.), *Interaktion und Herrschaft* (Konstanz 2004), 9–30: 23–24.

important part in everyday medical practice.[55] Our research shows that physicians often negotiated possible therapies with the men or women they treated. In this context both the kind of knowledge that the patients and relatives or confidants referred to and the degree of authority they allowed their physicians are of importance.[56] While the practice records were all written by male and – with the exception of Wachter – academically trained physicians, some of them nevertheless give insight into the stores of knowledge held by members of either gender and by patients from a wide range of social backgrounds.[57] The correspondence might contain far more detailed information on the concepts and ideas of patients and their milieu,[58] but the records provide additional information on what kind of knowledge physicians considered sufficiently relevant for recording, as in the case of Götz who marked such entries with the word *narrat*. More than a tenth of Götz's records quote such knowledge and in every other case it came from women, some of whom were even referred to as 'doctor'.[59] Not all of the recorded information – some of which also came from male or female servants – was related to the individual case histories.[60] Götz recorded a wide range of knowledge conveyed to him, ranging from statements relating to drugs and home remedies or reports of spectacular cures to observations regarding the sexual behaviour of inexpertly castrated pigs.[61] The patients and their families feature less prominently in Glaser's records although he, too, wrote down statements of third parties as late as 1763.[62] If one compares Bloesch, who tended to comment critically on his clients' interpretations ('as people used to think' or 'she thinks',[63]) and Götz,

55 Cf. KIM <http://www.kim-berlin.net/> (11 August 2014).

56 On authorization cf. Michael Stolberg, 'Formen und Strategien der Autorisierung in der frühneuzeitlichen Medizin', in Wulf Oesterreicher, Gerhard Regn and Winfried Schulze (eds), *Autorität der Form, Autorisierung, institutionelle Autorität* (Münster 2003), 205–18.

57 Cf. Marion Baschin, Elisabeth Dietrich-Daum and Iris Ritzmann, 'Who Consulted the Doctor? The Patients and their Motives' (in this volume).

58 Cf. Stolberg, *Experiencing Illness*, 2011; Ruisinger, *Patientenwege*, 2008.

59 UBE Ms. 1201/2. For more detail see Annemarie Kinzelbach, 'Women and Healthcare in Early ModernGerman Towns', *Renaissance Studies* 28 (2014), 619–38.

60 Cf. Kinzelbach *et al.*, '*Observationes*' (in this volume).

61 UBE Ms. 1201/2, pp. 24, 40, 49, 53, 57, 59, 67, 68, 71, 79, 100, 101, 127, 141, 150, 186, 215, 229, 233.

62 Institute for the History of Medicine (IGM) Charité, Glaser Ms. '*sagt die mutter es habe für die tochter vor 2 Jahren ein Toller hund...*,' (8) July 1763 – 'the mother says that for the daughter two years ago a rabid dog...', (8) July 1763; '*...convulsiones, es hat just seinen ersten hochzeit*[*tag*] *angefangen, liegt auch offt gar daran...*', (28) Juni 1763 – 'convulsions, it started just on his first wedding [day], is often also due to...', (28) June 1763.

63 Stadtarchiv (StA) Biel, 4 A 1, Tagebuch, vol. 18, 1840, 397, vol. 6, 1834, 161.

it appears that the interest of physicians in the knowledge of their clientele decreased sharply in the period under investigation, but Magirus, in the seventeenth century, hardly wrote down any of his patients' interpretations of illnesses.

As will be discussed below, components of humoral pathology generally constituted a link between physicians and their patients as well as between the record keeping physicians of the seventeenth to nineteenth century. This concept of humoralism differed, however, from that commonly misrepresented in older textbooks, which is simply based on the idea that illness is due to a 'bad mixture' of the four humours (blood, phlegm, yellow and black bile). In truth the concept encompassed, even in antiquity, the view that the balance of the humours depended on the 'concoction' and transformation of food in the organs, a process that might be congested or disturbed by influences arising in an individual's body, soul or environment.[64] More recent research shows, moreover, that early modern physicians often associated illness with toxins accumulated in particular parts of the body.[65] These unnatural and noxious substances needed to be diverted, discharged or 'dissipated'.[66]

Examination

In 1879 a medical practitioner described an important change in the examination of patients, from the physician's 'swift glance' to the 'physical' examination that involved the use of technical instruments.[67] Medical-historical research into examinations and the doctor-patient relationship has seen this development as a change from dialogue-based medicine, where the patient's subjective observations formed the centre of examination and diagnosis, to an

64 Cf. quoted in Wolfgang Eckart, *Geschichte der Medizin* (Berlin ^{2}1994), 60, 47, who also refers to the wider ancient concept; for more detail cf. Andrew Wear, *Knowledge and Practice in Early Modern English Medicine. 1550–1680* (New York 2000), 170–73; Beate Gundert, 'Humoralpathologie', in Karl-Heinz Leven (ed.), *Antike Medizin. Ein Lexikon* (München 2005), 436–41.

65 On epidemic disease cf. Wear, 2000, 275–79; Annemarie Kinzelbach, 'Infection, Contagion, and Public Health in Late Medieval and Early Modern German Imperial Towns', *Journal of the History of Medicine and Allied Sciences* 61 (2006), 369–89. We also thank Michael Stolberg for making available the relevant passage from his lecture manuscript.

66 UBE Ms. 1201/2, 6, 13, 25, 26, 46, 67, 75, 79, 99, 100, 104, 108, 109, 125, 113, 117, 504, 517.

67 Cf. Paul Niemeyer, 'Von ärztlicher Untersuchung', *Aerztliche Sprechstunden* 1, no. 3 (1879), 1–20: 1–2.

objectifying medicine dominated by the physician's interpretations.[68] This 'modern' medicine, it was said, had generated the 'scientific physician' as well as a new type of patient who was 'convinced of the epistemic superiority of the physical examination as the basis for reliable diagnosis'.[69]

A glance at the non-clinical medical practice belies these theses. Studies on a Canadian country doctor and a New York gynaecologist suggest that, up to the late nineteenth century, the conversation between practitioners and their patients constituted the most important part of the examination.[70] Michael Stolberg has shown, on the other hand, that, even in early modern times, physicians were not satisfied with the patients' narrative alone and had found a way of assessing processes inside the sick body by close examination of the patient's urine.[71] In his most recent work on the medical examination Stolberg strongly rejects the wide-spread claim that touching the patient was taboo in the early modern physician-patient relationship, mentioning multifarious, also physical, examination techniques in use at the time.[72]

At the first consultation the physician's experienced eye would take in the patient's complexion, the colour and irregularities of his skin and his stature.[73] Evidence of this visual medical inspection – as the most basic examination method – can be found in the practice records across the three centuries investigated by the Research Network. The second important, and certainly not only, component of the medical examination was the conversation with the

---

68 Cf. first Nicholas D. Jewson, 'The Disappearance of the Sick Man from the Medical Cosmology', 1770–1870, *Sociology* 10 (1976), 225–44; Barbara Duden, *The Woman Beneath the Skin. A Doctors Patients in Eighteenth-Century Germany* (Cambridge 1991); Malcolm Nicolson, 'The Art of Diagnosis. Medicine and the Five Senses', in William F. Bynum and Roy Porter (eds), *Companion Encyclopedia of the History of Medicine*, Vol. 2 (London 1993), 801–25: 806; Jens Lachmund, *Der abgehorchte Körper. Zur historischen Soziologie der medizinischen Untersuchung* (Opladen 1997); Volker Hess, *Der wohltemperierte Mensch. Wissenschaft und Alltag des Fiebermessens (1850–1900)* (Frankfurt am Main 2000); Mary E. Fissell, *Patients, Power, and the Poor in Eighteenth-Century* (Cambridge 2002).

69 Lachmund, 1997, 247.

70 Cf. Duffin, 1999; Kathleen E. Powderly, 'Patient Consent and Negotiation in the Brooklyn Gynaecological Practice of Alexander J.C. Skene. 1863–1900', *The Journal of Medicine and Philosophy* 25, no. 1 (2000), 12–27.

71 Cf. Stolberg, 2009.

72 Cf. Michael Stolberg, 'Möglichkeiten und Grenzen einer retrospektiven Diagnose', in Waltraud Pulz (ed.), *Zwischen Himmel und Erde. Körperliche Zeichen der Heiligkeit* (Stuttgart 2012), 209–27.

73 This observation technique is described in detail by Carl Ernst Bock (1809–1874), cf. Carl Ernst Bock, *Lehrbuch der pathologischen Anatomie und Diagnostik*, 2 vols (Leipzig 1853), vol. 2 (*Lehrbuch der Diagnostik*), 19–27.

patient, as is evident from the practice records of the early modern physicians Magirus, Götz and Glaser. Their descriptions of signs of illness suggest, moreover, that all three physicians derived their judgements from uroscopical findings. Magirus probably owned a uroscopy flask which he used for urine inspection; Glaser carried out urine examinations sporadically at the beginning of his medical practice and regularly once he had assumed office as town physician in 1763. In Götz' practice we find detailed descriptions of the colour, consistency and impurities of the urine, although it is not always clear whether the information was Götz' own or that of a third person. An early-nineteenth century inventory of the university policlinics overseen by Conrad Heinrich Fuchs lists a uroscopy flask. This and the fact that the case histories contain precise information on the amounts of urine obtained ('quantity of naturally voided urine slightly more than ½ pint',[74]) on its colour and impurities suggests that uroscopies were indeed carried out there. The Swiss practitioner Wachter, who was not university-trained, conducted uroscopies at the same time and described the colour of the urine in his notes. Bloesch, who practised in Biel, mostly measured the quantity of urine and occasionally described its consistency. Even in the late nineteenth century one finds a great variety of urine colours described in the records of the homeopath Bönninghausen. It can, however, not be established whether he carried out the examinations himself or whether the descriptions came from his patients. The fact that the findings are presented in such detail suggests, however, that Bönninghausen, at least occasionally, used uroscopy as a means of diagnosis. For the Tyrolese country physician Ottenthal uroscopy remained even in the late nineteenth century an important diagnostic measure. From the 1860s onwards Ottenthal used also a modern chemical procedure for urine examination (to test for protein, for instance). The urine samples were delivered by messenger[75] or obtained in his house. Up until the nineteenth century 'laypeople' strongly believed in the reliability of uroscopy.[76] Analysis of the various doctors' practices across three centuries supports the conclusion that the results of uroscopy were indispensable for diagnosis right up to the late nineteenth century.

74 Conrad Heinrich Fuchs (1803–1855), Krankheitsgeschichten, Staats- und Universitätsbibliothek Göttingen, Abteilung Handschriften und Alte Drucke (HSD SUB), Sign.: 80 Cod. MS. hist. nat. 61: I, vol. 4.

75 Flasks with the urine of children, whom Ottenthal usually never saw in person, were also brought to him. Cf. Alois Unterkircher, 'Ausser Konkurrenz? Strategien und Inanspruchnahme einer Südtiroler Landarztpraxis um 1890 bei Erkrankungen von Kindern', in Lina Gafner *et al.*, 2012, 158–77.

76 Cf. Stolberg, 2009, esp. 214.

Contrary to what researchers have assumed so far,[77] the examination of blood, of which we even find evidence in the early modern practices of Magirus, Götz and Glaser, was a further important measure for establishing diagnosis. When letting blood, Magirus observed the colour and the velocity with which the blood exited. When three cups were filled he would compare their content. He would wait until the serum had separated from the clot and then assess the weight of the former and the consistency of the latter. Glaser invented a special device for letting blood from the foot but his notes do not reveal whether he examined the blood himself.

Magirus stood out from other physicians because of his iatromathematical analysis of the state of health of – mostly wealthy – patients: this physician, who had devoted himself extensively to mathematical studies, calculated a birth chart (nativity) and analysed the influence of the contemporary planet constellation (revolution) on the health of individual patients. There was no need for patients to be personally present for the iatromathematical examination, since it could be done by letter. Other physicians also established and maintained contact with patients by correspondence. Throughout the three centuries physicians would, for instance, ask their patients to send stool samples, or they based their analysis of the consistency of the feces on the patients' written descriptions.

Feeling the patient's pulse was a common examination method even for early modern physicians.[78] Based on Galen's teachings, the procedure typically involved description of the quality of the pulse and determination of its relative velocity as 'faster than normal; slower than normal; neither fast nor slow, i.e. normal'.[79] In the notes of both Magirus and Götz one finds entries (such as *pulsus debilior et celerior* – a weaker and faster pulse) which suggest that they assessed the quality and velocity of the pulse. Although in physiological research, even in the sixteenth century, pulse frequencies were taken over a specific period of time, using hour glasses, the method did not immediately find its way into the doctors' practices. At the end of the seventeenth century the first watches with a third, second hand were produced. In the mid-eighteenth century the pulse rate was introduced in Dublin as part of clinical research. 'Monitoring the pulse with a watch' was not generally established in

77 Cf. Friedrich Lenhardt, *Blutschau. Untersuchungen zur Entwicklung der Hämatoskopie* (Hannover 1986).

78 On the importance of pulse diagnosis in early modern times cf. Stolberg, 2008. For a general history of pulse- taking cf. Werner Friedrich Kümmel, 'Der Puls und das Problem der Zeitmessung in der Geschichte der Medizin', *Medizinhistorisches Journal* 9 (1974), 1–22.

79 Kümmel, 1974, 3.

doctors' practices until the nineteenth century.[80] Ottenthal started to count the pulse in 1863, Bloesch and Bönninghausen at around the same time. Whether they used watches with a second hand or hour glasses is not known. Wachter, who practised until 1861, used to describe only the pulse quality ('strong pulse').

Contrary to the thesis of non-physical diagnosis in early modern times,[81] feeling the pulse, taking the temperature, as well as abdominal palpation, were among the examination methods used by Johann Magirus as early as the seventeenth century. Magirus related that with his patient Cupelmann, who suffered from abdominal pain, 'no lateral hardening was palpable' (*durities nulla in lateribus deprehensa*).[82] Götz also described findings that suggest abdominal palpation. In the policlinics palpation of parts of the body was a common examination routine in the early nineteenth century.

Determining a patient's temperature and body heat by touch was also still common usage in the policlinics. The Swiss physician Bloesch continued to feel his patients' temperature until he ceased practising in 1863. According to medical reference books and health guides, thermometers were used in doctors' practices as early as the 1870s,[83] while Ottenthal's practice records show that he only began taking his patients' temperature in the late 1890s. The fragile instrument might have been difficult to take around for country doctors, an assumption which is supported by the fact that the Canadian country physician began to use a thermometer very late indeed and only sporadically.[84]

Counting the pulse was considered antiquated as early as 1879. Physicians could distinguish themselves by employing the more modern method of auscultation, that is, listening to rather than palpating the heart.[85] This new and difficult method was not applied by Bloesch, Bönninghausen or Ottenthal, who preferred to confine themselves to the proven method of counting the pulse. There are exact descriptions of various heart sound qualities in Bönninghausen's records that suggest, however, that he used a stethoscope to monitor his patients' heartbeat.

80 Cf. Kümmel, 1974.

81 Cf. esp. Edward Shorter, *Bedside Manners. The Troubled History of Doctors and Patients* (New York 1985); Duden, 1991; Roy Porter, 'The Rise of Physical Examination', in William F. Bynum and Roy Porter (eds), *Medicine and the Five Senses* (Cambridge 2004), 179–97.

82 University library of Marburg (UB), Ms. 96, Diarium Johann Magirus, 207 [Zerbst, July 1651; Richter Cupelmann].

83 Cf. Volker Hess, *Der wohltemperierte Mensch. Wissenschaft und Alltag des Fiebermessens (1850–1900)* (Frankfurt am Main 2000); see also Niemeyer, 1879.

84 Cf. Duffin, 1999, 68.

85 Cf. Niemeyer, 1879.

Junior physicians in the policlinics of Würzburg and Göttingen learned to use stethoscopes as part of their medical training. The stethoscope was used primarily for diagnosing lung disease and to detect sounds in the chest such as 'cavernous respiration' or 'whistling'. Descriptions to that effect can also be found in Bloesch's notes.

Diagnosis

Few of the 'diagnoses' described in the practice notes of physicians from the mid-seventeenth to the late nineteenth century relate to clearly defined diseases. Nosological classification schemes and disease entities, while they existed and were used by physicians in categorizing specific syndromes and describing conditions, did not dominate the recordings of reasons for treatment. Pre-modern physicians appear to have found it neither necessary nor practicable to limit themselves to a single disease term when assessing a patient's condition and deriving a prognosis and therapies. Since they were not (yet) embedded in a national health insurance system, physicians were not compelled to resort to diagnostic terms in the modern sense[86] Only at the Würzburg policlinics, which followed the Natural History School of Medicine and pursued research and training interests, it was standard procedure to assign precise nosological entities to the ailments of patients.

Physicians, while generally influenced by humoralism, also subscribed to alternative concepts such as Broussais' theory of inflammation and the Brunonian theory of excitation,[87] both approaches that were derided as 'fashions' in the medical press.[88] As experienced practitioners they preferred to rely, however, on their proven empirical knowledge in their assessment of patients and were rather reluctant to embrace the new medical theories that entailed alternative models of symptom interpretation and abstract names for diseases. The new diagnostic concepts arising from the currently prevailing medical theories were chosen primarily by the town physicians and health officers, who, as well as being medical practitioners, acted as medical experts for the courts, the administration and the army. These office-holders were obliged

86 Norbert W. Paul, 'Diagnose und Prognose', in Stefan Schulz, Klaus Steigleder, Heiner Fangerau and Norbert W. Paul (eds), *Geschichte, Theorie und Ethik der Medizin* (Frankfurt am Main 2006), 143–52: 144–45.

87 Cf. Richter, 1813, XII; Friedrich August B. Puchelt, *Das System der Medicin: in Umrissen dargestellt* (Heidelberg 1826), XVI.

88 Richter, 1813, XII.

to adhere to the official diagnostic nomenclature to ensure that the results of their investigations could be allocated to general categories. Towards the end of the nineteenth century, the Southern Tyrolese physician Ottenthal, an assessor of military fitness, used the then newly established psychiatric diagnosis of neurasthenia to describe medical conditions.

The entries of the practice records provide brief summaries of the patients' ailments, describe their causes or concomitant circumstances and the physical or chemical diagnostic procedures applied. They therefore provide, in many cases, well-founded evidence for the fact that the establishing of a diagnosis was a complex cognitive process and also contain abundant information regarding the appearance of individual patients.

The ability to observe accurately and to evaluate and categorize the 'signs' of illness were seen as primary skills a physician had to master in order to understand a disease in all its complexity. The dialogue about the patient's condition, which always included the patient's subjective experience of illness, was at all times a crucial aspect of the diagnostic process. The specialist literature often warned that the patient narrative was 'subjective' and therefore misleading. Bloesch, the physician from Biel in Switzerland, also questioned the usefulness of patient knowledge. The patient narrative was nonetheless regarded as highly relevant and when patients were too sick or too young to articulate their experience this was seen as complicating the forming of a diagnosis.[89] The practice records that were evaluated demonstrate that the patients' descriptions of pain, lack of appetite or indigestion were indispensable for the diagnostic process. The entries brought together what the physicians observed and elicited from their patients with the information volunteered by these patients regarding their afflictions.

Due to the fact that the descriptions and names of illnesses were recorded in Latin and German, sometimes also in the local vernacular, the interpretations of lay-persons – or other physicians – have also been preserved in these medical records. The patients of the early modern physician Magirus referred to 'fluxes' when they described the physical imbalance they experienced. The records of the Göttingen policlinic include the statement of a female patient of the mid-nineteenth century who was suffering from a *carcinoma uteri.* Quoting her previous physician she alluded to her distended abdomen as a 'wench's belly'.[90]

89 Ernst Johann Wichmann, *Ideen zur Diagnostik, beobachtenden Ärzten mitgetheilet* (Hannover 1800), 23.

90 SUB Göttingen, HSD: 8 COD MS H NAT 61:I, Conrad Heinrich Fuchs (1803–1855), Krankengeschichten, Heft I, 1833/34, Nr. 3: Anna Biegner; Cf. Karen Nolte, 'Vom Umgang mit unheilbar Kranken und Sterbenden in "Kranken-Besuchs-Anstalten" zu Beginn des 19. Jahrhunderts', *Würzburger medizinhistorische Mitteilungen* 26 (2007), 28–52: 35–38.

In addition to the descriptions of the patients' physical and mental conditions, the entries also reveal the physicians' interest in the causes of illness. Any information regarding a patient's lifestyle and situation was consequently invaluable for the physician.[91] Allopathic as well as homeopathic doctors[92] would therefore inquire into their patients' occupational and domestic circumstances.[93] This means that, throughout the examination process, patients were seen as being – to a degree – responsible for their state of health. In addition, the physicians, up until the end of the nineteenth century, also noted the climate as an important factor in their search for the causes of illness.

The fact that a morbid condition was described as a series of symptoms does not mean that pre-modern physicians recorded only symptoms in their records in order to attempt to cure them individually.[94] They did not consider each symptomatic sign separately, but summarized them in constellations that were typical of particular illnesses, an approach that certainly required knowledge regarding the origin and causes of illness. The therapies aimed primarily at the presumed causes, which according to humoral tradition were seen as blockages in the functioning of organs or accumulations of pathogens that had to be discharged, dissipated or cleansed.[95] Physicians who embraced new medical theories were more likely to deviate from humoral medicine in determining the cause of a disease. The Nuremberg town physician Götz may have described his patients' complaints as organ-related but he did not believe that the affliction was caused by organ dysfunction. As a representative of pietistic medicine he would look for the true cause of illness in the patient's 'soul' or 'nature'. As a direct or indirect, though

---

91 Archives of the Institute and Museum of Medical History at the University of Zurich (MHIZ), Praxisjournal Gottfried Wachter (1776–1861), Ms H 17 – III (1808), 5; Cf. Karen Nolte, 'Schwindsucht – Krankheit, Gesundheit und Moral im frühen 19. Jahrhundert', *Medizin, Gesellschaft und Geschichte* 29 (2010), 47–70.

92 Marion Baschin, *Wer lässt sich von einem Homöopathen behandeln? Die Patienten des Clemens Maria Franz von Bönninghausen, 1785–1864* (Stuttgart 2010), 210–21.

93 Cf. Johann Wilhelm Conradi, *Einrichtung der Medicinischen Klinik im Academischen Hospitale zu Heidelberg* (Heidelberg 1820), 25–27. The domestic situation was described as a health hazard even by sixteenth century physicians. Annemarie Kinzelbach, *Gesundbleiben, Krankwerden, Armsein in der frühneuzeitlichen Gesellschaft. Gesunde und Kranke in den Reichsstädten Ueberlingen und Ulm, 1500–1700* (Stuttgart 1995), 194.

94 Cf. Carl-Gustav Schmalz, *Versuch einer medicinisch-chirurgischen Diagnostik in Tabellen oder Erkenntniß und Unterscheidung der innern und äußern Krankheiten mittelst Nebeneinanderstellung der ähnlichen Formen. Zweite, viel vermehrte und verbesserte Auflage* (Dresden 1812).

95 Götz, for instance, used 'flux-dissipating' balms, plasters, powders, drops, poultices, 'blood-purifying' drops UBE Ms. 1201/2, 6, 67, 77, 103, 105, 108, 121, 130, 504, 517.

not necessarily causal, influence in pathological processes, incidents such as falls were also recorded since they, in humoral terms, might occasion internal diseases.[96] A patient's state of mind was also considered as a potential cause of physical dysfunction.

Humoralism entailed an understanding of the body and mind as closely interacting entities rather than being strictly separated. Melancholy, for instance, was considered to be a state of mind caused by liver-related dyscrasia: too much bile accumulated in the blood and verifiably changed its consistency.[97] In their practice records the physicians of the seventeenth to nineteenth centuries recorded the interaction between their patients' physical complaints and their emotional state. Allopathic as well as homeopathic doctors saw emotions such as 'shock' or 'grief' as important factors in the origin and course of diseases.[98]

Around the year 1800 we discern a noticeable shift in the medical literature as regards the diagnostic process: the term 'diagnosis' was introduced as a progression from the traditional semiotic view. As the German medical professor Weber stated in 1847: 'The symptoms as such provide nothing but material for fantastical conjecture'.[99] The proponents of a 'scientific' approach thought that, in terms of accuracy, the new diagnostic system would eclipse the old semiotic system[100] because it combined the insights of pathological anatomy and physiology with physical examination.[101]

The practice notes do not substantiate the theory of a sudden change from semiotics to the new 'diagnostic approach' extolled in the contemporary medical publications. Diachronic comparison reveals instead that the traditional semiotic and the 'newer' diagnostic approaches co-existed far into the nineteenth century. The physicians, while holding on to traditional diagnostic procedures, also used the innovative diagnostic devices and techniques.

---

96 Cf. IGM Charité, Glaser, Ms., 31 March 1763.

97 MHIZ, Ms., H 17 – IV (1818); cf. Aline Steinbrecher, *Verrückte Welten. Wahnsinn und Gesellschaft im barocken Zeitalter* (Zürich 2006); Christina Vanja, 'Seele und Körper (Neuzeit)', in Peter Dinzelbach (ed.), *Europäische Mentalitätsgeschichte* (Stuttgart 1993), 200–07.

98 IGM Charité, Glaser, Ms., 31 December 1753; IGM Stuttgart, P 116, fol. 172.

99 Ferdinand Weber, *Kurze Bemerkungen über die Section der Leiche zu pathologischen Zwecken: ein Gruss an seine Herren Collegen in den Herzogthumern Schleswig-Holstein und Lauenburg, von Dr. F. Weber, Prosector an der Universitat in Kiel* (Kiel 1847), 9.

100 Bock, 1853, VI; Wichmann, 1800, 12, 32; Carl August W. Berends, *Vorlesungen über praktische Arzneiwissenschaft* (Berlin 1827), vol. 1 (*Semiotik*), 7–9.

101 Bock, 1853, 7; Weber, 1847, 7–9.

Therapy

The therapeutic practice was dominated by the diverting, discharging or balancing measures arising from the humoral view of the body and of illness. This applied for all the doctors' practices in the German-speaking realm, from the mid-seventeenth to the late nineteenth century, that were investigated within the Research Network. In those two and a half centuries the physicians diverged hardly at all from traditional and proven treatments, such as bloodletting, and they continued to prepare their medications from more or less the same stock of remedies.

Medical practitioners such as Bloesch, who was town physician and public health officer in the Swiss town of Biel – while they knew of new medical theories such as Brunonianism – appear not to have been influenced by them in their choice of therapies. Obvious therapeutic changes are only discernible in Bönninghausen's homeopathic practice and with the pietistic Nuremberg town physician Götz: both aligned their treatments, and especially their medication, to the pathologies that they saw as innovative in their respective life times. Bönninghausen rejected bloodletting, although it was very popular with his patients, and prescribed exclusively homeopathic medicines.[102] New pathological concepts along with their therapeutic implications were occasionally also integrated into the existing humoral system without calling its principles into question: the Nuremberg physician Götz, who gave his patients the new 'proprietary medicines' (*Carl'sche Pillen,* for instance) continued to prescribe the traditional therapies that mostly consisted in diverting or purging.[103]

In their choice of therapies the physicians were ultimately guided by their experience with the tolerability and effectiveness of interventions. Bloesch, who was well aware of the contemporary medical and diagnostic innovations, condemned the 'new dogmatic systematists',[104] insisting that a physician, when it came to treating patients, should be guided by 'empiricism' only.[105] Inquiry into the therapies applied in the Würzburg and Göttingen policlinics during the first half of the nineteenth century, revealed, on the other hand, that theory construction and empiricism were not necessarily mutually exclusive

102 Cf. Marion Baschin, 'The Patients' Choice. How and Why Sick People Used Homeopathy in 19th Century Münster', in Martin Dinges (ed.), *Medical Pluralism and Homoeopathy in India and Germany (1810–2010). A Comparison of Practices* (Stuttgart 2014), 149–65.

103 Cf. Kinzelbach *et al.*, '*Observationes*' (in this volume).

104 Cf. Boucherin, 2008, 153.

105 Cf. the subchapter 'Knowledge as prerequisite and objective' in this contribution.

and could well co-exist: enthused by the scientific impetus of the Natural History School of Medicine physicians hoped to develop, at the bedside, a pragmatic nosology, while their therapeutic concepts and practices still relied on the proven humoral methods of the past. Physicians not only adjusted their therapies to what was required in practice; they also assumed a pragmatic attitude towards the treatment methods chosen by the patients or suggested by other, non-academic, practitioners.

All practice notes show evidence that it was normal procedure for patients to make use of the multiple methods available on the extensive health market.[106] The university-trained physicians largely tolerated third-party or self-treatments, even if they were at variance with their own concepts. This applies even to the homeopath Bönninghausen, who had no choice but to accept the fact that his patients treated themselves or were treated by other practitioners using allopathic humoral methods. He certainly never dismissed patients because of their choices.

As is evident from Ottenthal's country practice in Southern Tyrol as well as from Goetz' Nuremberg practice, physicians managed to assume some kind of 'professional neutrality'[107] towards the 'lay healers'.[108] Analysis of their practice records suggests that the therapies offered by the licensed *doctores* of medicine and those of other groups blended seamlessly and co-existed in ways that seem to contradict the rhetoric of medical journalism. But the various approaches were not only based on the same humoral ideas they also used the same kinds of remedies.[109] Ottenthal reported that, before consulting him, many of his patients had taken medication they had prepared themselves or procured from itinerant merchants. Since there were but few apothecaries in Tyrol, Ottenthal supplied his patients from his own medicine chest. The university-trained physician produced his own salves based on traditional local recipes.[110]

University educated physicians, as well as using traditional methods of popular medicine, on occasion also applied magical practices. This we learn from the records of the Thuringian physician and executioner's son Glaser who used them against black magic. Glaser was also not opposed to his patients

106 Weikl *et al.*, 2012; Martin Dinges, 'Arztpraxen 1500–1900. Zum Stand der Forschung', in Dietrich-Daum *et al.*, *Arztpraxen*, 2008, 23–61: 25–26.

107 Oberhofer, 2008, 171; cf. Francisca Loetz, *Vom Kranken zum Patienten. 'Medikalisierung' und medizinische Vergesellschaftung am Beispiel Badens 1750–1850* (Stuttgart 1993), 99, 110, 227.

108 Cf. Weikl *et al.*, 2012.

109 Loetz, 1993, 113, 115.

110 Oberhofer, 2008, 174.

using amulets as long as these enhanced the healing process.[111] The examples quoted illustrate that there was no great discrepancy between the medical *doctores* and the so called 'lay healers' like Gottfried Wachter who was active in the Swiss canton of Thurgau. The gulf between academic medicine and the wider medical market can therefore only have emerged in the second half of the nineteenth century.

Diachronic comparison of the mid-seventeenth to late nineteenth century shows little change in the *materia medica* used by medical practitioners. Across that period the physicians, with the exception of the homeopath, drew on the same wide range of plant, animal and mineral substances for their medications, adding them either as powders to beverages, or preparing decoctions, pills or drops.

The – mostly diverting, dissolving or balancing – treatments applied in the practices investigated also hardly changed. Bloodletting, cupping and scarification, as well as applications of leeches, blistering plasters (with mustard for instance), 'Spanish fly' or fontanels remained the main treatments of choice from the seventeenth to the nineteenth century. Enemas and rinsing (with a syringe), using diverse oil mixtures or herbal solutions, also continued to be mainstays of medical therapy. External applications such as full baths or foot baths with ash, mustard or salt were recommended to support the healing process and alleviate ailments.

As well as generally established prescriptions, we find in the practice notes specific preparations or medicines developed by the individual physicians: Magirus prescribed his own specifics (*propria*) in Berlin and Götz gave proprietary medicines some of which were produced by pietistic manufacturers. Apart from herbal remedies, Magirus gave his patients also 'chymical' medicaments. The medicines which were based on the principle of contraries (*contraria contrariis*) were thought to have an effect on the inner *causa morbi* and stimulate the discharge of harmful matter. Magirus and Götz recorded their recommendations in the form of detailed prescriptions. The German terms used by both these early modern physicians to name their medicines implied the effect they aimed to achieve. Götz, for instance, prescribed *herzstärckendes Kraftwasser* (cardiotonic water), *schärfeabführendes Pulver* (antacid powder) and *blutreinigende Tropffen* (blood-cleansing drops).[112] Magirus, on the other hand, used Latin words, as in *spiritus antihypochondriacus*, *emplastrum ad*

111 Johann Friedrich Glaser, *Nützliches Verhalten bey der jetzo in Deutschland regierenden Fleckfieberseuche: um sie zu verhüten und wieder zu dämpfen* (Hildburghausen 1758), 117.

112 UBE Ms. 1201/2, 57, 99, 108.

hydropem (= dropsy plaster)[113] as well as German designations such as *Schläffsälblin* (salve for the temples), *Herzstärkendes Tränklein* (cardiotonic drink) and *Wasser zum Schwitzen* (water to induce sweating).[114] Fuchs' patient histories list the composition and all ingredients of medicines, usually in conjunction with indications regarding the desired therapeutic outcome.

The treating physicians were not always sure whether the chosen therapy would improve the patients' health or cure them. Treatments might also fail: the Nuremberg physician Götz recorded failed attempts and published them anonymously.[115] The entries of 'lay' and academic practitioners reveal that medicines and therapies were continually tested, adapted and changed.[116]

With medications and therapies one observes local or regional traits that also reflect the personal preferences of the prescribing physician. These preferences related either to particular products – such as the '*Carl'sche Pillen*' mentioned earlier, Dr Glaser's *Hirschhorngeist* (ammonium hydroxide) or the herbal wines of the healer Gottfried Wachter – or to particular therapeutic approaches as with Ottenthal's dietetics.

The lively discourse which arose in the eighteenth century on the therapeutic expediency of venesection[117] also reached the doctors' practices investigated here.[118] The practice of bloodletting might not have been equally important in all of them, but the popularity of the method remained almost undiminished despite the criticisms that were raised from the seventeenth right up to the late (!) nineteenth century.[119] Like other interventions that are classified as 'minor surgery', venesection tended to be conducted not by the physicians but by the practically trained surgeons or barber-surgeons. Numerous treatments – the tooth extractions that the Tyrolese physician Ottenthal carried out himself, for instance – prove, however, that the line that divided academic and non-academic practitioners was occasionally crossed.[120]

Ottenthal as well as Bloesch acted as obstetricians up until the second half of the nineteenth century. Their early modern colleague Götz, on the other hand, treated and advised pregnant and puerperal women but was never

113 UB Marburg, MS 96, 7, 16.

114 UB Marburg, MS 96, 11, 72, 33. All examples are from Magirus' Berlin practice, where he recorded details of his medications in his casebook.

115 Splinter, 'Medicus', 2011, 7–8.

116 MHIZ, Ms H 17 – II (1805), 31; IGM Charité, Glaser Ms., 27 July 1763.

117 Johann Samuel Carl and Georg Christian de Cilano, *Erfahrungs-Gründe von des Blutlassens wahren Gebrauch und Mißbrauch* (Flensburg 1739).

118 Glaser, *Verhalten*, 1758, 42.

119 Cf. Dinges, 2008, 48.

120 MHIZ, Ms. H 17 . XII (1843), 2.

present at the delivery.[121] Although the training and practice of midwifery was increasingly monitored due to the growing involvement of university-educated physicians in obstetrics, midwives did not forfeit their professional autonomy as a result.[122]

All case histories reveal evidence of patients treating themselves before or while being treated by an academic physician. Self-medications were often critically commented on but they were generally tolerated. In exceptional cases, as in pietistic medicine for instance, self-medication was part of the therapeutic concept – which is why the practice was encouraged by the Nuremberg physician Johann Christoph Götz.

Overall, the practice notes show that the wishes of patients were taken into account as long as the physicians did not consider the chosen therapies to be harmful. Patients clearly had a say in the choice of therapy, a fact that was discussed with ambivalence in the contemporary medical literature and seen as responsible for tensions in the doctor-patient relationship.[123]

Some patients demanded particular therapies so resolutely that the physicians were reluctant or unable to refuse them. Among these therapies were, in the case of Ottenthal, extreme purging, drastic pain relief as well as bloodletting, a measure that, as a traditionally proven cure, was frequently requested by patients across the centuries. One of the reasons why physicians gave in to the wishes of patients might have been that the success of their practice relied on patient compliance. A degree of tolerance and acceptance of the patients' therapeutic preferences was clearly part of the physician's *savoir faire*.[124] Götz – as well as the physicians in the policlinics – discontinued drugs if patients did not tolerate them or were put off by their taste. Children were treated with particular benevolence: several of the practitioners investigated, Magirus included, would sweeten medicines for children.

The various doctors were consulted by patients from all walks of life. The case histories suggest that therapy and medication *could* depend on a patient's social status: Magirus, for instance, not only spent more time on diagnosis with

121 Cf. Kinzelbach 2014.

122 Marina Hilber, 'Der Landarzt als Geburtshelfer – Dr. Franz von Ottenthal und der medizinische Markt in Südtirol (1860–1869)', in Lina Gafner *et al.*, 2012, 141–57; Philipp Klaas, 'Cäsar Adolf Bloesch – ein Geburtshelfer? Einblicke in seine ärztliche Praxis zwischen 1832 und 1850', in Lina Gafner *et al.*, 2012, 110–25; on the question of the 'deprofessionalization' of midwives in the nineteenth and twentieth century cf. Hans-Christoph Seidel, *Eine neue 'Kultur des Gebärens'. Die Medikalisierung von Geburt im 18. und 19. Jahrhundert in Deutschland* (Stuttgart 1998), 274–76.

123 Loetz, 1993, 100–01.

124 Cf. Ludwig Choulant, *Der junge Arzt am Krankenbette* (Leipzig 1823), 28–53.

rich patients he also prescribed more expensive medicines, some of them containing substances like emerald. For his poor patients Bloesch recommended mostly home remedies that were in line with their usual choice of self-treatment (baths, cupping, herbal teas etc.). In the Göttingen and Würzburg policlinics, physicians sometimes prescribed expensive ingredients such as liquorice which were not listed in the official pharmocopea for the treatment of poor patients, especially when patients were very sick.[125] This is the more surprising since the authors of medical writings explicitly recommended not adding such costly ingredients to prescriptions for poor patients.[126] This attitude may have been due to the fact that the policlinics depended on the goodwill of their patients. They were crucial for the practical medical training of the students and it was preferable to make sure that these patients did not turn to non-academic healers: again, compliance emerges as a crucial factor in therapy.

Healing, Recovery or Demise?

What therapists and patients expected from a chosen treatment cannot be as easily established as the two only symbols used by Götz since 1720 might suggest: he entered a red cross when a patient had died and the letter 'R' for patients who 'got healthy again'.[127]

The meaning of 'healthy' depended on how 'health' was defined at any given time. For Magirus, for instance, curing his patients meant removing their present complaints.[128] Likewise Götz entered 'R' for cured if the fever, for which he had been consulted, had actually subsided.[129] He achieved such relative cures in 1721 in about 46 per cent of his new patients; but this does not exclude the

125 Cf. Karen Nolte, 'Die medizinische Versorgung der Göttinger Stadtarmen durch das Poliklinische Institut der Universität', *Würzburger medizinhistorische Mitteilungen* 29 (2010), 131–57: 150.

126 Christoph Wilhelm Hufeland, 'Ankündigung des Königl. Poliklinischen Instituts auf der Universität zu Berlin nebst den Gesetzen desselben', *Journal der practischen Arzneykunde u. Wundarzneykunst* 31, no. 10 (1810), 3–56; Christoph Wilhelm Hufeland, *Armen-Pharmakopöe. Zugleich eine Auswahl bewährter Arzneimittel und Arzneiformeln, Fünfte vermehrte Auflage* (Berlin 1828).

127 Götz' translation of 'R', short for *reconvaluit*, UBE Ms. 1201/1, flyleaf after the frontispiece.

128 Sabine Schlegelmilch, *Die ärztliche Praxis des 17. Jahrhunderts – Johannes Magirus (1615–1697)* [scheduled for 2015].

129 UBE Ms. 1201/2, 2, 11.

possibility that he continued to treat the same persons for other ailments.[130] A number of factors may have led to the pronouncement of a treatment being successful: in his annual report of 1835 on the Würzburg policlinic Fuchs differentiated the categories 'cured' (721), 'improved' (56), 'referred' (19) (to other institutions) and 'deceased' (50). His definition of 'cured' was also based on subjective experience. Fuchs probably needed to be seen as particularly successful, not only among the 'poor'.[131] Wachter, who in 1805 reported a success rate (i.e. patients marked 'healthy' or 'cured') of almost 90 per cent might, as a 'lay healer', also have depended on a successful track record. Both practitioners followed the tradition of collecting evidence of their accomplishments in 'case histories'[132] as well as other demonstrations of success.[133]

Götz' entries, by contrast, were more nuanced. From his detailed notes and the up to 40 contacts per patient per year that he occasionally recorded, it is apparent[134] that his clients often only expected pain relief or other kinds of alleviation.[135] Most physicians provided long-term care for their patients, a fact that is particularly noticeable in Bloesch's practice: in 1842 almost half of his consultations were with patients who saw him more than thirty times in that year, with only a few of them being marked 'cured' or 'healthy'.[136] Evidence of therapies aiming at 'improvement' can also be found in the notes of Ottenthal, Bönninghausen and Glaser.

The second of Götz' symbols, a red cross, signified death – the end or other extreme of what could be expected. Up until 1863 we find the same symbol in Ottenthal's notes. In 1721, Götz marked 7 per cent of his patients with a red

130 UBE Ms. 1201/2. The success rate recorded ten years later by the practitioner Albrecht von Haller was noticeably lower, cf. Steinke *et al.*, 2007, 138.

131 Fuchs also registered a very high percentage (85%) of 'cured' patients three years later in the policlinic in Göttingen, cf. Nolte, 'Versorgung', 2010, 143.

132 Cf. Gianna Pomata, 'Sharing Cases. The Observationes in Early Modern Medicin', *Early Science and Medicine* 15 (2010), 193–236: 206; Marion Maria Ruisinger, 'Briefpraxis versus Besuchspraxis. Das Beispiel Lorenz Heister (1683–1758)', in Dietrich-Daum *et al.*, *Arztpraxen*, 2008, 65–77: 73.

133 On successful strategies of women cf. Gianna Pomata, 'Medicina delle monache. Pratiche terapeutiche nei monasteri femminili di Bologna in età moderna', in Gianna Pomata (ed.), *I monasteri femminili come centri di cultura fra Rinascimento e Barocco* (Roma 2005), 329–63: 335–36, 347; Annemarie Kinzelbach, 'Konstruktion und konkretes Handeln. Heilkundige Frauen im oberdeutschen Raum, 1450–1700', *Historische Anthropologie* 7 (1999), 165–90: 177–89.

134 Cf. Kinzelbach *et al.*, '*Observationes*' (in this volume).

135 UBE Ms 1201/2, 456, 487, 505.

136 Cf. Hubert Steinke, 'Arzt und Patient um 1850', in Peter Martig (ed.), *Berns moderne Zeit. Das 19. und 20. Jahrhundert neu entdeckt* (Bern 2011), 181.

cross.[137] At 6 per cent, the mortality rate recorded by Fuchs was also very low. Even in early modern times dying patients received palliative care from their physicians in the form of medical interventions they hoped would relieve symptoms or delay the course of illness. This *cura palliativa* was distinct from therapies that aimed at restoring a patient's health[138] and was, not least, a means of enabling patients to die 'appropriately' (in a religious sense).[139] Pain relief was therefore a minimum requirement of palliative care. The pietistic physician Johann Christoph Götz would treat and support dying patients in many ways, right to the end, and then record their 'appropriate', peaceful, passing away: 'At around nine o'clock he peacefully breathed his last'.[140] He also wrote reports about the physical, spiritual and moral state of patients who had died in his absence.[141] From around 1800 the medical provision for the dying, known as *euthanasia medica*, assumed a more central place in the professional self-image of university-trained physicians, as is apparent in the behaviour of the physicians at the policlinics in Würzburg and Göttingen. They did, however, not restrict their medical palliative care to pain relief but used tumour reducing drugs for patients with advanced cancer in the hope of easing their suffering and prolonging their lives.[142]

There is evidence, in some individual cases, of physicians' interest in diseases and their treatment continuing after the death of a patient and including the patient's entire family. Götz' notes contain detailed descriptions of post-mortems carried out with all family members present, including the children, servants, neighbours and friends of the deceased.[143] Such autopsies were part of the public funeral rites and a sign of a family's wealth.[144] Descriptions of autopsies carried out within the family circle are also present in Bloesch's journals, especially in the eight out of nine post-mortems that were paid for by the

137 UBE Ms. 1201/2.

138 Cf. Michael Stolberg, '"Cura palliativa". Begriff und Diskussion der palliativen Krankheitsbehandlung in der vormodernen Medizin (ca. 1500–1850)', *Medizinhistorisches Journal* 42 (2007), 7–29. On palliative care in early modern times cf. Michael Stolberg, *Die Geschichte der Palliativmedizin. Medizinische Sterbebegleitung von 1500 bis heute* (Frankfurt am Main 2011), esp. 21–66.

139 Cf. *ibid.*, 91–116. On the 'blessed passing away' in the pietistic bourgeoisie cf. Katharina Ernst, *Krankheit und Heilung. Die medikale Kultur württembergischer Pietisten im 18. Jahrhundert* (Stuttgart 2003), 201–18.

140 '*...ix circiter placide animam expiravit*', UBE Ms. 1201/2, 227.

141 UBE Ms. 1201/2, 56–57, 315–16; 1202/2, 488–89.

142 Cf. Nolte, 'Versorgung', 2010; Nolte, 2007.

143 UBE Ms. 1201/2, 469.

144 Cf. Ruisinger, *Patientenwege*, 2008, 208–10.

families. The number of officially commissioned and paid-for autopsies (23) shows, however, that public inquests into the cause of death predominated. The change in the way autopsies were recorded reflects a fundamental re-orientation in medical science.[145] The eighteenth-century *observatio*,[146] that involved the surgeons,[147] referred to individual corpses only, while the nineteenth century 'autopsy reports' were produced in series by the physician.[148] It excluded the families and, as with Fuchs for instance, served primarily the purpose of academic teaching.

Conclusion – Links between Theory and Practice

The expediency of theoretical knowledge was widely questioned as early as the eighteenth century. Still in 1866, Bloesch attached greater importance to personal observation.[149] Some existing studies suggest, however, that other medical practitioners believed that theory, science and practice were intimately linked.[150] But what little evidence research has found of how knowledge, science and medical practice influenced each other is based on very few examples.[151] The comparative inquiry into such questions carried out by our research network, as well as work on the resulting exhibition 'Praxiswelten' that strongly focused on particular practices and embodied knowledge,[152] have provided deeper insights that widen and qualify earlier research results.

The physicians' scientific publications and medical practice suggest that the division into periods proposed by historiography needs to be reconsidered.

145 Cf. Hess, *Semiotische Medizin*, 1993, 300.

146 StA Biel, 4 A 1, Acta 1730, 423–25, 461–63.

147 It is only apparent from the journal that the *sectio* was carried out by the surgeon. UBE Ms 1201/2, 469.

148 As medical authors had requested for more than a century, cf. Karin Stukenbrock, *'Der zerstückte Cörper'. Zur Sozialgeschichte der anatomischen Sektionen in der frühen Neuzeit (1650–1800)* (Stuttgart 2001), 123–49, 251–52.

149 Cf. Broman, 2003, 483; Cook, 2006, 432–34.

150 Thomas Schnalke, 'Praxis, Theorie und Wissenschaft im Selbstverständnis des städtischen Arztes Christoph Jacob Trew (1695–1769)', *Gesnerus* 52 (1995), 40–53: 41, 49–50.

151 Bresadola, 2011, 199, 212–16; John Crellin, 'Theory and Clinical Experience in Eighteenth-Century Extemporaneous Prescriptions – a Reciprocal Relationship', *Pharmacy in History* 48 (2006), 3–13: 8.

152 Cf. the illustrated catalogue of the exhibition: Isabel Atzl, Roland Helms, Stephanie Neuner and Ruth Schilling, *Praxiswelten. Zur Geschichte der Begegnung zwischen Arzt und Patient* (Ingolstadt 2013).

Whether acquired knowledge was retained or altered as a result of empirical experiences was by no means dependent on when a particular physician lived and took his notes. The holding on to a once adopted theoretical concept and its almost unaltered practical application over a period of time is as characteristic of the iatromathematician Magirus as of the homeopath Bönninghausen, even though the two were separated by a quarter of a millennium. While Bönninghausen and Ottenthal had a passive interest in scientific discourse because, as medical experts, they formulated the prevailing diagnoses, they did not actively participate in it. Whether their lack of interest can be attributed to personal reasons, such as being a member of the landed gentry, or whether it was the result of general medical and cultural developments in the second half of the nineteenth century[153] can only be established by comparative analysis of further practice records. But it is obvious how administrative requirements promoted the perception and implementation of new medical concepts.

The practice records reveal, in many instances, that forming a diagnosis was a multi-layered and complex cognitive process. The comparison of diagnoses and therapies shows, moreover, that the physicians were directly influenced by the scientific views they had adopted. Iatromathematical methods directly entered into the diagnoses established by Magirus[154] while Götz' therapies were informed by Stahlianism and Pietism. Bloesch admitted to having been influenced by the bedside training in Göttingen and Paris but he distanced himself from the publication methods of the Parisian clinicians.[155]

The example of Fuchs, however, illustrates that diagnoses were not just informed by a school of thought, in his case the Natural History School of Medicine. It was more a matter of diagnoses serving to teach the corresponding methods in medical school – once they had been developed at the bedside

153 Cf. Dietrich-Daum *et al.* (in this volume). Growing medical specialization (cf. Claudia Huerkamp, *Der Aufstieg der Ärzte im 19. Jahrhundert. Vom gelehrten Stand zum professionellen Experten. Das Beispiel Preussens* (Göttingen 1985), 177) may have constituted a hindrance, as much as the fundamental change of the communication culture from early modern times with their extensive correspondence networks (cf. Klaus-Dieter Herbst and Stefan Kratochwil (eds), *Kommunikation in der Frühen Neuzeit*, (Frankfurt am Main 2009)) to the nineteenth century when discussions tended to be confined to groups of 'experts' (cf. Eric J. Engstrom, Volker Hess und Ulrike Thoms (eds), *Figurationen des Experten. Ambivalenzen der wissenschaftlichen Expertise im ausgehenden 18. und frühen 19. Jahrhundert* (Frankfurt am Main 2005)).

154 Cf. Sabine Schlegelmilch, '"What a Magnificent Work a Good Physician is". The Medical Practice of Johannes Magirus (1615–1697)' (in this volume).

155 Cf. Lina Gafner, 'Administrative and Epistemic Aspects of Medical Practice: Caesar Adolf Bloesch (1804–1863)' (in this volume).

and brought into the scientific form of case histories. Fuchs, Götz and Bloesch were keen to generalize, through publication, the practical experience they had gained in their individual activities and the knowledge documented in their practice records, and to transport it from the medical practice to the world of the scholars. In other words, they aimed at discussing this knowledge as 'science' and, in doing so, establishing themselves regardless of location.[156]

Beyond that, it emerges with all the physicians that their recording of 'symptoms' was not a purely descriptive procedure, but required, even in early modern times, the consideration of causes and therefore implicit knowledge. Long before the shift towards scientific medicine occurred, objective examination methods had become an integral part of a physician's diagnostic practice. But new methods were added in the course of the nineteenth century, enabling physicians to gain insight into processes inside the body more independently of the patient. The mutual interpenetration of medical practice and medical and technical knowledge, facilitated by the industrial production of technical instruments, is reflected in the use of stethoscopes and thermometers. 'Uroscopy' was transformed into a chemico-technical examination of the urine consistency.

As little as diagnoses were established exclusively on the basis of the conversation between physician and patient in the seventeenth century, did conversations between doctor and patient lose their significance as a method of examination, even if nineteenth century medical authors strongly questioned that any useful information could be obtained from the 'subjective' narrative of patients.[157]

The tendency of physicians to hold on to their therapies illustrates how highly they valued empirical knowledge. Even though further and more detailed comparisons would be desirable in this respect, these tendencies can be explained by the fact that the patients' body of knowledge played an important part. Practices like bloodletting and certain medications show that physicians tended to consider the wishes also of poorer patients in order to achieve a high degree of compliance.

Historiography considers that, in early modern times, rich and educated patients and their relatives played an important part in defining what was acceptable as scientific innovation,[158] while a gap opened in the nineteenth

156 Cf. Kay Peter Jankrift *et al.* (in this volume).

157 Cf. Conradi, 1820, 17–27.

158 Cf. Stolberg, *Experiencing Illness*, 2011, 217; Domenico B. Meli, *Mechanism, Experiment, Disease. Marcello Malpighi and Seventeenth-Century Anatomy* (Baltimore 2011), 196; Bresadola, 2011, 195–98, 211.

century between the newly emerged medical knowledge of physicians and the wide-spread humoral views of their patients, which meant that the two no longer shared the same basis of knowledge.[159] Götz' interest in the views of his patients – which was in no way socially discriminating and even led him, at the beginning of the eighteenth century, to use edited parts of the records in his scientific publications – seems, initially, to confirm these theses to an extent, as does the opposite behaviour of Glaser in the second half of the eighteenth century and that of Bloesch in the 1830s and 1840s.

At the same time the records contradict the postulation that the gap between the knowledge of physicians and that of their clientele was a new phenomenon: this chasm existed even in the seventeenth century since we find hardly any narratives of patients or their relatives in Magirus' notes. The gaze into the body of a deceased patient during a dissection on the other hand, united relatives and 'family physicians' in Götz' early eighteenth century practice as much as it did in Bloesch's practice a century later when scientific anatomy began to lose its popularity.[160] As late as the second half of the nineteenth century, Ottenthal, who often treated patients from a geographic distance, relied on the narratives of patients or their messengers.[161] Their communication consequently required a shared store of knowledge.

---

159 Cf. Ruisinger, *Patientenwege*, 2008, 21.

160 Cf. Stukenbrock, 2001, 240–41, 247.

161 HM, 134/1852, 03.02.1852; HM, 325/1860, 5 March 1860.

CHAPTER 5

Medical Practice in Context: Religion, Family, Politics and Scientific Networks

Ruth Schilling and Kay Peter Jankrift

The road, the land, and the family had always defined and secured his place in the world.[1]

Introduction

The cultural, societal and political circumstances surrounding a doctor's practice decisively influence its structure and development.[2] Evidence of such influences can be found in the medical practice records even though one often has to read between the lines to discover them. While a patient's name, gender, age, illness and treatment do not, in themselves, say much about the physician's activities outside his medical practice, they provide the first clues for more

* We thank the following authors of the individual projects in the DFG Research Network 'Medical practices (1600–1900)', on whose partly still unpublished results this contribution is largely founded: *Sabine Schlegelmilch (J. Magirus), Kay Peter Jankrift/Annemarie Kinzelbach/Susanne Grosser/Marion Ruisinger (Chr. Götz), Ruth Schilling (J.F. Glaser), Stephanie Neuner/Karen Nolte (Ch. Fuchs), Katharina Weikl/Alois Unterkircher/Iris Ritzmann (G. Wachter), Lina Gafner/Philipp Klaas (C.A. Bloesch), Marion Baschin (F.P. von Bönninghausen), Marina Hilber/Eberhard Wolff/Alois Unterkircher (F. von Ottenthal).*

1 Jacalyn Duffin, *Langstaff. A Nineteenth-Century Medical Life* (Toronto 1993, repr. 1999), 10.

2 For examples cf. Mary Wilson Carpenter, *Health, Society and Medicine in Victorian England* (Santa Barbara [CA] 2010); Giorgio Cosmacini, *Storia della medicina e della sanità in Italia* (Roma ³1987); Mark S.R. Jenner, *Medicine and the Market in England and its Colonies, c. 1450–c.1850* (Basingstoke 2007); Robert Jütte, *Ärzte, Heiler und Patienten. Medizinischer Alltag in der frühen Neuzeit* (München 1991); Hilary Marland, *The Task of Healing. Medicine, Religion and Gender in England and the Netherlands, 1450–1800* (Rotterdam 1996); Roy Porter, *Disease, Medicine and Society in England* (Basingstoke 1989); Michael Stolberg, *Experiencing Illness and the Sick Body in Early Modern Europe* (Basingstoke 2011), 58–64.

 | DOI 10.1163/9789004303324_007

in-depth research in other written testimonies. Based on the academic physicians and one lay healer investigated as part of the Research Network 'Medical practices (1600–1900)' we will explore the mutual influence that existed between the medical practices and their cultural, political and societal contexts.

Recent historical research has shown that patients consciously compared the various kinds of healthcare available to them and chose what they deemed to be best, most affordable or most trustworthy.[3] This made it necessary for physicians to find ways of distinguishing themselves from their rivals, an endeavour that proved most successful when they had familial, religious or socio-political ties with the world of the patients. Such ties formed a solid foundation for a high degree of trust on both sides of the doctor-patient -relationship. Focusing on the medical market can, therefore, cast light on the relationship between the various healthcare providers and on the status the individual physicians or healers commanded in society. Both aspects are of interest: the influence of the context on the medical practice as well as the importance of the medical practice for the healers' general standing in society.[4] Under the title 'Penning Patient's Histories' a special issue of the journal 'Gesnerus' has investigated this dual relationship with regard to several projects within the DFG-Research Network 'Medical practices (1600–1900)'.[5] When it came to choosing a healer, trust was one of the criteria that remained constant across the centuries investigated.[6] The building of trust is, however, enhanced by factors which are difficult to quantify since they arise from the most diverse forms of community building and legitimation.[7] These forms can have domestic, religious, political or academic-scientific roots. It is precisely

3 Cf., with further information, Mary Lindemann, *Medicine and Society in Early Modern Europe* (Cambridge ²2010), 237–39; Marion Baschin, Elisabeth Dietrich-Daum and Iris Ritzmann: 'Who Consulted the Doctor? The Patients and their Motives' (in this volume); for case studies cf. Jütte, *Ärzte*, 1991, 89–131; Gianna Pomata, *Contracting a Cure. Patients, Healers, and the Law in Early Modern Bologna* (Baltimore 1998).

4 Cf. Robert Jütte, 'A Seventeenth-Century German Barber-Surgeon and His Patients', *Medical History* 33 (1989), 184–98: 191.

5 Cf. Lina Gafner, Iris Ritzmann and Katharina Weikl (eds), *Penning Patient's Histories – Doctors' Records and the Medical Market in the 18th and 19th Century* [special issue: *Gesnerus* 69, no. 1 (2012)].

6 Cf. Lina Gafner, Iris Ritzmann and Katharina Weikl, 'Gesundheitsmarkt in ärztlichen Praxisjournalen des 18. und 19. Jahrhunderts', introduction to Gafner *et al.*, 2012, 5–12: 10.

7 Cf. Ute Frevert, 'Vertrauen. Eine historische Spurensuche', in Ute Frevert (ed.), *Vertrauen. Historische Annäherungen* (Göttingen 2003), 7–66; on trust as a factor in scientific discourse cf. Franz Mauelshagen, 'Netzwerke des Vertrauens. Gelehrtenkorrespondenzen und wissenschaftlicher Austausch in der Frühen Neuzeit', in Ute Frevert (ed.), *Vertrauen. Historische Annäherungen* (Göttingen 2003), 119–51; Steven Shapin, *A Social History of Truth. Civility and Science in Seventeenth-Century England* (Chicago 2004).

this diversity of contexts – family, religion, official activities and scientific networks – that we will inquire into in order to explore their significance for medical practice. In summary, we need to ask about the continuity and disruptions in the significance of these contexts for the medical practice. Our starting hypothesis is that the growing dominance of the medical market by the group of academic physicians and their professionalization were closely linked. This dominance went hand in hand with the increasing power of the emerging state over the population.[8] While this relationship affected the corporate and political conditions of medical practice in Europe in various ways,[9] it created the foundation for the 'rise of the physicians' in the nineteenth century.[10] Studies of this development tend to look at the medical profession in general. Biographies such as Hugh Trevor-Roper's magisterial work on the Huguenot court physician Sir Theodore de Mayerne,[11] on the other hand, tend to portray in detail the interpenetration of professional legitimation, medical practice and confessional or domestic factors for individual physicians but do not analyse their results in the light of such long-term developments. Our Research Network makes it possible to look at a range of of individual practices and to ask how the individual practitioner responded to the trend towards placing the medical market under the control of academic physicians, especially if he relied on building and retaining the trust of his clientele in his local environment.

The Family as Social Capital

In the cases we investigated, the physician's family background played an important role in the interaction between his practice and society, and did so

8 Cf. Martin Dinges, '"Medicinische Policey" zwischen Heilkundigen und "Patienten" (1750–1830)', in Karl Härter (ed.), *Policey und frühneuzeitliche Gesellschaft* (Frankfurt am Main 2000), 263–95; Francisca Loetz, *Vom Kranken zum Patienten. 'Medikalisierung' und medizinische. Vergesellschaftung am Beispiel Badens 1750–1850* (Stuttgart 1993); Oliver Stenzel, *Medikale Differenzierung. Der Konflikt zwischen akademischer Medizin und Laienheilkunde im 18. Jahrhundert* (Heidelberg 2005); Michael Stolberg, 'Heilkundige. Professionalisierung und Medikalisierung', in Norbert Paul and Thomas Schlich (eds), *Medizingeschichte. Aufgaben, Probleme, Perspektiven* (Frankfurt am Main 1998), 69–86.

9 Laurence Brockliss and Colin Jones, *The Medical World of Early Modern France* (Oxford 1997), 227–29, 290; Harold J. Cook, *The Decline of the Old Medical Regime in Stuart London* (Ithaca 1986), 47–56.

10 Thomas Broman, 'Rethinking Professionalisation. Theory, Practice, and Professional Ideology in Eighteenth-Century German Medicine', *The Journal of Modern History* 67, no. 4 (1995), 835–72; Claudia Huerkamp, *Der Aufstieg der Ärzte im 19. Jahrhundert. Vom gelehrten Stand zum professionellen Experten. Das Beispiel Preussens* (Göttingen 1985).

11 Hugh Trevor-Roper, *Europe's Physician. The Various Life of Sir Theodore de Mayerne* (New Haven 2006).

consistently for centuries. The family ties generally provided the basic social capital available to the physician on entering the medical market.[12] This capital gained in value when a physician, after finishing medical school, settled in his hometown where everyone knew his family; and it could be further augmented in specific ways through planned marriages and, above all, by finding access to existing networks via contacts with scholars, influential patients or commitments outside the medical practice, for instance in associations, societies, the parish or local politics. While some patients may have sought out a particular practice because of the physician's religious or political orientation, others might have avoided it because they did not share those particular views. A look at some of the medical practices investigated within the Research Network can illustrate such processes.

The Swiss physician Caesar Adolf Bloesch, for instance, returned to his home town of Biel after finishing his studies in 1827.[13] His father, also a physician, had died too early for his son to benefit from his clientele. But the young Bloesch nevertheless possessed a certain amount of social capital: his family had been established in Biel for generations and his mother, the mayor's daughter, was placed in charge of the boarding facilities at the local grammar school after her father's demise. One of his two brothers was active on the political stage while the other made his fortune as an industrialist. Still in his first year of practice, after his return to Biel, Bloesch married Elisa Pugnet, the daughter of the Napoleonic military surgeon Jean-François-Xavier Pugnet, from whose reputation as a poor-physician Bloesch undoubtedly benefited. His first patients were, accordingly, members of the Biel bourgeoisie. As a result of the smallpox vaccinations Bloesch's clientele grew considerably and was joined by members of the working classes who were now prepared to give him their trust. Bloesch also sought to increase his social capital by becoming a member of the 'Casino-Society' in 1830.

We see a similar trend a century earlier with the Nuremberg physician Johann Christoph Götz who also grew up fatherless, but came, like Bloesch, from a reputed and well-established family. It was only a few weeks after Götz had started practicing in his hometown, following the conclusion of his medical studies and after his admittance to the *Collegium Med*icum, that he married, on 29 May 1713, Margaretha Catharina Spitz, the daughter of the theology

12 The use of the terms 'social', 'cultural', 'symbolic' and 'economic' capital is based on the fundamental deliberations of Pierre Bourdieu, cf. Pierre Bourdieu, 'The Forms of Capital', in John G. Richardson (ed.), *Handbook of Theory and Research for the Sociology of Education* (New York 1986), 241–60.

13 Cf. Lina Gafner, 'Administrative and Epistemic Aspects of Medical Practice: Caesar Adolf Bloesch (1804–1863)' (in this volume).

professor Felix Spitz (1641–1716). Her brother Georg would, a few years later, establish the contact between Götz and his most eminent patient, Count Ernst von Metternich.

There were other up-and-coming physicians who benefited from their family ties right from the start because they took over the practice of a father or other relative. As a result, they had from the beginning, a regular client base even if not all the patients of their predecessors decided to consult them. The Münster homeopath Dr Friedrich von Bönninghausen (1828–1910), for instance, took over his father's practice in 1864. The practice of Johann Friedrich Glaser in eighteenth century Suhl constitutes a special case since he presumably took on a thriving medical practice from his uncle, an executioner.

While medical careers were not always predetermined by family tradition one still observes that physicians, up until the end of the nineteenth century, tended to come from families who possessed social standing, influence and occasionally also money. Conrad Heinrich Fuchs (1803–1855), from 1836 professor of pathology in Würzburg and director of the policlinic there, was not born into a family of physicians, but his father had held the elevated position of court advocate. The family of Caesar Adolf Bloesch was certainly also wealthy, as were the families of the Nuremberg physician Johann Christoph Götz and of the aristocrat Franz von Ottenthal. Studying was, afterall, an expensive undertaking. The same applies to the other physicians whose activities were investigated within the Research Network. Even the family of Johann Friedrich Glaser owned extensive property, which was however of little use in the days before the descendants of public executioners gained the right to aspire to a higher status in society. They only gained this right when the occupation of executioner which, for centuries, had been regarded as dishonourable, was finally officially declared respectable. The only exception among the physicians and practices that were examined was the non-academic practitioner Gottfried Wachter whose family background was in farming. But even in his case, the family tradition in healing is likely to have benefitted his son's (unfinished?) academic studies of medicine.

There is also evidence that, in almost all cases, the place of origin played a part in the establishment of the practices. Johannes Magirus, born in Joachimsthal in 1615, was one of few among the physicians investigated who did not settle down in his birthplace. He changed his location repeatedly before finally establishing himself as a professor in Marburg. The Swiss physician Caesar Adolf Bloesch, while conducting his studies in Göttingen, Paris and Berlin, returned to his hometown of Biel once these studies were finished. The example of Johannes Magirus illustrates that it could take years to set up a practice in a new environment. He had to work particularly hard to establish a medical reputation. It took him no less than four years to find some kind of access to the existing structures in Berlin. It is therefore hardly astonishing that most physicians chose to settle in familiar territory.

Vita Activa, Religio and Praxis Medica

Religious beliefs and any active involvement in the network of a faith community or – on a smaller scale – in the local parish were usually tied to the family background. Some patients may have deliberately chosen a particular practice because they belonged to the same parish as their physician and appreciated his commitment to a cause they also believed in. It needs to be considered, however, that from the mid-seventeenth to the nineteenth century the religious structures in which the physicians moved were determined by the political situation.

In the age of religious schism Catholicism forfeited its dominant position in Europe, while religion and sovereignty remained inseparably connected.[14] Article 17 of the 1555 Augsburg Settlement gave Protestants, as well as Roman Catholics within its ambit, access to political power, even though this regulation applied originally only to Lutherans. While this concession had long-term consequences, in particular for the various territories of the German Reich, it was also adopted in other parts of Europe. In keeping with the motto *cuius regio, eius religio*, the creed of the territorial lord determined from then on the faith of all his subjects, with the exception of the Jews. Before the importance of religious faith began to decrease as part of the secularization process, the beliefs of healers would affect their careers and the success of their practice.

This applied to an even greater extent to physicians like Johannes Magirus who not only acted as town physicians, but also served at court. Magirus' career was inseparable from his reformed faith. Beginning with his studies at the *Leucorea* – the alma mater of Martin Luther (1483–1546) – through Berlin and Zerbst in the Duchy of Anhalt, which had suffered severely from the ravages of the Thirty Years' War, to his position as professor in Marburg at the first reformed university of the Reich, he worked exclusively in environments with protestant and, in this case, reformed structures. The protestant creed of Johann Christoph Götz, the early eighteenth century Nuremberg physician, also played an important part in his career, as is apparent from his regular church attendance on Sundays of which we read in his records, as well as from

14 As one example from the extensive literature cf. Axel Gotthard, *Der Augsburger Religionsfriede* (Münster 2006).

the noticeable pietistic influences in his medical practice.[15] The majority of his patients belonged to the same parish. And the persons of higher rank who sought Götz' medical services – often as part of an extensive consiliary correspondence some of which is enclosed with his practice journals – were equally rooted in the protestant faith. They included the Margrave of Bayreuth-Ansbach and Count Ernst von Metternich, envoy of Brandenburg-Ansbach to the Perpetual Diet.[16] Although Nuremberg had passed its economic heyday by that time, the city retained a strongly institutionalized medical system as well as a wealthy and reputable bourgeoisie. These bourgeois citizens and their servants made up a considerable share of Götz' clientele.

In the early nineteenth century the political commitment of physicians gained importance, as we see from the cases examined, while the significance of religious affiliation dropped significantly. Only in rural areas did religion remain an issue, at least from a societal point of view, as we see from the example of Gottfried Wachter. He was borough councillor and church treasurer, but religious aspects no longer affected his medical practice. Wachter operated in a confessionally mixed environment and, as a reformist, he also treated catholic patients.

In the Swiss town of Biel, where Caesar Adolf Bloesch was town physician from 1853, people's political views had also gained greater importance than their confession. Due to his family background Bloesch was able to contribute to the political life of the town as president of the municipality and *Bürgergemeinde* (citizens' community) and become active in poor relief and on the school board. It is apparent from the development of his practice that this commitment affected the composition of his clientele.[17]

It was different in the homeopathic practice of Friedrich von Bönninghausen because he was active in a Catholic environment. The relationship between the old-established Westphalians and the ruling Prussians was not entirely without conflict.[18] Bönninghausen, who did not actively engage in politics or church matters, was primarily consulted by patients who chose him for his

15 Cf. Annemarie Kinzelbach, Susanne Grosser, Kay Peter Jankrift and Marion Ruisinger, '*Observationes et Curationes Nurimbergenses*'. 'The Medical Practice of Johann Christoph Götz (1688–1733)' (in this volume); Christa Habrich, *Untersuchungen zur pietistischen Medizin und ihrer Ausprägung bei Johann Samuel Carl (1677–1757) und seinem Kreis* (Habilitationsschrift, Universität München 1981).

16 Susanne Friedrich, *Drehscheibe Regensburg. Das Informations- und Kommunikationssystem des Immerwährenden Reichstags um 1700* (Berlin 2007), 299–301. Peter Fuchs, 'Metternich', in *Neue Deutsche Biographie*, vol. 17 (Berlin 1994), 232–35.

17 See the contribution by Gafner in this volume.

18 Jens Metzdorf, *Münster wird preußisch. Von der fürstbischöflichen Residenzstadt zur Provinzialhauptstadt* (Münster 1992).

homeopathic orientation.[19] That he was also a member of the established Catholic elite in Münster does not seem to have influenced their choice.

Even if references to church attendance and the importance of religious affiliation occasionally shine through, especially in the practice journals of Johann Christoph Götz, the records do not generally refer to a patient's faith. Before the process of secularization set in, this might have been due to the fact that the territorial lord in question determined the faith and all Christian subjects followed suit. It was therefore, outside communes with mixed confessions, not necessary for physicians to record their patients' religious affiliation. Their faith would be obvious unless they were foreign travellers who fell ill *en route* and ended up in the practice by accident. After the secularization the question of confession became largely meaningless and there was no longer a reason for practitioners to record a patient's religious affiliation in their practice journals.

Jewish patients were always an exception to this rule. Johann Christoph Götz, for instance, would add comments such as *judea Fürtensis,* a Jewish woman from Fürth.[20] The Suhl town physician Johann Friedrich Glaser also referred expressly to Jewish patients in his notes, as he did on 4 July 1763 when he was consulted by Hirsch Katz from Bauerbach.[21] The Biel physician Caesar Adolf Bloesch occasionally included the designation *Hebrew*.[22] Why Jewish patients consulted Christian physicians is a question that the journal entries have so far not been able to answer. While the activities of Jewish medical practitioners and the numerous polemical comments against their treatment of Christian patients have been the subject of specific investigations, there are no studies as yet on Jews as patients of Christian physicians.[23] The Jerusalem

19 Marion Baschin, *Wer lässt sich von einem Homöopathen behandeln? Die Patienten des Clemens Maria Franz von Bönninghausen, 1785–1864* (Stuttgart 2010).

20 Universitätsbibliothek Erlangen-Nürnberg, Handschriftensammlung, Ms. 1202/2, p. 52. Entry of Sunday 10 February 1726.
Cf. Annemarie Kinzelbach *et al.*, '*Observationes*' (in this volume).

21 Institut für Geschichte der Medizin (Charité Berlin), Johann Friedrich Glaser: Medicinisches Register auf das Jahr Christi 1750 (Praxistagebuch 1750–1763), Ms., s.p., cf. also the entries of 31 March, 27 July, 4 September, 25 September, 25 October and 6 December 1763.

22 Stadtarchiv Biel, 4 A 1, Tagebuch, vol. 54, 1862, 436. I warmly thank our colleague Philipp Klass in Bern for this information.

23 Kay Peter Jankrift is working on an essay on this topic. On the role of Jewish physicians and the wide-spread polemics against their activities cf., for instance, John M. Efron, *Medicine and the German Jews. A History* (New Haven 2001); Kay Peter Jankrift, 'Eigenes und Fremdes. Zur sozialen Stellung jüdischer Gelehrter und ihrer Rolle für die Vermittlung von Wissen im Mittelalter', in Kay Peter Jankrift and Florian Steger (eds), *Gesundheit – Krankheit. Kulturtransfer medizinischen Wissens von der Spätantike bis in die Frühe Neuzeit* (Köln 2004), 139–54; Nicoline Hortzitz, *Der Judenarzt. Historische und sprachliche Untersuchungen zur Diskriminierung eines Berufsstands in der frühen Neuzeit* (Heidelberg 1994).

Talmud stipulates that no Jew should settle in a town that has no physician.[24] The text does not explicitly specify that it has to be a Jewish physician, but the other Mishna tractates – especially *Avodah Zarah* (idolatry) – make this seem likely.[25] Despite these religious guidelines some practice records include evidence of the treatment of Jewish patients. But their number is so too small to have made a noticeable difference to a doctor's practice. It is also unlikely that physicians went to any particular lengths to recruit Jewish patients, even if they would not have refused to treat them because of their religion.

Physicians in Administration

From an early stage physicians were included in the municipal health services and at court.[26] Their efforts to provide disease control and safeguard the sovereign's health helped to retain the basic framework that gave the authorities in medieval and early modern times political legitimacy. These physicians wanted to benefit from their key positions and were keen to claim superiority over their competitors. On a normative and discursive level the close connection between medicine and state-building has often been described.[27] But little or no attention has so far been given to a possible link between the medical practice and political functions.[28] Although some authors assume that public office had a positive effect on a physician's establishment on the medical market, this is difficult to prove on the basis of the sources available.[29]

---

24 Talmūd Jerūšalmī, Ende Kiddušin. The Babylonian Talmud restricts this interdiction to scholars (b. Sanh. 17b). For a full Hebrew edition of the Jerusalem Talmud visit http://www.mechon-mamre.org/b/r/ro.htm (accessed on 22 August 2014).

25 *'Avoda Sara: Götzendienst*, ed. by Michael Krupp (Ein Kerem 2002).

26 Enno Bünz, 'Leibärzte', in Werner Paravicini (ed.), *Höfe und Residenzen im spätmittelalterlichen Reich. Bilder und Begriffe* (Ostfildern 2005), 156–57; Vivian Nutton, 'Introduction' to *idem* (ed.), *Medicine at the Courts of Europe, 1500–1837* (London 1990), 1–14; Andrew W. Russell (ed.), *The Town and State Physician in Europe from the Middle Ages to the Enlightenment* (Wolfenbüttel 1981).

27 Without any claim to completeness: Christian Barthel, *Medizinische Polizey und medizinische Aufklärung. Aspekte des öffentlichen Gesundheitsdiskurses im 18. Jahrhundert* (Frankfurt am Main 1989); Torsten Grumbach, *Kurmainzer Medicinalpolicey 1650–1803. Eine Darstellung entlang der landesherrlichen Verordnungen* (Frankfurt am Main 2006); Bettina Wahrig and Werner Sohn (eds), *Zwischen Aufklärung, Policey und Verwaltung. Zur Genese des Medizinalwesens 1750–1850* (Wiesbaden 2003).

28 It is not mentioned, for instance, in the comprehensive and relevant anthology by Jürgen Helm and Renate Wilson (eds), *Medical Theory and Therapeutic Practice in the Eighteenth Century. A Transatlantic Perspective* (Stuttgart 2008).

29 Anne Digby, *Making a Medical Living. Doctors and Patients in the English Market for Medicine 1720–1911* (Cambridge 1994), 117–21.

Magirus as well as Glaser acted as town physicians at certain periods in their careers. Götz was a member of the Nuremberg Collegium Medicum. This institution consisted of sworn physicians and was in charge of the kind of policing that, in other towns, was carried out by individual town physicians.[30] The comparison of Magirus, Götz and Glaser has shown that neither a uniform definition nor a diachronic line of development can be established for the office of town physician.[31] It was more that the rights and duties that went with the office of town physician offered each of the three physicians a framework that was compatible with their individual scientific networks and discourses, but particularly also with their family background. Magirus' office in Zerbst meant that he was involved in municipal poor-relief and that he could teach at the *Gymnasium Illustre* which was under ducal patronage.[32] As town physician he was able to combine medical teaching and medical practice. Götz' membership in the Collegium Medicum provided above all the welcome opportunity to gain distinction among Nuremberg's medical scholars. Glaser was probably the one who most needed the official appointment since it helped him – the son of an executioner – to attain an honourable position in the society of Suhl and to participate, due to the utilitarian thinking associated with this office, in the republic of scholars.

From these three cases it can be concluded that, up until the eighteenth century, the office came without a clearly defined 'job description'. While the official duties were specified through oath and appointment, the normative framework was often applied according to pragmatic considerations and depended on the local structures and the situation of each physician.[33] This relative openness is reflected in the practice diaries. Because of Götz' scientific interests, there is a close connection between his case histories, practice diary and printed publications.[34] In Glaser's case there is only limited evidence of a link between his work

30 Cf. Susan Splinter '"Ein jeder der Stadt Nürnberg bestelter Medicus und Physicus ordinarius soll geloben…". Medizinale Strukturen Nürnbergs zu Beginn des 18. Jahrhunderts', *Würzburger medizinhistorische Mitteilungen* 30 (2011), 334–49.

31 Ruth Schilling, Sabine Schlegelmilch and Susan Splinter, 'Stadtarzt oder Arzt in der Stadt? Drei Ärzte der Frühen Neuzeit und ihr Verständnis des städtischen Amtes,' *Medizinhistorisches Journal* 46 (2011), 99–133.

32 Cf. Sabine Schlegelmilch, '"What a Magnificent Work a Good Physician is". The Medical Practice of Johannes Magirus (1615–1697)' (in this volume); Joachim Castan, *Hochschulwesen und reformierte Konfessionalisierung. Das Gymnasium Illustre des Fürstentums Anhalt in Zerbst* (Halle/Saale 1999) 144, 184.

33 Schilling *et al.*, 2011, 124–26.

34 Cf. Susan Splinter, '"Der Aufrichtige Medicus". Eine Zeitschrift des Nürnberger Arztes Johann Christoph Götz (1688–1733) als Vorläufer des "Commercium Litterarium"', *Jahrbuch für Kommunikationsgeschichte* 13 (2011), 5–15.

as town physician and the *Medicinische Register*,[35] while the direct connection between his career as town physician and his publication activities is obvious. In 1758, the year when Glaser was sworn in as official town physician of Suhl, he published treatises on typhus[36] and on a blood-measuring device he had invented.[37] Two years earlier he had commended himself to the town council as an expert on fire protection, a topic that was of great interest to the council since, some years earlier, a fire had laid waste to the entire town.[38] Fire protection and disease control were among the tasks of officially appointed physicians and both gave Glaser a greater chance to distinguish himself as a useful member of society than his medical activities could have done. The situation was entirely different for Magirus and Goetz, where the connection between medical legitimation and clientele was far less complex. For both of them, the medical practice constituted an important building block in their careers as scholars.[39] The overall picture that arises for the early modern period is that of a relatively high degree of freedom that the physicians had in their official and medical activities and their participation in scientific discourse. With all three physicians we see two different correlations: one is between the physician as scientist and his medical practice, the other has to do with the social advantages a physician could gain if he proved to be a 'useful' member of his community.

Gottfried Wachter holds an ambiguous position among the healers examined in this context.[40] He was not academically trained and, due to the medical laws of the Canton of Thurgau where he was active, he became caught up in the control measures of the authorities and the newly emerging medical

35 This refers to the uroscopy that was only regularly registered after he became town physician in 1758.

36 Johann Friedrich Glaser, *Nützliches Verhalten bey der jetzo in Deutschland regierenden Fleckfieberseuche: um sie zu verhüten und wieder zu dämpfen* (Hildburghausen 1758).

37 Johann Friedrich Glaser, *Beschreibung einer neuerfundenen Blutwaage und Blutmeßgeschirrs: womit man beym Hand- und Fußaderlassen das Blut, ob es schon unter das dabey gebräuchlichen Wasser läuft, dennoch sobald Loth zu Loth richtig und sehr nützlich wägen und messen, auch sonst durch einen angewiesenen Vortheil unter Wasser abgesondert allein auffangen kann* (Hildburghausen 1758).

38 Johann Friedrich Glaser, *Nützliche und durch die Erfahrung bewährte Vorschläge, bey heftigen und geschwinden Feuersbrünsten Häuser und Mobilien sicher zu retten: nebst einer gründlichen Anweisung, grosse und gefährliche Feuersbrünste zu verhüten, und was man bey Auf- und Ausbauung der Häuser und Aufbehaltung der Mobilien, auf eine nutzbare Weise, zu beobachten habe* (Hildburghausen ^{3}1764).

39 Cf. Schilling *et al.*, 2011, 111, 114–17.

40 Alois Unterkircher and Iris Ritzmann, 'Unlicensed Practice: A Lay Healer in Rural Switzerland' (in this volume).

administration. As a result he faced certain regulations with which he had to comply. His *Operation Büchli* were composed as a result of this monitoring. The new political and administrative conditions made the situation more difficult for Gottfried Wachter than it had been for his father who had still been able to pursue his sideline as 'peasant doctor' without interference. But in the long term these changes brought the Wachter family closer to academic medicine, when Gottfried Wachter's son continued the family's healing tradition by going to medical school. As a lay healer Gottfried Wachter was disadvantaged because he had to have his medical skills monitored and scrutinized by the central medical authority.[41] He enhanced his social standing in the local community by taking on various official functions. As church treasurer he carried out accounting tasks, as borough councillor he was in charge of traffic and trade, agricultural and moral policing. Wachter's close social ties with the community are apparent from the fact that, in 1822, he treated the wife of the mayor in Hugelshofen.[42] Surprisingly, there was no such close correlation between patients and social networks where Wachter's activities as military surgeon were concerned. When he acted as assistant military surgeon in 1813 and 1827, he had to attend the regular field exercises of the contingent troops,[43] but he neither gained more patients nor carried out more surgical interventions as a result.[44] While their official functions opened up new career prospects for the three early modern physicians, this did not happen in Wachter's case. His official duties secured the social position he had already established for himself, but they had no notable effect on his medical practice.

From 1854 onwards Cäsar Adolf Bloesch was town physician of Biel.[45] He donated the rather modest remuneration he received for this office to the workhouse where, for the most part, orphans were kept, educated and employed. In the Biel town council he devoted himself to poor relief and the school board. In addition to these official tasks Bloesch also had political interests and was active in various associations in his home town: in 1830 he joined the liberally-minded 'Casino-Society' and in 1832 the liberal *Schutzverein* (protection association) in Biel. The official medical appointments granted by the communes in the canton of Bern also seem to have been a result of his societal commitments and of

41 Cf. Katharina Weikl and Iris Ritzmann, 'Ein Auslaufmodell. Spuren einer ländlichen Heilerpraxis im Zeitalter der Akademisierung', in Gafner *et al.*, 2012, 76–94.

42 Archives of the Institute and Museum of Medical History at the University of Zurich (MHIZ) MS H 17 – V, 11 (1822, Fall Nr. 43).

43 Albert Schoop, *Geschichte der Thurgauer Miliz* (Frauenfeld 1948), 186.

44 Only one soldier is mentioned in Wachter's record: MHIZ MS H 17–III, 19.

45 Gafner (in this volume).

the reputation he had acquired as a politically and socially active member of the local community. His reputation in society certainly helped Bloesch to build up a regular clientele that included members of all social classes.

With Franz von Ottenthal we find a similar mix of official medical activities and socio-political commitments: from 1852 he acted as court cashier, from 1866 as borough councillor and administrator of the school fund.[46] From 1861 he represented his local community in the Tyrolese parliament, where, six years later, he was admitted to a group of liberally minded aristocratic landowners. Between 1861 and 1888 Ottenthal acted repeatedly as community and forensic physician in the communities of the Tauferer Valley.[47] The catchment area of his medical practice was quite extensive, in geographical as well as social terms.[48]

Even though Bloesch and Ottenthal belonged to the group of academically trained physicians, what they had in common with Wachter was that their involvement with administrative tasks enhanced their standing locally and had therefore a positive effect on their medical practice, whilst not affecting its quality.

Throughout his medical career Bloesch acted as a forensic expert. The fact that he included his expert reports with his journals suggests that he regarded them as part of his medical practice. As an expert he acted by order of the authorities, but he also issued certificates and post-mortems on behalf of private persons. Lina Gafner has demonstrated that the certificates he issued helped him to recruit and retain patients.[49] In Ottenthal's case the only expert reports available are those he issued in his role as community or forensic physician.[50] This role involved only two kinds of commissions: requests that fell within his remit as medical expert such as conducting post-mortems,[51] assessing military fitness or establishing a person's fitness to work or mental state;[52]

46 Elisabeth Dietrich-Daum, Marina Hilber and Eberhard Wolff, 'An Alpine Doctor's Private Practice: Franz von Ottenthal (1847–1899)' (in this volume).

47 Elena Taddei published a monograph on the biography of Franz von Ottenthal: Elena Taddei, *Franz von Ottenthal. Arzt und Tiroler Landtagsabgeordneter (1818–1899)* (Wien 2010).

48 Cf. Andreas Oberhofer, 'Eine Landarztpraxis im 19. Jahrhundert am Beispiel der Ordination des Dr. Franz von Ottenthal (1818–1899)', in Elisabeth Dietrich-Daum, Martin Dinges, Robert Jütte, and Christine Roilo (eds), *Arztpraxen im Vergleich. 18.–20. Jahrhundert* (Innsbruck 2008), 167–91: 178–88.

49 Lina Gafner, 'Ärztliche Zeugnisse auf dem medizinischen Markt. Die professionalisierte Zeugenschaft im wachsenden Verwaltungsapparat des frühen 19. Jahrhunderts, am Beispiel einer Bieler Arztpraxis', in Gafner *et al.*, 2012, 95–109.

50 Taddei, 2010, 145–52.

51 *Ibid.*, 153–54.

52 *Ibid.*, 146.

and scientific requests from the Academy of Sciences in Vienna. This Academy asked Ottenthal in 1861 to register all 'cretins' in his catchment area, a request that he rejected referring to his lack of time and financial resources.[53] Ottenthal, in other words, refused to scientifically evaluate findings he had established in his position as forensic physician. As with Bloesch, Ottenthal's official and political commitment was not aimed at opening up new, scientific or social, areas of activity, but played a part in consolidating the position he held in society.

In Conrad Heinrich Fuchs' case, the policlinics in Würzburg and Göttingen, of which he was director, were not additional, but fundamental to his medical activity.[54] The patient histories he asked his students to record were of the greatest scientific importance for him, a fact that is apparent from the form of these histories.[55] The combination of scientific interests and poor-relief led to tensions between the physician and the town administration: while the town elders of Göttingen were not prepared to support their town physician's primarily scientifically inspired interest in an increased healthcare budget, Fuchs' active support of an improved poor-relief system enhanced his standing with the people of Göttingen.[56] The case histories collected under Fuchs' aegis nevertheless reflect a medical practice that was informed by scientific interest and in his environment he found ideal conditions for such aspirations.

While Bloesch, Ottenthal and Wachter had another focus than Fuchs, there is evidence with all of them of some kind of relationship between their official medical function and their private medical practice. The homeopath Friedrich von Bönninghausen, on the other hand, was not interested in any public health office. This was not, or not entirely, due to the medical specialization he represented. There are known cases of homeopaths who worked with health insurance companies[57] or served as forensic physicians.[58] Bönninghausen's reluctance to take up a public office may have had to do with the fact that most

53 *Ibid.*, 149–50.

54 Stephanie Neuner and Karen Nolte, 'Bedside Training and Healthcare for the Poor in the Würzburg and Göttingen Policlinics in the First Half of the Nineteenth Century' (in this volume).

55 Volker Hess and Sabine Schlegelmilch, '*Cornucopia officinae medicae*: Medical Practice Records and their Origins' (in this volume).

56 Neuner *et al.* (in this volume).

57 The two lists, from Münster, of physicians who worked with insurance companies that have been preserved mention two homeopaths: Stadtarchiv Münster, Stadtregistratur Fach 28, Nr. 129: Vereinbarung zwischen den Ärzten und den Krankenkassen 1905–1910, 149b–150.

58 Michael Stolberg, *Die Geschichte der Homöopathie in Bayern (1800–1914)* (Heidelberg 1999), 29.

public medical positions in Münster were already filled.[59] We can also assume that Bönninghausen was, due to his sound social and financial position, not dependent on subsidizing his private practice.

Was Bönninghausen an exception? Or did his behaviour represent the endpoint of an intra-professional differentiation between state-appointed physicians and private practitioners? This question cannot be answered on the basis of the sources available, but it can help to further categorize the physicians of his time and of the time before, that were investigated by the Research Network. If one compares the two groups of physicians – those of the seventeenth/eighteenth and those of the nineteenth century – one notices a shift in emphasis between official function and medical practice. While both are central to the understanding of the social as well as medical behaviour in the early modern case studies, the nineteenth century practices of Wachter, Bloesch and Ottenthal can be understood without their official medical activities. The situation is entirely different for Fuchs because of his scientific emphasis. This shows that it would be wrong to assume that administrative activities generally lost significance in medical practice. But it also suggests a distinction between the roles of physicians as practitioners and scientists that is already noticeable with the three seventeenth-eighteenth century practitioners but only surfaces fully in the nineteenth century.

Scientific Networks and Medical Practice

Their academic studies gave early modern physicians an enhanced status within their profession and in society.[60] University education can be seen as an initiation into the practices and rituals of a scholarly culture that saw itself as universal rather than specialized.[61] The academic journey from university to university, if possible in conjunction with a visit abroad, was part of this 'code of conduct' that not only the early modern physician Magirus adhered to, but also his nineteenth-century colleague Caesar Adolf Bloesch.

By attending medical school physicians not only gained a particular status, they were also included in a particular scientific current. They became part of

59 Cf. Marion Baschin, 'Warum besucht man einen Homöopathen? Eine Praxis in Münster im 19. Jahrhundert', in Gafner *et al*, 2012, 126–40.

60 Marian Füssel, *Gelehrtenkultur als symbolische Praxis. Rang, Ritual und Konflikt an der Universität der Frühen Neuzeit* (Darmstadt 2006), 109–17.

61 Alf Lüdtke and Reiner Prass, 'Einleitung. Gelehrtenleben. Wissenschaftspraxis in der Neuzeit', in Alf Lüdtke and Reiner Prass (eds), *Gelehrtenleben. Wissenschaftspraxis in der Neuzeit* (Köln 2008), 1–29.

a European culture of communication.[62] Magirus, for instance, contributed to the transfer of Cartesian ideas and Dutch models of medical practice to the town of Zerbst in central Germany.[63] The scientific networks in which Magirus and Götz were active were strongly informed by the reformation or pietistic thinking. The religious factor no longer played a part in the scientific networking of Johann Friedrich Glaser in the mid-eighteenth century. He made use of his official position as town physician and medical officer to distinguish himself, regionally and further afield, as an expert who was knowledgeable outside medicine in areas such as fire protection or the promotion of fruit-growing.[64]

While all three early modern physicians were scientifically active, this applied to only two of the nineteenth century practitioners. As a type, Caesar Adolf Bloesch needs to be included with the early modern physicians: in addition to his pronounced reflection on his position within medical science[65] he composed, like Johann Friedrich Glaser, non-medical works, for instance a 900-page chronicle of his hometown of Biel.[66] The patient histories written down under the aegis of Conrad Heinrich Fuchs were directly linked with his scientific interest in a nosology. Fuchs is the only representative of the type of expert who concentrates exclusively on medical issues, an approach that led to his difficulties with the town's administration.[67]

With Wachter, Bönninghausen and Ottenthal, scientific contexts play hardly any part. While this is not surprising in Wachter's case, since he was not academically trained, it needs to be explained for Bönninghausen and Ottenthal. Why didn't Bönninghausen publish a justification for his decision to prescribe the high-potency homeopathic remedies that were controversial even then?[68] Why didn't Ottenthal publish case histories from the Taufen Ahrn Valley and why didn't he express his criticism of the Vienna Health Authorities in

62 Interesting attempts at reconstructing this universalities can be found in cf. Jean Boutier, 'Compétence internationale, émergence d'une "profession" et circulation des savoirs. Le tuteur aristocratique dans l'Angleterre du XVIIe siècle', in Maria Pia Paoli (ed.), *Saperi a confronto nell' Europa dei secoli XIII–XIX* (Pisa 2009), 151–80; Patrizia Delpiano, 'Precettori e allievi nel Settecento. La circolazione di saperi in Italia e in Europa', in Maria Pia Paoli (ed.), *Saperi a confronto nell' Europa dei secoli XIII–XIX* (Pisa 2009), 181–96.

63 Schlegelmilch (in this volume).

64 Ruth Schilling, 'Social Mobility and Medical practice: Johann Friedrich Glaser (1707–1789)' (in this volume).

65 Gafner (in this volume).

66 *Ibid.*

67 Neuner *et al.* (in this volume).

68 Marion Baschin, 'A Special Kind of Practice? The Homeopath Friedrich von Bönninghausen (1828–1910)' (in this volume).

writing?[69] With either of them, two aspects may have influenced their decision not to become distinguished members of the scientific community of their time: one was the great importance their medical practice had for their own self-image as physicians, the other their pronounced awareness of the aristocratic roots that distinguished them from the other physicians examined by the Research Network. While the medical practice, a scientific-scholarly self-image, local relations and family were still equally important for Magirus, Götz, Glaser and Bloesch, it was the combination of medical practice and an aristocratic milieu that prevailed with Bönninghausen and Ottenthal.

Results

Comparison of the medical practices investigated within the Project Network reveals differences and similarities in the relationship between a physician's position in the medical market and the wider context of this position. It also illuminates how the influential factors developed across the centuries. The importance of the family background for a practitioner's success in the medical market emerges as a constant that maintained its importance throughout the entire period under investigation, often in conjunction with the family's economic capital, which was essential for the more-or-less expensive academic training away from home and the setting up of a practice. The continuingly important role played by the family suggests that the decisive changes to the medical practice only occurred in the late nineteenth and in the twentieth century: only then did it develop from the former all-round activity in the physician's semi-private environment to an occupation cushioned by professional insurance associations.[70]

The importance a physician's religious confession had for his medical practice and for the composition of his clientele faded even earlier. While we still observe, with Magirus as well as Götz, a strong mutual influence between clienteles from a homogeneous confessional environment and the physicians' personal integration in parish matters as a means of increasing their social capital, the religious question was of little importance for Glaser and even less so for Wachter in their recruitment of a particular clientele. For the medical practices of Bloesch, Bönninghausen and Ottenthal confession was of no concern.

While changes and continuity can be established for the spheres of religion and family in all examples, especially in diachronic comparison, the relationship between official administrative activities and medical practice reveals a

69 Dietrich-Daum *et al.* (in this volume).

70 Cf. also Elisabeth Dietrich-Daum, 'Arztpraxen im Vergleich. 18.–20. Jahrhundert. Der Vergleich des Unvergleichbaren?', in Dietrich-Daum *et al.*, *Arztpraxen*, 2008, 9–22: 9–11.

different, and surprising, picture, especially with regard to the question as to how the early modern period compares to the nineteenth century. As a first working hypothesis, two types of medical spheres of activity can be distinguished for the nineteenth century: one where the social aspect implied in the medical activity is intimately connected with a socio-political commitment to the *bonum commune*, and one where the physician used the social welfare framework to add a scientific dimension to his medical practice. In a second step, parallels emerge with the early modern examples of medical practice investigated within the Research Network: again, we find two types depending on how physicians saw their practice: one where the encounter with the patient is seen as part of an activity that is rooted in the local environment, and a second one, where this activity is detached from the everyday world and incorporated into the sphere of scholarly *observationes*. Scientific activity and service for the good of the community therefore emerge as two facets of medical practice that are partly intimately connected and mutually enhancing, but can also come into conflict with one another. This ambivalence remained as a constant of medical practice from early modern times into the nineteenth century.

Analysis of the interrelation between medical practices and their social or cultural contexts, based on the case studies of the Research Network 'Medical Practice', has shown on the whole that broad developments such as 'medicalization' and 'professionalization' need to be differentiated, also – and in particular – by looking more closely at the physicians as agents. The investigation of examples from early modern times as well as from the 'long' nineteenth century in particular has accentuated changes more clearly than individual studies could have done: for instance the loss of the physician's universality. But it has also uncovered new connections between medical practice and administrative or scientific medical activities that can be found in the seventeenth/eighteenth as well as in the nineteenth century.

PART 2

CHAPTER 6

'What a Magnificent Work a Good Physician is': The Medical Practice of Johannes Magirus (1615–1697)

Sabine Schlegelmilch

Introduction

Practice records reflect the personality of their author. This is certainly true for the *Diarium Medicum* of Johannes Magirus. Because it was recorded in two different places, this particular casebook shows, at a glance, how strong the mutual influence is between the physician's personal situation and his record-taking. Of a total of 540 pages filled with notes from everyday practice[1] less than half (the first 197 pages) were written in Berlin where Magirus had his first practice from 1641 until the end of 1650. The greater part (342 pages) consists of notes from his five-year stint (1651–1656) as town physician of Zerbst, a town north of Leipzig.[2] Because of this division into two parts, the document is remarkable, as it gives insight into Magirus' armamentarium while raising the question whether a comparison of the two surgeries might yield patterns of activity that were informed by the respective religious, political or social contexts. Magirus allocates the individual patient's entries to a range of categories such as occupation (*cancellarius*), name (*die Kootin* = Koot's wife) or descriptive attributions (*a man from Roßlaw*).[3] The index relies on subjective criteria and cannot be construed as complete. As a result, Magirus' particular logging system does not allow for conclusive inferences to the

* All references in this contribution that are not supported by a source will be dealt with in a monograph scheduled to be published in 2016: *Die ärztliche Praxis des 17. Jahrhunderts – Johannes Magirus (1615–1697)*.

1 As well as a twenty-four-page index, the *Diarium* includes 620 pages, the last sixty of which are, however, filled with notes for Magirus' first history lecture in Marburg.

2 On Magirus' biography cf. Sabine Schlegelmilch, 'Johannes Magirus: Stadtarzt in Zerbst (1651–1656)', *Mitteilungen des Vereins für Anhaltische Geschichte* 20 (2011), 9–30: 12–16; Ruth Schilling, Sabine Schlegelmilch and Susan Splinter, 'Stadtarzt oder Arzt in der Stadt? Drei Ärzte der Frühen Neuzeit und ihr Verständnis des städtischen Amtes', *Medizinhistorisches Journal* 46 (2011), 99–133: 104–05.

3 It is impossible for the reader today to establish whether or not a patient mentioned by name once might be identical with another patient listed under his or her occupation.

 | DOI 10.1163/9789004303324_008

patients' social background, but on the whole it does grant insight into the physician's activities and their dependence on factors outside the practice.

The Physician's Practice

In order to understand the importance of each of his practices in their respective location, one needs to know that Johannes Magirus – unlike many of his contemporaries – never wished to work, or be perceived, as a medical practitioner only. His objectives, of which his practice was always a vital part, were informed by two aspects associated with his early years. The first of these was his home background: his father, Tobias Magirus (1586–1652), who would gain a degree of prominence with his comprehensive publications while Johannes still was a teenager, had received a professorship at the newly reformed and restructured university in Frankfurt on the river Oder. He also maintained intimate connections with the Brandenburg electoral court where he and his brother (Johannes' uncle Joachim, a theologian and professor at the same university) were responsible for the religious education of the young electoral Prince, Friedrich Wilhelm. To his son – who signed his early publications 'Johannes Magirus T.M.F.'. (*Tobiae Magiri Filius*) – Tobias Magirus represented the personification of an ideal, participating in academic honours while at the same time holding a prominent position at the reformed royal courts (where pointing out one's own usefulness was of great importance). In addition to these early influences, Magirus' horizon was broadened when he went on a study trip to the Netherlands, Southern England and Northern France from 1637 to 1640. On his journeys he came across two developments that were only just emerging and that would have been entirely new for a student coming from the German speaking realm. The first of these was early modern rationalism as promulgated by René Descartes. His mechanistic view of the world was discussed at the medical school in Leiden and immediately found a staunch supporter in the young Magirus who already had a profound interest in mathematics (see below).[4] The second was the method of bedside teaching that had only just (1636) been introduced at the Caecilia Hospital in Leiden involving the acquisition of diagnostic and therapeutic skills by means of witnessing and taking part

4 As early as 1636, Descartes' mechanistic doctrine was taught in the Netherlands and soon afterwards defended in public disputations: cf. Roger French, 'Harvey in Holland: Circulation and the Calvinists', in *idem* and Andrew Wear (eds), *The Medical Revolution of the Seventeenth Century* (Cambridge 1989), 46–86: 54, 65. Magirus heard lectures in Leiden 1639–1640 by Johannes Walaeus who later endorsed Harvey's theory of the blood circulation: cf. French, 1989, 55.

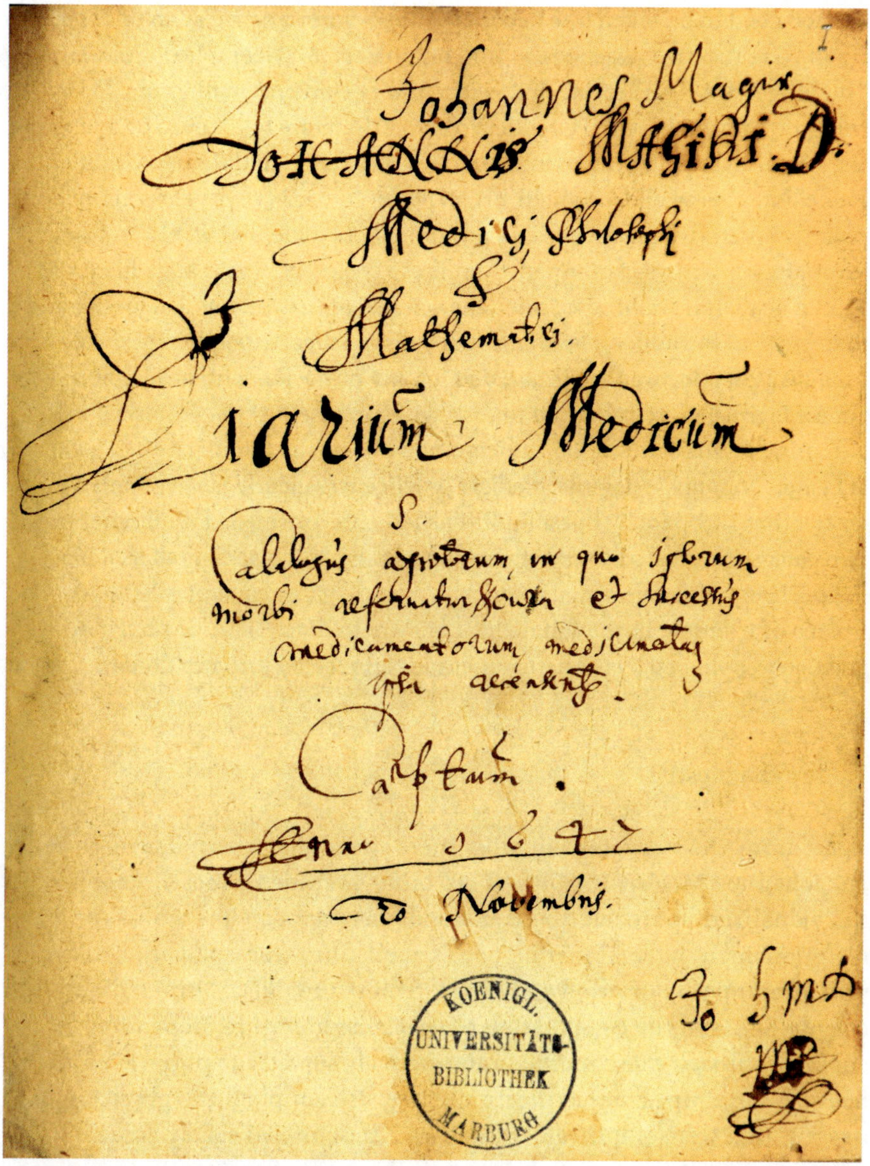

FIGURE 6.1 *Title of Johannes Magirus'* Diarium Medicum.
UNIVERSITY LIBRARY MARBURG, MS. 96.

in actual treatment.[5] These influences and experiences – which can only be very briefly outlined here – helped to shape Magirus' self-image as *medicus, philosophicus & mathematicus* (see Figure 6.1).

5 Cf. Tim Huisman, *The Finger of God. Anatomical Practice in 17th-Century Leiden* (Leiden 2009), 128–32.

They need to be taken into account if one wants to arrive at an understanding of Magirus' two practices as building blocks of a complex, multi-layered structure: like his father, Magirus wished to become a member of the society around a reformed Court. His endeavour was therefore not only to be a provider of medical services, but also to be seen as a competent and therefore influential councillor. This would heighten his standing in society while paving his way towards a lucrative practice with a prominent clientele that would enhance his reputation as a physician. Apart from these socio-economic aims he also harboured academic ambitions: since the science of the time no longer defined a good *medicus* by his book-learning but by his practical experience (*experientia*), a medical practice was a *sine qua non* for any physician with academic aspirations. Magirus, moreover, tried to distinguish himself in the empirical field by proposing for his own students the method of bedside training he had acquired in Leiden. Johannes Magirus' medical practice was clearly never simply a place for treating sick people, but a multi-faceted enterprise:[6] as for all other physicians alike, it provided a way of earning a living, but also figured as a place to establish and maintain social contacts, a place to gain academic distinction and a place of training. In the following, I will try to illustrate how such a doctor's practice was shaped by the actual methods applied by its practitioner.

Diagnostics

Apart from the conversation with the patient that every diagnosis relies on, Johannes Magirus employed six other methods. Since he applied an individualized approach, these methods may appear in each of the case-histories singly, in combination or not at all. They were: palpation, observation, examination of urine and blood, feeling the pulse, iatromathematical calculations and the consultation of specialist literature. Depending on the chosen approach these methods alternatively served the purpose of identification, monitoring or justification. They also supplied the physician with insights that were empirically gained and therefore, allegedly, more reliable than the patients' own verbal account, who were often in pain or in search of an explanation for their condition.[7] In contrast to the popular image,

6 Cf. Kay Peter Jankrift and Ruth Schilling, 'Medical Practice in Context. Religion, Family, Politics and Scientific Networks' (in this volume).

7 Patients were not necessarily prepared for such a conversation: cf. Robert Jütte, *Ärzte, Heiler und Patienten. Medizinischer Alltag in der frühen Neuzeit* (München 1991), 108; cf. also the example of the patient Ryder who stated: 'But when I came to talk with him I was at a loss what to say', quoted in: Andrew Wear, 'Medical Practice in Late Seventeenth- and Early Eighteenth-Century England: Continuity and Union', in French *et al.* (eds), 1989, 311.

early modern physicians thought that accurate examination necessitated physical contact with the patient.[8] As with physicians today, they would primarily look for (and find) the symptoms they considered to be relevant according to the prevailing school of thinking.[9] According to Galenic medicine there were two aspects to illness: firstly, the humours were vitiated due to a particular *causa* and secondly, these humours were insufficiently excreted. It therefore was equally important to look for concentrations of humours and to examine the excreted matter. When the son of a swordsmith pointed out that he had 'hard scorbutic tumours on his feet',[10] Magirus insisted on examining the feet, since patients' statements were often not corroborated by the findings of the physician. Another patient, by name of Cupelmann, complained about a pain in his sides (*conqueritur de doloribus laterum*) although Magirus could not feel any lateral indurations (or hardened humours): *durities nulla in lateribus deprehensa*.[11] This was good news since it meant that the humours were still in flow: they were real 'fluxes', as the patients liked to point out.[12] But these fluids might nonetheless be corrupted or produce heat – another sign that needed to be established by touch: when he felt a *facies calida* (hot face) the physician knew that there were hot, burning humours projecting their heat onto the skin. In some rare cases, Magirus was instantly able to confirm the connection between outer and inner heat, as when, with a patient called Kresse, he established (by dipping in a finger) 'hot and serous humours, streaming from the feet – which means the blood is hot and burnt'.[13]

The skin, on to which the humours streaming through the body projected their condition, was an essential diagnostic medium, not only for observing temperature changes. Visible skin alterations were regarded as manifestations of underlying processes which the physician could manipulate so that the harmful matter was eliminated through the skin by means of sudorific agents or measures such as bloodletting, scarification or blistering plasters. But first the condition needed to be diagnosed through exact observation.

8 Cf. Michael Stolberg, 'Examining the Body c. 1500–1750', in Sarah Toulalan *et al.* (eds), *The Routledge History of Sex and the Body, 1500 to the Present* (London 2013), 91–120.

9 Cf. Annemarie Kinzelbach, Stephanie Neuner and Karen Nolte 'Medicine in Practice: Knowledge, Diagnosis and Therapy' (in this volume).

10 Universitätsbibliothek (hereinafter: UB) Marburg, Ms. 96, 51: 'has *tumores scorbuticos duros pedum*'.

11 UB Marburg, Ms. 96, 271.

12 On the phenomenon of fluxes cf. Michael Stolberg, *Experiencing Illness and the Sick Body in Early Modern Europe* (Basingstoke 2011), 95–97.

13 UB Marburg, Ms 96, 111: '*humores calidi et serosi e pedibus effluentes, indicat sanguinem calidum et adustum*'.

If Magirus lapsed into using the German language in his writing, for instance when describing someone's hands as 'peeling off like a glove' (*die hände plastern sich ab als handschuch*) or pointed out that a woman had 'blisters the size of hazelnuts' (*blasen wie haselnüsse*),[14] we need to bear in mind that the physicians of his time – unlike their colleagues today who can refer to illustrated dermatology books – only had recourse to descriptive texts, mostly without any illustrations. These descriptions also reflected the immediate environment of their authors, as we see from Daniel Sennert who stated that scorbutic blotches, in the early stages of the disease, were 'not unlike fleabites'.[15]

The obvious dissimilarity between the frequency with which Magirus applied the other diagnostic procedures in his practices in Berlin and Zerbst returns us to the question raised in the introduction to this contribution: do the methods used by a physician depend on where he uses them? If we compare Magirus' two surgeries we arrive at the following picture:[16]

	Berlin practice (1648)	**Zerbst practice (1653)**
	97 patients/ 184 contacts	**100 patients / 266 contacts**
uroscopy	17	35
haematoscopy	—	6
feeling the pulse	—	3
iatromathematical analysis	—	13
consulting specialist	11	16
literature	(once; 3 authors in all)	(frequ.; 30 authors in all)

14 UB Marburg, Ms 96, 39; 130.

15 Daniel Sennert, *Opera omnia in quatuor tomos divisa*, tomus tertius (Lyon 1656), 516: '*illis quae a pulicum morsibus relinquuntur non absimiles*'. The English physician Thomas Sydenham used the same comparison, which was easily accessible to all patients cf. Volker Hess, *Von der semiotischen zur diagnostischen Medizin. Die Entstehung der klinischen Methode zwischen 1750 und 1850* (Husum 1993), 29.

16 1648 and 1653 were the only years when Magirus worked only as a medical practitioner without pursuing any other professional activities (travelling etc.).

Here it can be noted that two of the diagnostic methods under investigation were never used in the Berlin practice, but are adopted after Magirus' relocation to Zerbst, indicating that his practice changed markedly in the space of only five years. With regard to the examination of the blood (haematoscopy), one might conclude that Magirus was influenced by a local surgeon[17] – if the Berlin diary had not already included the procedure, but as a self-experiment carried out when Magirus was a young physician rather than as a diagnostic tool. As it is, Magirus must have been interested in the method even before he moved to Zerbst.[18] Why this discrepancy between the two surgeries? To answer this question we need to consider three factors:

1. *His newly obtained autonomy in practising, devoid of any interference from senior and/or more renowned colleagues.* In the nine years (1641–1650) of Magirus' Berlin practice (which was also his first), his situation grew increasingly difficult. He had moved to a town which had been destroyed by war and was, medically, inadequately supplied (at least from the point of view of the academic physicians). Magirus was received with open arms by the private physicians of the Prince Elector and immediately admitted as a partner in their practice.[19] But this was exactly what, in the course of time, caused him problems, for in Berlin Johannes Magirus – as we can see from the cases recorded in the *Diarium* – never treated truly important patients on his own: he usually was dispatched to look after minister's wives when they caught a cold, obese vicars from the surrounding parishes, poor patients who vanished after one treatment (like the 'poor woman with four children from Müncheberg whose husband had been shot by the blind Valtin') and chronically ill patients such as the 'manic man from Spandau'. What had initially been so important for

17 Haematoscopy as an early modern practice is mentioned occasionally but has not been investigated in more depth. Cf. the author's monograph (see note 1).

18 UB Marburg, Ms. 96, 94.

19 As early as 1642, Magirus (as the youngest of a team of doctors!) was permitted, following the death of Margrave Ernst of Brandenburg, to compose the official post-mortem report for the court and the Königsberg Medical Faculty: Geheimes Staatsarchiv Preußischer Kulturbesitz, BPH. Rep 32, V 46, 13a–91a: Akte über die Krankheit und den Tod des Markgrafen Ernst von Brandenburg; Staatsbibliothek Berlin, Ms. boruss. qu. 81: Kurtzes, jedoch gründliches Gutachten über Ihre Fürstlichen Gnaden Herrn Ernst Markgrafen zu Brandenburg und Herzogen zu Jägerndorffs Krankheit und todt und derselben Ursache durch Joannes [!] Magirus.

Magirus' career, that is, to be seen with his prominent and experienced colleague Martin Weise on house calls or at mass,[20] became a hindrance. Confined as he was by the narrow society of the reformed Court and its reliance on the appointed court physicians Magirus was unable to develop a style of his own. This changed when he arrived in Zerbst: here he was the one town physician, appointed by the town council, and the only learned physician around. His patients came from all classes and he alone decided on their treatment.

2. *His rise to the position of academic teacher.* In addition to his post as town physician, which he assumed in the spring of 1651, Magirus was appointed professor of mathematics at Easter 1652 and professor of physics in October of the same year: he lectured on *physica* and thus on the foundations of medicine at the *Gymnasium Illustre* in Zerbst.[21] Considering his mathematical approach to medicine – he had presented mathematical-medical courses in Berlin (see below)[22] – combining the two chairs seemed a logical step that allowed him to teach his students theory as well as practical skills at the bedside.[23]
3. *The necessity for self-promotion that came with the office and the professorship*. In order to be convincing as an academic authority, it certainly was important for Johannes Magirus to establish a good reputation as town physician and, to this end, he probably was prepared to meet expectations that went beyond the usual call of duty. The fact that the number of uroscopies nearly doubled may have been a result of this endeavour, since the common people in particular would expect the urine flask ritual.[24] And because Magirus advocated a practical approach to teaching,[25] it was important for him to practise what he taught. All the standard medical textbooks he discussed in his lessons referred to the diagnosis of *urina & pulsus*: in Zerbst we see evidence of Magirus'

20 Cf. Sabine Schlegelmilch, 'Vom Nutzen des Nebensächlichen. Paratexte in den Kalendern des Johannes Magirus', in Klaus-Dieter Herbst (ed.), *Astronomie – Literatur – Volksaufklärung. Der Schreibkalender der Frühen Neuzeit mit seinen Text- und Bildbeigaben* (Bremen 2012), 393–411: 396–97.

21 Cf. Schlegelmilch, 2011, 21–22.

22 Cf. Schilling *et al.*, 2011, 109.

23 On a student entry in the *Diarium* cf. Schlegelmilch, 2011, 24.

24 Cf. Michael Stolberg, *Die Harnschau. Eine Kultur- und Alltagsgeschichte* (Köln 2009), 125–31, 203–12.

25 Cf. Schilling *et al.*, 2011, 109–11.

taking the pulse (several instances), a procedure he had not previously applied. Similarly, the sudden rise of iatromathematical treatments and recourse to the specialist literature may reflect his teaching activity: Magirus presented his cases in the lessons to illustrate what he taught[26] and lent authority to his own work by referring to the luminaries in the field of medicine. The procedure of pondering over successively filled containers with blood gained from a patient by venesection, and comparing the blood's appearance[27] as well as that of holding a urine flask against the light again and again (Magirus tended to repeat the procedure up to eight times per case to monitor treatment), the concentration exhibited in feeling the pulse and the industrious leafing through books and handling of compasses allowed him to display his diagnostic competence in the presence of patients[28] much more effectively than any silent contemplation at the bedside ever could.

Case Study: Johann Köppen Jr., Fever and Worms

The following case, based on Magirus' own notes, which is exceptionally lengthy and elaborate for a child patient, illustrates what has just been described above. The young patient was the son of Johann Köppen, who was not only syndic of Zerbst, but also privy councillor to Prince John VI of Anhalt-Zerbst and, since 1637 (i.e. 14 years before Magirus' arrival) professor of law at the *Gymnasium Illustre* – an influential man, one can assume, in the town, at the school and at Court. The case history begins with two columns on page 311 of the *Diarium* (see Figure. 6.2, left side):[29]

26 This is apparent from Magirus' announcements of his lectures where he not only vowed to 'communicate' the cases of prominent persons but to present to his students patients who were poor or who lived outside town and whom he treated free of charge – an early policlinical concept.

27 On the technique of collecting blood in various bowls (Magirus used three) cf. Friedrich Lenhardt, 'Zur Ikonographie der Blutschau', *Medizinhistorisches Journal* 17 (1982), 63–77: 65–72.

28 On this aspect of the construction of the physician's authority cf. Michael Stolberg, 'Formen und Strategien der Autorisierung in der frühneuzeitlichen Medizin', in Wulf Oesterreicher, Gerhard Regn and Winfried Schulze (eds), *Autorität der Form, Autorisierung, institutionelle Autorität* (Münster 2003), 205–18: 214–15.

29 UB Marburg, Ms. 96, 311. Latin text translated by the author.

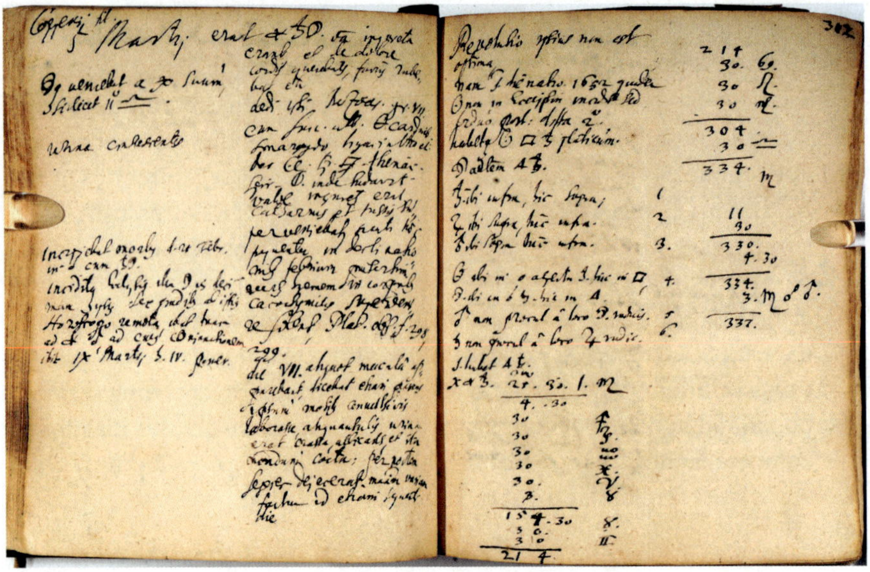

FIGURE 6.2 *The case of Johann Köppen, Jr., in Johannes Magirus'* Diarium Medicum.
UNIVERSITY LIBRARY MARBURG, MS. 96, 311–2.

Coppenii fil[ius]

5 March

The Moon (☽) reached its sextile (✱), i.e. 2° autumnal point (♎).

Sextile aspect (✱) of Saturn (♄) and Moon (☽), he was generally restless and complained about heart-ache, face was flushed etc.

Urine thickening

I gave him 7 grains of bezoar with white amber, nettle salt, hyazinth emerald and oil [...], tartar, theriac,

spirit of nitre; after that he developed a sweat, he was very restless, catarrh and cough occurred as happens frequently when fever subsides, mainly in winter, in cacochymical bodies:

The illness began on 28 February, during a conjunction (☌) of Saturn (♄) and Moon (☽).

And this eclipse of the Moon (☽) entered the 10th house, Six degrees from his horoscope; it then entered into the sextile (✱) with Mars (♂), and will enter conjunction with it on 9 March, at the fourth hour of the afternoon.

Plat[ter]observ[ationes] p. 298 299.

A few blotches appeared on the 7th day.

His father said he had also suffered a few spasms. The urine was thick, whitish and not concocted.

He excreted various matters six times in the night.

The same occurred on the next day.

The subsequent page (312; see Figure. 6.2, right side) shows astronomical calculations, beginning with the statement that the child's revolution was not very good because the Sun deviated from the nativity of 1652 by two degrees. This is followed by information about the aspects of planets and their positions, and (listed in addition tables) calculations about the shift of the 12 houses of the horoscope in relation to the zodiacal houses. Magirus continued his notes on page 313:

> The urine was simply thicker than ordinarily which is not very ominous. Signs of failed concoction do not necessarily presage death. They point to long drawn-out rather than fatal illness: Zacutus Lusitanus, Book 4, p. 816. On the 8th day of his illness he excreted a worm in the night and from midday another one crawled out of his mouth. He slept well. I prescribed grass water, scordium with spirit of vitriol and tartar, theriak & scordium syrup. I gave orders to rub his pulse and the soles of his feet with a balm. On the 8th day the usual fever convulsions were absent by the grace and mercy of God.
>
> [An additional marginal note contains extracts in unfinished sentences (Mercurialis, f. 757) to the effect: long and deep sleep is good for children as experience and reason show].

Even before Magirus recorded the patient's symptoms, he noted that at the beginning of the treatment Saturn (the planet with the most portentous influence on human health) and the Moon (due to its proximity to Earth the most influential planet, also with negative connotations) were in the sextile (angle of 60° between two planets). This generally beneficial position was in this case, however, determined by the negative nature of the two planets, a fact used to explain the poor state of the child. The rest of the case history is interspersed with iatromathematical calculations which need not be explained in detail here. Basically, they indicate that at the beginning and during the course of the illness, planets with negative qualities exerted their influence on the child. This is followed by prognoses regarding the negative/positive constellations to be expected for the subsequent days. Importantly, for us, there were two professors at this child's sickbed and one of them, the professor of mathematics, provided his colleague with calculations regarding the illness of his child, demonstrating his professional competence as both physician and mathematician. For Magirus, the planetary movement he had established constituted a clear explanation for the illness and he would now fight the manifestation of that illness within the body – the *cacochymical* state (vitiated humours) – with *chymical* substances (that his wealthy colleague could well afford): minerals, theriac, various distillates. He demonstrated the effects of this medication to his academic colleague by inspecting the urine (three times) and underpinning his iatromathematical diagnosis (cacochymical state) and course of therapy by referring to authoritative publications (Platter's *Observationes*, Lusitanus' *De medicorum Principum Historia*), supplemented by seemingly banal truisms (Mercurialis: sleep is good for sick children).[30] As well as his own observations (the child developing blotches), Magirus also considered those of the father who deputized for the physician when he was called away. How often that happened is difficult to establish, as with all cases recorded in the *Diarium*. The fact that there were three uroscopies, two prescriptions, a single observation of blotches and the listening to the father's report suggests that the doctor called three times.

Therapeutic Practices

Therapies were tailored to the individual patient. That they were not necessarily dependent on the symptoms is apparent from the case of young Köppen who received expensive *chymica* while other children with the same diagnosis (fever and worms) would be given preparations made of only a few, mostly herbal, substances.[31] The treatment seems wholly out of proportion, especially since the statements regarding revolution and nativity seem to suggest that the child was only one year old.[32] It is probably justifiable to say that the physician's activity was aimed as much at the father as at the son, in

30 In this case Magirus consulted the following reference books: Felix Platter, *Observationum in hominis affectibus plerisque, corpori et animo, functionum laesione, dolore, aliave molestia et vitio incommodantibus libri tres* (Basel 1614); Zacutus Lusitanus, *De medicorum principum historia liber tertius, in quo medicinales omnes medicorum principum historiae de uteri, genitalium, & inferiorum partium affectibus, describuntur & copendiose explanantur* (Amsterdam 1637); Hieronymus Mercurialis, *De puerorum morbis tractatus* (unknown edition).

31 As purges Magirus chose mechoacanna/jalap for children, and for adults primarily senna leaves and rhubarb. He usually added sweeteners to children's medicines.

32 'Nativity' refers to the planetary constellations at the time of birth; 'revolution' to planets returning to that constellation: cf. Hans Gaab, *Der Altdorfer Mathematik- und Physikdozent Abdias Trew (1597–1669). Astronom, Astrologe und Kalendermacher* (Frankfurt/Main 2011), 278–79. The nativity was clearly established in 1652; a year had passed since then, the Sun (Θ) returned after 365 days (see date in the case history), but obviously with ominous deviations. If the information regarding heart-ache given at the beginning of the treatment was not an interpretation of the father, the child must have been old enough to verbalise such an observation.

the hope of securing the recognition of the one and the health of the other. With children at least, Magirus refrained from surgical interventions while fever in an adult patient would most likely have entailed bloodletting as a cooling therapy.

The case presented may, nevertheless, convey a distorted picture of Magirus' medical practice. He was a physician who, on the whole, was modest in his use of medication and surgery. His surgical interventions were restricted to bloodletting (*venaesectio*) and dry, as well as bloody, cupping (*cucurbitales / scarificationes*).[33] He would judge from the patients' remaining physical strength what kind of therapy was required ('The false doctors also neglect the strength of patients, whichever way it goes…'),[34] since he saw it as the task of the physician to support the healing powers of nature (*natura medicatrix*). Looking at all the long-term cases we can discern a system in the therapies Magirus applied: phases of medical intervention alternated with phases of rest that were only disrupted in emergency situations. He recorded the phases of rest as meticulously as the other medical measures. For a woman patient by name of Seesen he wrote: 'when she had rested for two days and gained new strength' (*cum per biduum quiesceret et vires recolligeret*),[35] clearly considering that allowing nature to play its part was an essential element of the therapy ('when nature wants to rid itself of a harmful thing, as in the paroxysms of fever, the true physician will leave nature to take its course').[36] Which is why, with young Köppen, Magirus regarded sleep as a crucial aspect of the treatment.

Magirus prepared the medicines for each patient individually. While still in Berlin, he would consult his older and more experienced colleague Weise, which is not to say that he took on the latter's prescription uncritically: 'Dr Weis [sic] prescribed a stomachic electuary for her. I took some of it to try it myself and felt pain and a rumbling in the left *hypochondrium* – so it is true: sugar and all things sweet are harmful for diseases of the spleen'.[37] In other cases, especially in his Zerbst practice where he had no colleagues he could consult, he tended to use prescriptions that he found in the specialist literature, as long as the case he treated was similar to the one described in the book.

---

33 The use e.g. of fontanels and ropes of hair to produce pus are not mentioned.

34 Johannes Magirus, *Prognosticon, das ist: Astrologische Mutmaßung vom Gewitter, Frucht- und Unfruchtbarkeit, Gesund-oderKranckheiten, Krieg und Friedens des* MDCXLIX. *Jahres nach Christi Geburt* (Leipzig [1648]), fol. C2v.

35 UB Marburg, Ms. 96, 215.

36 Magirus, *Prognosticon*, 1648, fol. C1v.

37 UB Marburg, Ms. 96, 71: '*D.D. Weis praescripserat Electuarium stomachum confortans. Ego de ipso sumebam aliquid degustandi gratia; sed dolores et rugitus in sinistro hypochondrio percepi, itaque verum est omnino sacharum et dulcia omnia in splenis morbis noxia*'.

He always recorded the patient's response and changed the prescription if undesired side-effects occurred – as for instance in the case of a spinster called Weiler: 'Because it contained musk and is sweet it did not agree with her, so I replaced it'.[38] An objective observer, he registered successes ('with the plaster all pain disappeared') as well as failures ('took half of it, but vomited twice, had no success'),[39] compared patients ('it is the plaster I also prescribed for Schwenk's son'),[40] and by doing so, built up his own store of *nostra*, that is, medicines he considered to be special and generally effective.[41] Although these procedures seem 'modern' and scientific, we must not lose sight of the fact that they were still based on metaphysical foundations: Magirus was convinced that the success or failure of a physician was ultimately decided by God's omnipotence. The case histories which challenged his medical competence often end with the qualifying formula L.D.O.M – *Laus Deo Optimo Maximo*.

Conclusion

As evident from Magirus' case-book, the development of a physician's practice has to be seen as an never-ending process. It is shaped by the accumulation of social and professional approaches and the specific ways they affect each other in different contexts.[42] Our inquiry into only two surgeries of one and the same physician, whose training and career embraced almost a century, shows that there is no such thing as the definitive seventeenth century medical practice.

> It smacks of a commonplace to say that a man carries through life some vestiges of his early training and education. If his own aptitudes coincide with the interests of those who taught him, particularly during the formative years of his university education, it is likely that his whole intellectual development will be colored by his teaching and that his maturity will be the fullest working out from the general principles and habits of thought and work learned at this time.[43]

38 *Ibid.*, 55.

39 *Ibid.*, 268: '*ex emplastro omnis dolor evanuit*'; see also 115.

40 *Ibid.*, 103: '*emplastrum est illud, quod Schvencii fililo praescripseram*'.

41 On the *nostra* cf. Wear, 1989, 306–09.

42 From a praxeological point of view surgeries were 'typed, routinized and socially 'comprehensible' clusters of activities': cf. Andreas Reckwitz, 'Grundelemente einer Theorie sozialer Praktiken. Eine sozialtheoretische Perspektive', *Zeitschrift für Soziologie* 32 no. 4 (2003), 282–301: 289.

43 Gweneth Whitteridge, *William Harvey and the Circulation of Blood* (London 1971), 3.

This statement put forward by Gweneth Witteridge to describe a much more prominent contemporary – William Harvey (1578–1657) – also applies to Johannes Magirus. His lifelong aspiration to implement and apply the knowledge acquired during his studies resulted in the simultaneous application of seemingly contradictory and not contemporary practices[44] – such as astrology and rationalist science – in his medical routine. Magirus was barely twelve years old when he was introduced to the foundations of mathematics by Benjamin Ursinus, the Brandenburg court mathematician and a friend of Magirus' father. Ursinus, a maker of calendars, steered Magirus towards iatromathematics, a science of the sixteenth century that was already being questioned in the seventeenth century and gradually vanished from medical practice from 1650 onwards, the time when the *Diarium* was composed. Magirus continued to defend the approach throughout his life: in his astronomical calendars he claimed – quoting a great number of sources – that no-one could be a true physician without consulting the stars – a view he will probably have come across as a student at Wittenberg:[45]

> Those who do not know astronomy I will warn that they do not obey Hippocrates who recommended it, saying...that every medicus ought to know about the rising and setting of the most prominent stars in his place of practice; and Galenus insists that a medicus must be able to cast a nativity and use it with his patients....[46]

Magirus' words reveal that the iatromathematical treatment of his young patient Köppen was, in his eyes, also part of an orthodox Galenism and Hippocratism. This insisting on a Galenic-Hippocratic practice and combining it with a variety of other elements constitutes an approach that he shared with all physicians of the seventeenth century: they might have broken away from some aspects of the ancient doctrines previously defended with humanist

44 This is evidence that the much-cited 'fundamental experience' of the 'non-synchronicity of the synchronous', described by Reinhart Koselleck as arising from the experiential horizon of early modern times, also applied to medical surgeries: cf. Reinhart Koselleck, '"Neuzeit". Zur Semantik moderner Bewegungsbegriffe', in *idem* (ed.), *Vergangene Zukunft* (Frankfurt/Main 1989), 300–47: 323–25.

45 Following the teachings of Philipp Melanchthon, a staunch believer in astrology, Wittenberg was the centre for astrological training: cf. Claudia Brosseder, *Im Bann der Sterne. Caspar Peucer, Philipp Melanchthon und andere Wittenberger Astrologen* (Berlin 2004), 11–17.

46 Johannes Magirus, *Prognosticon Astrologicum auff das Jahr nach der gnadenreichen Geburt unseres Herrn und Seligmachers MDCXLVI* (Altenburg [1645]), fol. C1a.

zeal, but they continued to call on their authority to justify their own actions.[47] We observe the same tendency with Magirus' prescriptions: by remaining within the boundaries of Galenic medicine, he followed his Wittenberg teacher Daniel Sennert, who had succeeded, by clever reasoning in his treatise *De Chymicorum cum Aristotelicis et Galenicis consensu et dissensu* (1619), to make chymical (i.e. Paracelsian) medicines palatable to his Galenic colleagues. Magirus considered himself on the right side in the on-going debate about the genuine *chymiater*: '[he] is a bad doctor in these days, too, …who does not understand *chymica* and uses chymical medicaments when they are indicated'.[48] He certainly prescribed the new medicines for his patients. While non-academic practitioners who used these *chymica* were still abused as *empirici*, the same empiricism had grown to be an accepted category of *experientia* and was seen as a hallmark of scientific practice.[49] Magirus' self-experiments (haematoscopy, drug testing), and his exact observation of patients' reactions to particular medicines, served as a means to acquire practical knowledge by experience: his entire medical activity is imbued by this scientific spirit.

There was another characteristic quality in Magirus that marked him as a man of his time. The 'prominent…position of mathematics in seventeenth-century science'[50] deeply influenced his thinking. He saw himself primarily as a *mathematicus*, a fact that was apparent in his objective, unemotional way of looking at the world.[51] He had already promulgated the expediency of the mathematical sciences for the medical practice in his lecture courses for medical students: geometry, he said, taught the positions and proportions of the limbs and the structure of the locomotor system; arithmetic, among many other uses, taught the dosage of medicines and how to calculate the duration of a pregnancy; optics helped to judge eye diseases; geography helped dealing with conditions that were specific to particular regions – and astronomy, of course, revealed the true *causae morborum* which then manifested in the

47 Cf. Volker Hess and Sabine Schlegelmilch, '*Cornucopia officinae medicae*: Medical Practice Records and their Origins' (in this volume).

48 Cf. Schlegelmilch, 2013, 194.

49 On the similarity of the medical practices of academics and empirics cf. Wear, 1989, 304–12.

50 Eduard J. Dijksterhuis, *The Mechanization of the World Picture: Pythagoras to Newton* (Princeton 1986), 403–04.

51 In an irate letter, that unfortunately caused him to repeatedly slip in his mastery of Latin and syntax, Magirus seemed to express a warning that even mathematicians might put a spoke in other people's wheel, meaning that they could also be emotional schemers if provoked: cf. Schlegelmilch, 2011, 29.

body.[52] Magirus' view of the body marked him as a Cartesian. He was convinced that 'Descartes was right in thinking that *mechanicas regulas easdem esse ac regulas naturae*'.[53]

But how did all this affect Magirus' work as reflected in his *Diarium*? He never uses images of geometry, hydraulics or mechanics there. And one would not expect him to, because all of these statements are part of a mere theoretical approach that is distinct from the development of new medical practices.[54] Yet one can discern a particular style that distinguishes Magirus' notes from those of contemporary physicians. Magirus noticeably eschewed emotive language in his writing: he never mentioned his feelings or those of his patients. Emotions were as irrelevant as the purely subjective speculations of patients regarding the external causes of their illnesses, which were not recorded either: the whole field of the six *res non naturales*, usually considered to be fundamental to early modern medicine, hardly makes an appearance.[55] Magirus' notes are strictly symptom-oriented. His interest was clearly limited to the complaints patients described to him and to what he observed during the consultation. He would look back, comparing ('has the same illness as before but has boils now as well'),[56] and ahead prognosticating, and in doing so he assimilated the symptoms that appeared at various times into the history of an illness rather than assuming that each complaint was a new disease, as other physicians were still wont to do at that time.

The conspicuous lack of feeling the pulse in Magirus' practice might have been due to the fact that the method only supplied relative and not definite results. Magirus clearly preferred presentable, i.e. visible or calculable, facts: the consistency of blood and urine, alterations of the skin (as a result of his prescriptions), the information gained from constellation charts, evidence from the specialist literature. He focused strongly on producing evidence for the patients in order to be perceived as a physician and therefore comprehensible

52 Johannes Magirus, *Epistolica declaratio collegii medici mathematici et practici* (Berlin 1646), fol. A2r–v; on this cf. also Schilling *et al.*, 2011, 109.

53 '...the rules of mechanics are the same as those of nature': cf. Johannes Magirus, *Alt und Neu Prognosticon Astrologicum, das ist Vernünfftige und aus dem Himmelslauff genommene Mutmaßung vom Gewitter, Frucht- und Unfruchtbarkeit, Gesund-oder Kranckheit, Krieg, Frieden und anderen Welthändeln des* MDCLV. *Jahres nach Christ unseres Herrn Geburt* (Nürnberg 1654), fol. A3r; cf. Schilling *et al.*, 2011, 109 n. 44.

54 Cf. Wear, 1989, 304: 'Practitioners rationalized about their remedies and made them appear appropriate and attractive to buyers... The language in which this was expressed could vary according to the practitioner's theoretical standpoint'.

55 On the 'six non-naturals' cf. Wear, 1989, 300, 312; Stolberg, 2011, 85–87.

56 UB Marburg, Ms. 96, 132.

in his actions. Therefore, it is not surprising that his calendar texts from 1649 – the 'true mirror of genuine and false physicians' – offered readers ways of identifying professional physicians, as well as charlatans, on the basis of the practices they applied. 'Nobody knows better than the sick what a magnificent art the art of medicine is: what a magnificent work a good physician is nobody knows better than the sick who ever had the misfortune to fall into evil hands…'.[57] Patients, in other words, would be able to recognize a good doctor – such as Magirus – from the methods he applied.

57 Johannes Magirus, *Alter und Neuer Schreibkalender auf das Jahr nach Christi unseres Erlösers und Seligmachers Geburt* MDCIL (*Leipzig* [*1648*]), fol. A2r.

CHAPTER 7

Observationes et Curationes Nurimbergenses: The Medical Practice of Johann Christoph Götz (1688–1733)

Annemarie Kinzelbach, Susanne Grosser, Kay Peter Jankrift and Marion Ruisinger

'Having been called away from the church, I hear that no progress has been made,' Johann Christoph Götz wrote in his journal on Sunday, 24 February 1726.[1] Margaretha Seum's condition bore no delay. The physician had already been called a day earlier to assist the young woman with her difficult birth (*partus difficilis*). Now she was still in labour and the pain had become so unbearable that Götz was summoned from the church service to her bedside. It would not be his only consultation on that Sunday.

The practice journals reflect in detail the everyday activities of Johann Christoph Götz (1688–1733). Although the physician did not record all patient contacts, his notes, which are written almost entirely in Latin, reveal on which days of the year he was consulted, how many patients he treated and how often he called on them or they on him.[2] Insertions in German represent literal renditions of the patients' statements regarding their state of health. As a rule, Götz also recorded the patient's age, place of residence and occupation. He noted down his diagnosis in the margins, adding a cross when a patient had died. The patient journals contain notes on payment, too, which were transferred to a separate account book; but this has not, unfortunately, been preserved.

Information on prescriptions is also included in the journals. The names of the pharmacies which received the prescriptions are added in a different handwriting, probably by Götz' wife Margaretha Catharina. As a result of the

1 '*e templo vocatus audio, nihil hactenus profectum esse*'. Universitätsbibliothek Erlangen-Nürnberg, Manuscript Collection (hereafter: UBE) Ms. 1202/2, 72.

2 Ruth Schilling, Sabine Schlegelmilch and Susan Splinter, 'Stadtarzt oder Arzt in der Stadt? Drei Ärzte der Frühen Neuzeit und ihr Verständnis des städtischen Amtes,' *Medizinhistorisches Journal* 46 (2011), 99–133: 113; for a general characterization cf. Volker Hess and Sabine Schlegelmilch, '*Cornucopia officinae medicae*: Medical Practice Records and Their Origins' (in this volume).

 | DOI 10.1163/9789004303324_009

Medical Act of 1652, Nuremberg had a special regulation which specified that, for each doctor's practice, a notebook had to be placed in the pharmacies so that the preparations and dispensation of medicines could be recorded. The newer edition of this Act, from 1700, which was in force in Götz' time, included an amendment of this requirement.[3] Some of these notebooks have been preserved in the University Library in Erlangen, in the German National Museum and in the Nuremberg Town Archives,[4] but those of Johann Christoph Götz are not among them.

There are seven practice journals from the period between 1716 and 1726;[5] the volumes from the early years (the practice started in 1713) are missing, as are those of the final years up until 1733, the year of Götz' death. But the expanding size of the extant volumes suggests a steady increase in patient contacts. While there are no patient letters enclosed in the years 1716, 1717 and 1718, those bound in at the end of the other volumes provide evidence of his medical correspondence.[6] On the whole, the journals portray Götz as a medical practitioner and aspiring 'corresponding physician'. His correspondence would play an important part in other respects towards the end of his life.[7]

---

3 Verneuerte Gesetz und Ordnung Eines Hoch Edlen und Hochweisen Raths, des Heiligen Reichs Stadt Nürnberg, Dem Collegio Medico, den Apotheckern, und anderen Angehörigen daselbst gegeben (Nürnberg 1700).

4 Germanisches Nationalmuseum Nürnberg, Historisches Archiv, Pharmaceutica 14–16; UBE Ms. 1186–1188 u.a., Stadtarchiv Nürnberg, N E9 (III), NW 233; we thank Bettina Blessing, Stuttgart, for pointing out the *Rezeptierbücher* in the Nuremberg city archives (Stadtarchiv Nürnberg).

5 UBE Ms. 1200/1 (1716); 1200/2 (1717); 1200/3 (1718); 1201/1 (1720); 1201/2 (1721); 1202/1 (1725); 1202/2 (1726).

6 Cf. Marion Maria Ruisinger, 'Briefpraxis versus Besuchspraxis. Das Beispiel Lorenz Heister (1683–1758),' in Elisabeth Dietrich-Daum, Martin Dinges, Robert Jütte and Christine Roilo (eds), *Arztpraxen im Vergleich. 18.–20. Jahrhundert* (Innsbruck 2008), 65–77; Kay-Peter Jankrift, Annemarie Kinzelbach and Marion Ruisinger, 'Ernst von Metternich (1656–1727). Ein patientenzentrierter Einblick in den medizinischen Markt um 1720,' in Lina Gafner, Iris Ritzmann and Katharina Weikl (eds), *Penning Patient's Histories – Doctors' Records and the Medical Market in the 18th and 19th Century* [special issue: *Gesnerus* 69, no. 1 (2012)], 12–35.

7 Tilman T.R. Rau, *Das Commercium Litterarium. Die erste medizinische Wochenschrift in Deutschland und die Anfänge des medizinischen Journalismus* (Bremen 2009); Jankrift *et al.*, 2012; Susan Splinter, '"Der Aufrichtige Medicus". Eine Zeitschrift des Nürnberger Arztes Johann Christoph Götz (1688–1733) als Vorläufer des "Commercium Litterarium",' *Jahrbuch für Kommunikationsgeschichte* 13 (2011), 5–15; Schilling *et al.*, 2011, 114; Thomas Schnalke, *Medizin im Brief. Der städtische Arzt des 18. Jahrhunderts im Spiegel seiner Korrespondenz* (Stuttgart 1997), 224.

Johann Christoph Götz

Johann Christoph Götz was born in Nuremberg on 8 March 1688. He was baptized the next day in the Church of St Lawrence and was given the name of his father who had died several months before his birth.[8] His mother, Anna Elisabetha (born 1653), was the daughter of the merchant Carl Albrecht and came from a respected Nuremberg family. She had sufficient means to afford a good education for her youngest son. After attending the grammar school of St Lawrence, the *Gymnasium Aegidianum* in Nuremberg and the *Gymnasium Casimirianum* in Coburg, Johann Christoph Götz began his medical studies in Altdorf in 1708. Having gained his doctorate, he embarked on the customary study-tour, which also led him to Halle where he heard the lectures of Georg Ernst Stahl (1660–1734), Friedrich Hoffmann (1660–1742) and Michael Alberti (1682–1757).[9] The university lectures and the religious and spiritual ambience in Halle made a lasting impression on the young physician. The pietism and teachings of Georg Ernst Stahl deeply influenced his medical work when – having finished his academic peregrination and having been accepted as a sworn doctor of medicine by the Nuremberg *Collegium Medicum* – he began to practise in his hometown on 4 April 1713.[10] This Stahlian influence is reflected in his practice journals, both in occasional references to religious rituals and in his prescriptions.[11]

A few weeks after his admission to the *Collegium Medicum*, on 29 May 1713, Götz married Margaretha Catharina Spitz, the daughter of theology professor Felix Spitz (1641–1716). Some years later her brother Erhart would introduce Götz to the Count Ernst von Metternich, who repeatedly sought advice by letter from the Nuremberg physician regarding the treatment of bladder stones.[12] Three children were born to Johann Christoph Götz and Margaretha Catharina Spitz. When their youngest son, Ernst Felix Christoph, was baptized on 30 November 1716, Georg Ernst Stahl stood as godfather.

In addition to his medical practice, Götz devoted much time to scientific study. The monthly meetings of the *Collegium Medicum* provided opportunity for discussion with colleagues and for continued professional training.

---

8 The biographical notes on Johann Christoph Götzes are based on Susan Splinter's research during the first phase of the project. Splinter, 'Medicus,' 2011; Schilling *et al.*, 2011, 112–18.

9 Leopoldina-Archiv MNr. 391 Götz, Johann Christoph, 5.

10 Susan Splinter '"Ein jeder der Stadt Nürnberg bestelter Medicus und Physicus ordinarius soll geloben…". Medizinale Strukturen Nürnbergs zu Beginn des 18. Jahrhunderts,' *Würzburger medizinhistorische Mitteilungen* 30 (2011), 334–49.

11 E.g. UBE Ms. 1202/2, 72; Habrich 1981.

12 Jankrift *et al.*, 2012.

The *Collegium* had a small library, a *theatrum anatomicum* and its own botanical gardens. In 1726, Götz released the first issue of the weekly journal *Der Aufrichtige Medicus* (The sincere physician).[13] By deliberately deciding to publish in German rather than Latin (then the *lingua franca* of scholars) he intended to open up debate on topical medical questions to interested laypersons. But the experiment failed and publication ceased in the second year. In 1730, Götz, Christoph Jacob Trew – who had joined the *Collegium* in 1720 and would go on to assume a central role among the scholars of Nuremberg – and other physicians founded a society which published the *Commercium litterarium,* the world's first medical weekly.[14] But Götz was not to witness the progress of this new media project for long. He died in 1733 at the early age of 45.

Nuremberg

At the time when Götz was active in Nuremberg the city had passed its heyday. Scholars such as Johannes Regiomontanus (1436–1476) had once regarded the free imperial city as the centre of Europe.[15] After the Thirty Years' War it lost much of its political significance and was weighed down by economic problems. The population decreased steadily from around 50,000 in 1622 to 25,126 in 1806 when the city was ceded to Bavaria.[16] This did not prevent the anatomist and surgeon Lorenz Heister (1683–1758) from considering Nuremberg to be the loveliest city in Germany.[17] There were a number of reasons why, as late as the

---

13 Splinter, 'Medicus,' 2011.

14 On Christoph Jacob Trew and his Nuremberg milieu cf. Thomas Schnalke, *Natur im Bild. Anatomie und Botanik in der Sammlung des Nürnberger Arztes Christoph Jacob Trew* (Erlangen 1995); on the *Commercium litterarium* cf. Rau, 2009.

15 '*quasi centrum europae*,' quoted in Siegmund Günther, 'Müller, Johannes,' *Allgemeine Deutsche Biographie* 22 (1885), 564–81.

16 Rudolf Endres: 'Bevölkerungsentwicklung,' in: Michael Diefenbacher and Rudolf Endres (eds), *Stadtlexikon Nürnberg* (Nürnberg 1999) (http://online-service2.nuernberg.de/stadtarchiv/dok_start.fau?prj=biblio&dm=Stadtlexikon).

17 '... *Nurimbergam, pulcherrimam fere Germaniae urbem*'. Lorenz Heister, in a letter to Giovanni Battista Morgagni, 20 December 1748 (Biblioteca Comunale di Forlì, collezione Morgagni, vol. 29, 111r). On the attraction of Nuremberg for eighteenth century scholars cf. Marion Maria Ruisinger, 'Die Medizinische Fakultät Altdorf im frühen 18. Jahrhundert – eine Wiege der wissenschaftlichen Chirurgie,' in Hanns Christof Brennecke, Dirk Niefanger and Werner Wilhelm Schnabel (eds), *Akademie und Universität Altdorf. Studien zur Hochschulgeschichte Nürnbergs* (Köln 2011), 207–23.

FIGURE 7.1 *Nuremberg around 1750. Coloured copper engraving by Tobias Conrad Lotter, Augsburg 1740/50.*
GERMANISCHES NATIONALMUSEUM, KUPFERSTICHKABINETT, KAPSEL 1039B, SP. 6374.

eighteenth century, it was still attractive to physicians who harboured scientific ambitions. It offered, for instance, the opportunity for on-going study and learning outside the university and boasted excellent copperplate engravers and printers, overseas mercantile relations that facilitated the exchange of correspondence, books and natural produce and, not least, a highly structured municipal healthcare system which had been governed by pharmaceutical and medical laws since the sixteenth century. The *Dispensatorium* of Valerius Cordus (1515–1544), published by the city council in 1592, is one of the oldest official pharmacopoeias in existence.[18] In 1592 a medical act came into effect that included the institution of a *Collegium Medicum* of which all academically trained medical practitioners in Nuremberg and the professors of medicine in

18 Christoph Friedrich, 'Pharmakopöen. Spiegel der pharmazeutischen Technologie,' *Pharmazeutische Zeitung* 145 (2000), 2122–127.

Altdorf were members.[19] The fact that the city was well supplied with pharmacies was another attraction. In Götz' time there were two pharmacies in the district of St Lawrence where he lived and five more in the Sebald district across the river Pegnitz.[20]

Patients

The journals reveal that Götz, with his years of practice, managed to gain the trust of a growing number of patients and that by the 1720s his practice was well established. The number of Götz' patients and of physician-patient contacts tripled from 91 patients and 404 contacts in 1717 to 283 patients and 1216 contacts in 1721. Patient numbers remained high in later years.[21] These figures do, however, conceal a diversity of contacts that is only revealed by an analysis of individual entries and patients.

Indirect contact was not restricted to consultations by letter. Almost every tenth first contact was not established by the patients themselves, but through male or female servants, messengers, relatives, employers or neighbours who asked for advice or medicines, as in the case of the Nuremberg lawyer Dr Wolfgang Hieronymus Herold, who sent his servant to ask for a particular soothing powder.[22]

Götz dispensed 'proprietary medicines' (i.e. medicines of mostly unknown composition sold under a specific name) even where no consultation is recorded. Twenty-eight such cases are documented for 1721, most of them featuring

19 Philipp L. Wittwer, *Entwurf einer Geschichte des Kollegiums der Aerzte in der Reichsstadt Nürnberg* (Nürnberg 1792).

20 Collegium Pharmaceuticum Nürnberg (ed.), *Versuch einer Geschichte des Apothekenwesens in der freyen Reichsstadt Nürnberg* (Nürnberg 1792).

21 UBE Ms. 1200/2, 1201/2, 1202/2. These figures seem low compared to those of the nineteenth-century Nuremberg practice of Johann Osterhausen, cf. Martin Dinges, 'Arztpraxen 1500–1900. Zum Stand der Forschung,' in Dietrich-Daum *et al.*, *Arztpraxen*, 2008, 23–61: 39; the clearly lower figures of Albrecht von Haller in Bern from the 1730s suggest that Götz had established himself as a physician by the 1720s; cf. Hubert Steinke, 'Der junge Arzt und seine Patienten. Albrecht von Hallers Praxis in Bern 1731–1736,' in Dietrich-Daum *et al.*, *Arztpraxen*, 2008, 79–86: 80; Philipp Klaas, Hubert Steinke and Alois Unterkircher, 'Daily Business: the Organization and Finances of Doctors' Practices' (in this volume).

22 *Herold, Wolf Hyeronimus. Herold, Advocatus huius Rei publicae ordinarius, per famulum petens accepit Pulv. Carl. Temp.* UBE Ms. 1201/2, 11; Dr. jur. Wolfgang Hieronymus Herold, StaN (Stadtarchiv Nürnberg) E 1/585. Sample 1721 UBE Ms. 1201/2, 9, 11, 18, 53, 83, 127, 137, 141, 147, 191, 194, 234, 270, 276, 297, 307, 318, 373, 386, 391, 414–15, 455, 484, 495.

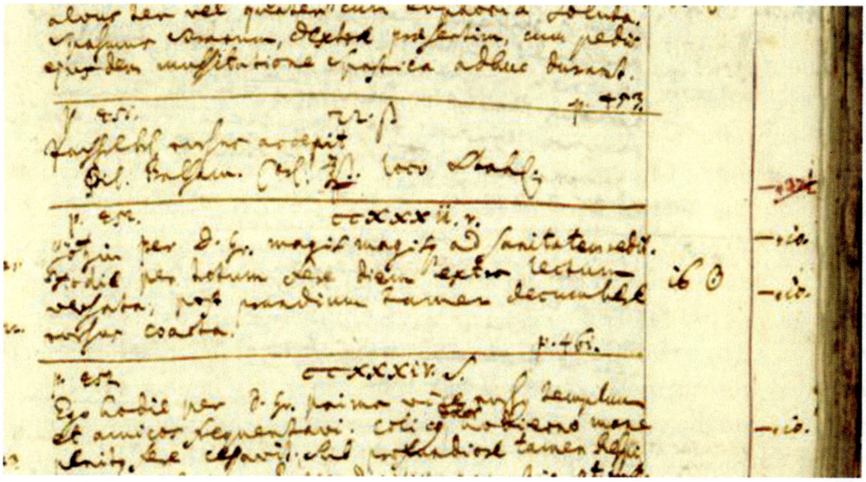

FIGURE 7.2 *Extract from Götz's practice journal of 1721.*
UNIVERSITY LIBRARY ERLANGEN, MS. 1201/2, P. 453.

remedies by Johann Samuel Carl.[23] Pachhelbel, the organist at Saint Sebald, and other patients received these medicines although they were not registered patients at the time.[24] Götz obviously regarded such contacts as different enough to mark them separately in his indexing system: he used titles with Arabic numerals, in this case *22* and *22. ß*, instead of Latin letters as in the subsequent entry: *ccxxxii. r.* (Figure 7.2).

Götz had on average three (direct or indirect) contacts per case of treatment. In some cases treatment was more intense, as with the 55-year-old merchant Leonhard Wegleiter. Götz called on him several times a day, giving

23 UBE Ms. 1201/2, 56, 60, 77, 82, 91, 101, 192, 198, 206, 212, 215, 230, 243, 297, 306, 373, 386, 418, 426, 446, 451, 453, 455, 464, 471, 472, 517; there are only two cases where he dispensed different medicines: a Bachhelbel solution and Stahl's pills, 297, 373. On Carl cf. Christa Habrich, 'Untersuchungen zur pietistischen Medizin und ihrer Ausprägung bei Johann Samuel Carl (1677–1757) und seinem Kreis' (Habilitationsschrift, University of Munich 1981); *eadem*, 'Johann Samuel Carl (1677–1757) und die Philadelphische Ärztegemeinschaft,' in Hartmut Lehmann, Hans-Jürgen Schrader and Heinz Schilling (eds), *Jansenismus, Quietismus, Pietismus. Im Auftrag der Historischen Kommission zur Erforschung des Pietismus* (Göttingen 2002), 272–89.

24 '*... Pachhelbel. Pachhelbel, organist bey S. Sebald petens accepit Pil. Carl. Balsam; loco Stahl*'., comparable entry three days later, UBE Ms. 1201/2, 451, 453. Götz did not mention a first name but one organist of St Sebald was Wilhelm Hieronymus Bachhelbel (1686–1764), son of the famous organist and composer Johann Bachhelbel. Stadtlexikon Nürnberg online (15 June 2012).

direct treatment as well as exchanging information via messengers. The treatment of Wegleiter, which took place from early June until late December 1721, occasioned 43 entries, some of which sum up several doctor's calls per day or consultations stretching over two days.[25] This is apparent in the first entry of Sunday, 1 June 1721, when Götz recorded the patient's history followed by the diagnosis: colic of the kidneys and the hemorrhoids.[26] According to Götz, the patient, who had been suffering from nausea and vomiting after meals, called him at supper time. He was by then experiencing testicular pain and dizziness and neither of these symptoms could have been caused by a wrong diet or by excitement. Götz recorded that the patient informed him of the time of his last venesection, the amount of blood extracted and the consistency of the blood, and that he immediately administered Carl's soothing powder. Götz possibly went on to the *Heilig-Geist-Apotheke* to ask for the preparation of an 'antinephritic' medicine.[27] Other contacts on that day were included under the same entry. A week later he summarized the results pertaining to the period from Saturday to Tuesday: medicines dispensed by his wife and the pharmacist during his absence of several days, two personal calls and several indirect contacts that entailed diverse prescriptions and dietary recommendations.[28]

Further random checks reveal more summaries of this kind.[29] Any conclusions as to the intensity of treatment therefore require in-depth study, but even then the phrasing does not always reveal with certainty whether individual steps of treatment were recommended by Götz or whether they had been reported to him. The entries on Felicitas Bartschmid, the 24-year-old daughter of a button-maker in the neighbourhood, whom Götz treated for several weeks for a 'daily fever with cough,'[30] make it difficult to discern how much

25 UBE Ms. 1201/2, 187, 190, 192, 197, 204, 206, 211, 213, 225, 263, 284, 295, 303, 309, 310, 312, 313, 317, 322, 331, 346, 347, 364, 375, 382, 386, 394, 400, 403, 406, 410, 414, 417, 420, 423, 425, 426, 428, 432, 434, 436, 454, 506.

26 *Colica Nephritico-Haemorrhoidalis*, UBE Ms. 1201/2, 187–89.

27 *Antinephritica, sunt medicamenta morbis renum medentia*, i.e. medicines for the treatment of renal disease, cf. Blanckaert 1748, 66.

28 UBE Ms. 1201/2, 192–94.

29 On Wegleiter UBE Ms. 1201/2, 309, 375, 394–95, 403, 406, 410, 423; other examples: Ms. 1201/2, 4–5, 13–15, 16–17, 25, 27–28.

30 *Febris Quotidiana cum Tussi*, cf. Blanckaert, Steven/Schulze, Heinrich: *Lexicon medicum. Ioann. Henrici Schulzii Opera Insigniter Auctum Et Emendatum, Nunc Denuo Recognitum Variisque Accessionibus Locupletatum Curante Mich. Gottl. Agnethlero. Praefationem Praemisit Andreas Elias Büchner* (Halle 1748) (http://www.uni-mannheim.de/mateo/camenaref/blanckaert.html), 317, 684.

he himself added to the treatment, since it also involved other members of the *Collegium Medicum*. It seems that he only learnt from others the young woman's death six months after his last entry on her.[31]

It is mostly due to the fact that the journal has not been preserved in full that the patients' places of residence are only recorded in twenty to thirty per cent of entries. In a comment from 1720, Götz refers to detailed information on a patient in an earlier volume: *Poemerin, Cunigunda, Joh. Jacobi uxor. Poemerin, cuius Historia n. MMCCXXii et sequ. habetur*. But the last preceding volume that is preserved dates from 1718 and ends with number *MCCXiX*.[32] Many patients can be identified as residents of Nuremberg because they were relatives or servants of local people. With another ten per cent of cases the relevant conclusions can be derived from archival searches. Overall, it is fair to say that around three quarters of the patients whose location can be identified lived in Nuremberg. But Götz was also consulted by patients from neighbouring towns and villages, even across political or confessional borders, as is apparent from a case of Jewish patients from the town of Fürth. Occasionally Götz would also travel to see a patient, such as the lieutenant's wife, a member of the eminent aristocratic family of Schmettau,[33] who was staying on the estate of General von Barner, more than fifty kilometres from Nuremberg.[34] This was, however, an exception since Götz' patients as a rule came to see him in his house in Nuremberg or exchanged letters with him. We find no indications of any kind of patronage.[35] Götz was a family physician and 'family,' in early modern parlance, included the entire

31 1721: 15 entries in January alone, Ms. 1201/2, 2, 5, 9, 11, 16, 19, 21, 22, 24, 29, 31, 36, 38, 40, treatment up until 17 February 1721; note on 31 August 1721 that the patient had died, 53, 314. There is mention of *D.D. Gockelius* and *Rost*, Dr Christoph Ludwig Göckel, hospital physician and Dr Johann Carl Rost, two members of the Collegium Medicum in Nuremberg, cf. Wittwer, 1792, 29, 36–37.

32 Ms. 1201/1, 2; 1200/3, 714.

33 Barner and Schmettau cf. Johann Heinrich Zedler, *Grosses vollständiges Universal-Lexicon aller Wissenschafften und Künste, Welche bißhero durch menschlichen Verstand und Witz erfunden und verbessert worden* (Halle/ Leipzig 1732–1754) (http://www.zedler-lexikon.de, 11 August 2014), sup. 3, 54–55, vol. 35, 337–39.

34 UBE Ms. 1201/2, 2, 9, 11, 12, 13, 18, 23, 29, 191, 192, 204, 270, 307, 318, 484; Ms. 1202/2, 52–54, 74, 212, 251–52, 308, 313, 369, 450, 498, 499; Ms. 1201/2, 191–92.

35 This did not seem to apply to Haller in Bern, cf. Hubert Steinke and Urs Boschung, 'Nützliche Medizin. Theorie und Praxis bei Albrecht von Haller,' in André Holenstein, Martin Stuber and Gerrendina Gerber-Visser (eds), *Nützliche Wissenschaft und Ökonomie im Ancien Régime. Akteure, Themen, Kommunikationsformen* (Heidelberg 2007), 133–47: 145–46.

household. In sixty-three per cent of new contacts he treated several members of the same family or household.[36]

Analysis of the patients' occupations, as far as this is possible on the basis of the journals, shows that a quarter of patients belonged to the households of Nuremberg craftsmen (28%). A similar proportion (25%) came from the families of lawyers, clerks, priests, teachers or artisans. These figures alone reveal that Götz did not only treat members of the city's upper classes. The high percentage of male servants, apprentices and day-labourers (14%) and of female servants, maids and cooks (27%) shows, moreover, that less established and mostly young men and women turned to Götz or were sent by their employers. The entries on his 30-year-old cook Helena Schönknecht and her 56-year-old mother Margareta, who lived in Fürth, illustrate that Götz' role as a much-consulted 'family physician' was not restricted to the upper classes.[37] The case of the cook is also proof that people who knew Götz personally – in this case an employee – consulted him as a physician and recommended him to relatives.

Gender distribution shows, with 48% to 59%, a slight predominance of women.[38] This result qualifies the thesis that the female body was only medicalized in the nineteenth century,[39] since even in the early eighteenth century, numerous women in Nuremberg consulted a physician specifically for women's problems. Of the 509 cases in the three years that were statistically evaluated, the 57 complaints relating to the female body constitute the third most frequent cause of consultation after 'fever' (60) and venesection (58).[40] Even though Götz' journal reveals that childbirth was a specifically female matter involving only midwives, female healers, relatives or neighbours, pregnant women and nursing women in childbed often consulted Götz for themselves and their babies, especially when complications arose.[41]

36 In 1721 these were 189 out of 301 new contacts, UBE Ms. 1201/2.

37 UBE Ms. 1201/2, 18, 23, 32, 33, 37, 105, 268, 274, 275, 514, 517. Not only the case studies reveal a broad social distribution, cf. Matthias Blanarsch, 'Die Arzt-Patienten-Beziehung zu Beginn des 18. Jahrhunderts, untersucht anhand Johann Storchs Kasuistik zu Molenschwangerschaften,' *Medizin, Gesellschaft und Geschichte* 28 (2009), 121–52: 126–27; for diachronic comparisons cf. Marion Baschin, Elisabeth Dietrich-Daum and Iris Ritzmann, 'Who Consulted the Doctor? The Patients and their Motives' (in this volume).

38 Statistical evaluation of the years 1717, 1721, 1726, UBE Ms. 1200/2; 1201/2; 1202/2.

39 Cf. the criticism in Michael Stolberg, 'The Monthly Malady. A History of Premenstrual Suffering,' *Medical History* 44 (2000), 301–22: 320.

40 UBE Ms. 1200/2; 1201/2; 1202/2.

41 The patterns of interaction between women and Götz were very complex, cf., among others, Annemarie Kinzelbach, 'Women and Healthcare in Early Modern German Towns,'

Knowledge and Communication

Götz' activities as a physician required wide medical, but also patient-specific, knowledge. He therefore had to continue to develop what he had learned as a student. In this process, indirect knowledge acquisition was as important for him as gaining direct insight through reading and observation. A wealth of examples of such indirect ways of gaining knowledge can be found in the practice journals, since Götz not only recorded multiple kinds of knowledge he had found access to, but always mentioned the respective sources, channels and contexts of the underlying communication.

Starting with the immediate communication between Götz and his patients, it is noticeable that while he generally kept his journal in Latin, he wrote down the complaints described by his patients word for word in German, especially when these reports were vividly illustrated with metaphors or similes, as in the case of Barbara Catharina Geider who likened the pain in the left side of her head to the sensation of having 'a red-hot wire pulled through it,'[42] or that of the butcher Peter Wölfel who related that, when he had coughed, a sound had escaped from his ear 'as if three rockets were fired from it, one after the other'.[43] Götz also recorded other illness-related observations reported by his patients, such as the statement of Catharina Susanna Merckel that she had tolerable and less tolerable days.[44]

With new patients Götz would also write down personal information next to their current complaints, including age and occupation, previous illnesses, medication taken so far, lifestyle and dietary habits as well as the use of venesection. He elicited this information, using a list of specific questions which has been preserved in two cases.[45]

Communication between Götz and his patients was not limited to history-taking, but extended to the question of treatment. Götz noted when patients

---

Renaissance Studies 28 (2014), 619–38.

42 UBE Ms. 1201/2, 124. Götz' pietistic colleague Carl also frequently included German phrases in his notes, cf. Johann Samuel Carl, *Specimen Historiae Medicae Ex Solidae Experientiae Documentis, Maxime Vero Monimentis* [!] *Stahlianis, In Syllabum Aphoristicum Redactum. Ut Ita Ex Morborum Circumstantiis Perpetuis Essentialibus Et Extraessentialibus Universa Oeconomia Medicinae Internae, Indeque Per Se Fluens Vera Et Constans Pathologica Et Therapeutica Consideratio Patescat* (Halle 1719) (http://reader.digitale-sammlungen.de/de/fs1/object/display/bsb10054187_00005.html), 47–49, 56–59, 65, 88, 91, 94.

43 UBE Ms. 1201/2, 177.

44 UBE Ms. 1202/2, 243: '*Hodiernum tamen minus, quam hesternum, narrat tolerabilem, et observasse vult dies molestos et tolerabiles alternari*'.

45 Ms. 1202/1 (1725), 102; Appendix F.

flatly refused to use therapies he suggested or did not comply with them. He did not respond with criticism and apparently accepted that his clients had their own views on treatment which they often resolutely expressed.[46] The journals, in reflecting the physician-patient relationship mostly as an encounter between equals, confirm that the position of patients tended to be stronger in early modern days than it is today.[47]

But Götz did not gain all his information about patients exclusively from direct communication with them. The journals reveal that third persons played an important part, even at the bedside.[48] Information was supplied particularly by relatives (spouses, parents, children, siblings and so on) or by servants belonging to the household. Female servants especially, who took on the role of sick-nurses, often kept the physician posted. It was a maid who kept Götz informed about the bowel movements of the 2-year-old Erhard Metzger.[49] Servants also reported to the physician behind the patient's back, as is shown by the case of Johann Christian Schleicher who was treated for *hypochondriacum malum*. The cook he sent as a messenger to Götz told the physician all about his employer's temper tantrums.[50]

The journal entries document, moreover, the indirect and not necessarily case-related sharing of information about the treatment of illnesses. Notes about individual patients often include references to the letters of medical colleagues if Götz made use of therapeutic considerations contained in these letters.[51] His wide network of communication is apparent both from his correspondence and from the *Commercium litterarium*.[52] Götz also recorded facts worth knowing that were not associated with particular cases. As with his notes

46 Cf. UBE Ms. 1202/2, 245–46. The patient Sabina Elisabetha Wegleiter refused the use of enemas: '*Clysteres ob Haemorrhoides, quas fere nunquam non habet, admittere recusat*'.

47 Cf. Michael Stolberg, *Experiencing Illness and the Sick Body in Early Modern Europe* (Basingstoke 2011), 64.

48 In his journals Götz often added *narrat* as an indication that the information in question was from a third person.

49 UBE Ms. 1201/2, 72: '*Deiectiones autem materiae teterrimae tres ab ancilla meridie, cum adirem, narrabantur*'.

50 UBE Ms. 1201/2, 36: '*Iras multiplices simul subesse clam narrat coquus nuncius*'.

51 UBE Ms. 1201/2, 40.

52 On Götz' correspondence in Trew's collection of letters cf. Eleonore Schmidt-Herrling, *Die Briefsammlung des Nürnberger Arztes Christoph Jacob Trew (1695–1769) in der Universitätsbibliothek Erlangen* (Erlangen 1940), 225. The *Commercium litterarium* includes references to other correspondents of Götz, for instance in *Commercium litterarium ad rei medicae et scientiae naturalis incrementum institutum quo quicquid novissime observatum agitatum scriptum vel peractum est succinte dilucideque exponitur* (1731), 355–56; (1734), 1–4.

on the dispensation of proprietary medicines, he used Arabic rather than Roman numerals for such records. Such information might go back to conversations with medical colleagues or pharmacists in Nuremberg[53] or to medical lay-people. References to 'miracle cures' or 'panaceas' are quite frequent, as in the case of an old cobbler who attributed his longevity to a miraculous root.[54] This kind of information was relevant for Götz, not least since better situated patients were also interested in such wondrous remedies and he could enhance his status among the advisers of noble patients by providing the appropriate remedies.[55] On the whole, the practice journals confirm the importance of multiple communication processes for a regular and continuous increase of practice-relevant knowledge at various levels.

The Range of Illnesses

The first entry of each case of illness in the practice journal usually contains a marginal note in Latin which served as an index entry and facilitated orientation (Figure 7.3). Since the ink colour and stroke width of these annotations often differ from the main entries, it can be assumed that Götz added them later. The terms he used can be divided into three groups depending on whether they are based (1) on the patient's own perception, (2) on changes which could be perceived from the outside or (3) on a secondary diagnostic interpretation. The annotation 'venesection' constitutes a special case.

Most of the index terms used by Götz are based on the patients' own perspective, adding the appropriate technical terms to the narratives provided by the patients in their native tongue. Götz most frequently noted localized pain in the head (*cephalgia*) or heart (*cardialgia*), toothache (*odontalgia*), earache (*otalgia*) or pain when passing water (*stranguria, ardor urinae*); also respiratory disorders (*asthma, dyspnoe*), indigestion (*oppletio primarum viarum, alvus obstructa, anorexia*) and menstrual problems (*mensium ataxia, mensium cessatio*). These subjectively experienced complaints formed the 'causes for consultation'[56] that prompted patients to make contact with Götz.

---

53 UBE Ms. 1200/3, 463; Ms. 1200/3, 598.

54 UBE Ms. 1202/1, 12–13: '*Vierthaler, ultra octuaginta annos agens sutor, …longaevitatem, et quasi tenacitatem, suae vitae attribuit cuidam radici, quam…ab eremita quodam…acceperat*'.

55 Cf. Jankrift *et al.*, 2012.

56 'The cause of consultation is what prompts patients to see a general practitioner': Frank H. Mader and Herbert Weißgerber, *Allgemeinmedizin und Praxis. Anleitung in Diagnostik und Therapie* (Berlin ²1995), 16.

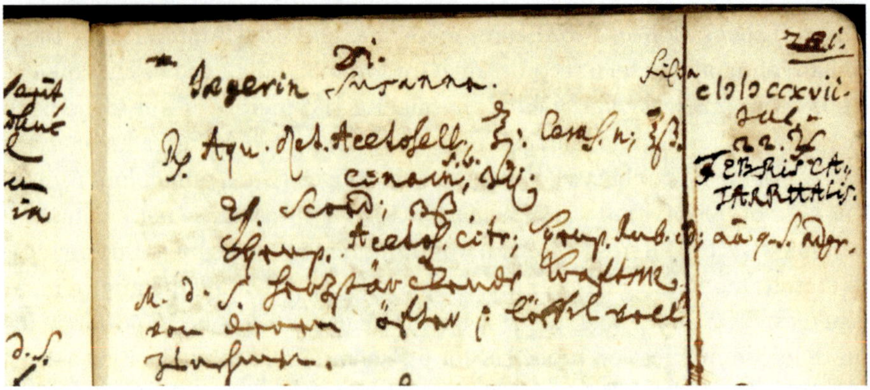

FIGURE 7.3 *Götz's annotation of 'Febris catarrhalis' in the margin of the entry on Susanna Jaeger's case in the practice journal of 1717.*
UNIVERSITY LIBRARY ERLANGEN, 12FIG. /2, P. 281.

The index terms of the second category refer to the external appearance of the patient's body and denote outwardly perceptible changes which were perceptible from the outside, often relating to the skin, such as skin marks (*naevi materni, serpigo*), cradle cap (*crusta lactea*), holy fire (*erysipelas*), smallpox (*variolae*), scabies or other forms of rash (*exanthemata*). Other visible changes included swellings (*tumores*) of knee, ankle, foot or the whole body (*tumor totius corporis*) and afflictions of the eye (*ophthalmia*) or the bed of nails (*paronychia*). Other phenomena that were noticeable by bystanders included those associated with pregnancy and childbed (*abortus, fluor albus gravidae, partus difficilis, puerperii varia symptomata*). Non-visible, but audible causes for treatment, were hoarseness and cough (*tussis*). Somewhere between 'internal' and 'external' were vomiting (*vomitus*), diarrhea (*diarrhoea*), bleeding (*haemorrhagia*) and coughing blood (*haemoptysis*).

The index terms of the third category result from Götz' processing of information – gained from the inner and outer perspectives – into diagnoses in the traditional sense. Cases referred to as fever (*febris*) are particularly frequent. In the three years analysed there are more than thirty different descriptions of fever that differ either in their aetiology and pathogenesis (*f. biliosa / rheumatica / ex iracundia*), in the course they took (*f. incipiens, f. in prima herba suffocata / suppressa / complicata*) or in their frequency of occurrence (*f. quotidiana / continua / tertiana / quartana*).

There are very few cases where Götz associated particular conditions with yellow bile (*cholera*) or black bile (*melancholia*). He never referred to an overabundance of blood (*plethora*) or, in iatromechanic terms, to the congestion or stagnation of humours. This is all the more astonishing since Götz frequently prescribed medicines that he hoped would cleanse the blood, 'dissipate flux' or 'eliminate sharpness'. There is probably a pragmatic explanation for the fact that he based his aetiological interpretations and the therapeutic concepts he derived from them on an advanced form of humoralism, but refrained from using it in his index. If the journal was to serve as a reference book for the treatment of new patients, the index terms, in order to facilitate the fast finding of suitable cases of reference, had to relate to the causes for consultation.

We do not find much evidence of injuries or other treatments involving surgical intervention. In surgical or obstetric cases, Götz' task was restricted to preparatory medical measures and post-operative care, while the operation or delivery itself fell within the remit of the surgeons or midwives. One exceptional case was that of the gatekeeper's wife *beym Hallerthörlein* (at the Haller Gate) who presented herself to Götz – ten days after having been beaten by her husband – with severe bruising on her left breast.[57] In this case Götz himself took on the treatment, prescribing a herbal infusion, laxative pills and a 'dissipating' plaster. Maybe the gatekeeper's wife, who was called Göz, was a relative and therefore preferred to consult him rather than the surgeon.

Venesection

While Götz clearly did not offer surgical interventions we find many instances of venesection in the practice journal. On taking a patient's history, he would generally ask about former bloodletting habits, as is apparent from the questionnaire he sent to a patient who did not consult him in person.[58] The term bloodletting (*venaesectio*) is also prominent among the case index terms. Analysis of the distribution of these entries reveal a definite increase: in the years 1716, 1717 and 1718 not a single case is annotated 'VSio,' while there are 10 and 17 respectively in 1720 and 1721, and 34 and 43 entries respectively in 1725 and 1726. In interpreting these findings it is important to distinguish between preventive and curative bloodletting.[59]

There are only a few cases where the venesection recorded by Götz was part of a treatment. One of them is that of the 24-year-old sick farm labourer Paul

57 '*abhinc X diebus iam a marito verberibus valde male habita,*' Ms. 1201/2 (1721), 50.

58 Ms. 1202/1 (1725), Appenix F.

59 Marion Maria Ruisinger, 'The Circulation of the Blood and Venesection. On the Relation between Medical Theory and Practice in the Early Eighteenth Century,' in Jürgen Helm and Renate Wilson (eds), *Medical Theory and Therapeutic Practice in the Eighteenth Century. A Transatlantic Perspective* (Stuttgart 2008), 37–59.

Dummler, who was at risk of developing an acute fever. Götz advised that he should immediately have the venesection that had been delayed for three weeks.[60] As a rule, the cases recorded constitute preventive measures, a finding that is corroborated by the seasonal surge of applications in the weeks close to the equinoxes – a time particularly recommended for bloodletting. 71 out of 104 venesections recorded were carried out in the spring (March, April), a further 18 in the autumn (September, October). The preventive character of these venesections is also apparent from the fact that married couples and household members had their veins opened on the same day and by the same surgeon.[61] The regular annual venesection for 'full-blooded' persons was seen as a crucial health-promoting measure. One such patient was the 31-year-old baker Johann Georg Bauer who, being 'puffy, robust, sanguine-phlegmatic,' had become used to annual venesections and monthly bloody cupping.[62]

Venesections were not carried out by academically trained physicians. The Nuremberg Medical Act of 1700 stated that it was not 'customary…for *Medici* to act in *Chirurgicis casibus* and apply fontanels, venesection and similar,' since these interventions fell to trained 'barbers and surgeons'.[63] This corresponds with Götz' notes which state that bloodletting had been carried out by a *chirurgus tonsor* or a *chirurgus balneator.* He was generally not present for these procedures. Any exceptions were stated explicitly, as in the case of his brother, Johann Philipp, who was bled on 21 April 1721 by a barber called Igel under the supervision of Götz and another physician.[64] The venesection of Clara Regina Elisabetha Gäntner was certainly exceptional: on 4 April 1725, she met the surgeon Wittwer in Götz' house (*in mea domo*) where she was bled from her right foot.[65]

If preventive venesection was a measure that could be carried out without the involvement or presence of a physician, one must ask why it is mentioned increasingly in Götz' practice notes – especially as his accounts suggest that he

60 *'Febrim timens Acutam suasi ut VSio, quam per tres iam Septimanas, frustra nobis alia monentibus, de die in diem meticulosus distulerat, sine ulteriori temporis administraretur dispendio,'* Ms. 1202/1, 102.

61 Like the council clerk Andreas Merkel and his wife Catharina Susanna who were bled on 28 March 1721 by a barber called Geißler. Ms. 1201/2, 104.

62 *'spongiosus, robustus, sanguineo-phlegmaticus, VSnibus annuis et scarificationibus menstruis assuetus,'* Ms. 1201/2, 153.

63 Rat der Stadt Nürnberg: *Verneuerte Gesetz und Ordnung Eines Hoch Edlen und Hochweisen Raths, des Heiligen Reichs Stadt Nürnberg, Dem Collegio Medico, den Apotheckern, und anderen Angehörigen daselbst gegeben* (Nürnberg 1700), § 33.

64 *'praesentibus me et D.D. Wincklero,'* Ms. 1201/2, 133.

65 *'Gäntnerin a balneatore Wittwer in mea domo vena Pedis dextri secta,'* Ms. 1202/1, 72.

must have been involved in a way that warranted payment. The reason was the same as in his obstetric cases: he was not actively involved in the intervention, but acted as a medical consultant.

With bloodletting there were two possibilities for Götz to be involved: he could be asked several days in advance to prepare the intervention by giving internal medicines[66] or he might call on the patient towards the evening (*vesperam*) on the day of the intervention – which usually took place between 11 a.m. and noon – in order to inspect the blood. In his entries he repeatedly referred to the first bowl (*patella*(*m*) *prima*(*m*)), which indicates that the venesection was carried out in stages and that the blood that came out first was caught in a separate bowl and kept for later inspection. The amount of blood extracted was also recorded. If we disregard the quantities mentioned in prescriptions, these are the only regular quantitative statements that occur in the practice journal. They show that 6 to 7 ounces of blood were usually extracted in preventive venesection and around 10 ounces in therapeutic bloodletting. Based on the Nuremberg ounce of 29.8 grams[67] one arrives at quantities of 180 to 300 grams of blood in each case.

This information conceals a potential source of error that Götz himself alluded to. By opening a vein in the elbow the blood could be collected almost without any loss, allowing for exact measurements. It was much more common, however, to open a vein on the bridge of the foot or on the hand.[68] In these cases the barber would encourage the blood-flow by using a warm foot or hand bath which meant that the blood was mixed with water and could not be measured. As a result more blood than necessary might be let and the patient might faint.[69] This lack of precision was vexing for the consultant physician because it meant that an important factor in the assessment of the venesection – and ultimately of the barber in question – was absent.[70]

66 For his brother Götz prescribed a digestive powder and laxative pills four days before he was bled, Ms. 1201/2, 130, 133.

67 Diefenbacher *et al.*, 1999, 1246.

68 Götz usually recorded the part of the body used for bloodletting: for instance: *VSio Brachii* or *VSio Pedis*. The Nuremberg Medical Act of 1700 specified guidelines for the remuneration of barbers: 10 to 15 kreutzers for bloodletting from the arm, 15 to 20 kreutzers from the feet or hands – an indication that venesection on the arm was seen as technically less challenging.

69 As in the case of Sybilla Albrecht who swooned when she was bled from the hand (*lypothymia*).

70 In 1758 the Suhl town physician Johann Friedrich Glaser presented the blood scales he had invented to solve precisely this problem. Ruth Schilling, 'Social Mobility and Medical Practice: Johann Friedrich Glaser (1707–1789)' (in this volume).

Götz tended neither to initiate nor carry out the preventive venesections. The fact that the index term 'VS' appeared increasingly in his journals can be seen as an indication that he was becoming ever more established as a physician in Nuremberg and was being chosen by a growing number of citizens as medical supervisor for their annual venesections.

Post-mortems

The last service Götz could afford his patients was to take part in the autopsy once they had died. The practice journal includes several autopsy reports,[71] the most extensive being that of 23 November 1721 on his daughter Catharina Susanna. She died at the age of seven, four weeks after contracting scarlet fever with subsequent arthritic symptoms.[72] Götz was not satisfied, in her case, with reporting the medical findings, but described the external circumstances of the autopsy: the dissection was carried out by an apprentice surgeon in the presence of his master, under the supervision of Götz and three other physicians. Friends of the family (a couple and another friend with her daughter), Mrs Götz with their maid and cook, were also present as was a *Seelfrau* (woman providing pastoral care). It was not scientific interest that brought the attendants of this post-mortem together, but their grief about the deceased child and their hope to learn more about her disease and the cause of her death.

Conclusion

Johann Christoph Götz' activities outside the everyday medical routine leave no doubt that he saw himself as a physician and a scholar. When his experiment of making current medical issues and discussions accessible to a wider public by presenting them in the journal *Der Aufrichtige Medicus* failed, he started editing the *Commercium litterarium*. How these editorial activities affected his medical practice remains an open question. The practice journals

71 Ms. 1200/2, 443–45; 1201/2, 469–70. Götz published more postmortems in 1730 in the Journal of the Leopoldina: Johann C. Götz, 'Observationes Physico-Medicae. Observatio CLXXXI-Observatio CCVII,' in *Nova Acta Physico-Medica Academiae Caesareae Leopoldino-Carolinae Naturae Curiosorum Exhibentia Ephemerides sive Observationes Historiae et Experimenta a Celeberrimis Germaniae et Exterrarum Regionum Viris Habita et Commmunicata Singulari Studio Collecta* 2 (1730), 423–63.

72 '*Febris scarlatina cum sequentibus symtomatibus arthriticis*,' Ms. 1201/2 (1721), 429.

from 1730 up until Götz' death in 1733, had they been preserved, might have revealed whether the number of his patient contacts dropped because of his having to devote more time to scholarly pursuits or whether his growing popularity resulted in a wider and therefore different clientele. We are left with the picture of a striving physician who was well on his way to securing a place for himself in the world of scholars when his premature death, at the age of 45, put an end to his career as a scholar and practitioner.

CHAPTER 8

Social Mobility and Medical Practice: Johann Friedrich Glaser (1707–1789)

Ruth Schilling

> To the passement maker[1] of Römhild (1)[2] 12 gran[3] (powder) mix (2) elixir (of the day) 4 times 20 gran: angry bile after eating, obstructed sweat and shudder in stomach obstructio hepatis vomits violently this has been festering for a long time, shudders tearing in the back.~~paid~~[4]

This entry from the Medical Register of Johann Friedrich Glaser documents the visit of the 'passement maker' from Römhild who attended Glaser's practice on 22 January 1753.[5] Thirteen years' worth of such entries have been preserved – bound posthumously – in the Institute for the History of Medicine at the Charité in Berlin.[6] Entries from four separate years were taken from this source, transcribed and evaluated by means of a database.[7] These entries cast light on the daily practice of a man who, in 1758, was appointed town physician

* This paper is based on previous studies by Volker Hess; cf. Volker Hess, 'Ärztlicher Alltag in Thüringen um 1750. Auswertung eines Praxistagebuchs', in Elisabeth Dietrich-Daum, Martin Dinges, Robert Jütte and Christine Roilo (eds), *Arztpraxen im Vergleich. 18.–20. Jahrhundert* (Innsbruck 2008) 87–107; Andrea Thümmler, 'Rekonstruktion des Alltags eines thüringischen Arztes im 18. Jahrhundert anhand seines Praxistagebuchs 1750–1763' (Med. diss., University of Berlin, 2004); student assistant Malte Senska made the entries into the database and procured cards and statistics.

1 Posamentierer, manufacturer of passements: decorative trimmings on clothes.

2 The numbers indicate the prescription.

3 Abbreviations and pharmaceutical symbols are spelt out in brackets.

4 *deletur* (is deleted) means that the patient had paid.

5 Institut für Geschichte der Medizin (Charité Berlin), Johann Friedrich Glaser: *Medicinisches Register auf das Jahr Christi 1750* (Praxistagebuch 1750–1763), Ms: '(22) *dem poßenmentirer von Römhild,* (1) 12 *g*(*ran*) (*Pulver*) *m*(*isch*) (2) *elex*(*ier*) (*des Tages*) 4 *m*(*al*) 20 *g*(*ran*)*: nach dem eßen erzürnt gall verlegener schweiß und Schauer im magen obstr*(*uctio*) *hepat*(*is*) *bricht sich stark dieses hat Schon lang geschwärt, Schauer reißen im rück*(*en*) ~~*del*~~*.*

6 The book was privately owned by Klaus Stark Voss, MD. The Institute for the History of Medicine (Freie Universität Berlin) borrowed and filmed it in 1981 and purchased it in 1988/89. I am indebted to Johanna Bleker for this information.

7 The total number of evaluated consultations is 3668.

 | DOI 10.1163/9789004303324_010

and health officer in the German town of Suhl. Glaser's case is extraordinary even if one considers the greater social mobility of the early modern period.[8] He was born in 1707 in Wasungen. His father and one of his brothers were executioners and his mother came from a dynasty of executioners and surgeons who resided in the Hessian town of Schmalkalden.[9] Glaser is one of many academically trained physicians from executioner families who have been the subject of sporadic investigation.[10] The question we need to ask is whether, and how, this background affected the practice of medicine and how it manifests in Glaser's 'Medical Register'.

The Location of the Practice

After practising medicine for a number of years in his birth town of Wasungen, Glaser relocated to Suhl in 1738.[11] A brother of his father's, Johann Heinrich Glaser, was executioner there in the early decades of the eighteenth century. This uncle's medical activities are documented in a license granted by the Duke of Saxe-Hildburghausen in 1710 and in an entry in the Römhild church

8 On the concept of social mobility and its application in the investigation of executioners' families cf. Gisela Wilbertz, 'Zur sozialen und geographischen Mobilität einer Scharfrichterfamilie in der Frühen Neuzeit: Die Clauss (Clausen, Claessen, Clages) aus Lemgo', *Beiträge zur westfälischen Familienforschung* 53 (1995), 253–322; 54 (1997), 183–256; 55 (1998), 204–44.

9 Cf. Leopoldina, Archiv, Matrikelmappe Sign. 630, Johann Friedrich Glaser: *Curriculum Vitae* vom 28. Januar 1760 (unpaginiert).

10 Cf. Petra Pechaček, *Scharfrichter und Wasenmeister in der Landgrafschaft Hessen-Kassel in der Frühen Neuzeit* (Frankfurt am Main 2003), 233, 240–41, 304–05; Johann Glenzdorf and Fritz Treichel, *Henker, Schinder und arme Sünder*, 2 vols (Bad Münder am Deister 1970), vol. 2, 415–74; Jutta Nowosadtko, *Scharfrichter und Abdecker. Der Alltag zweier 'unehrlicher Berufe' in der Frühen Neuzeit* (Paderborn 1994), 310–30; Hubert Weckbach, 'Ein Scharfrichter wird Arzt. Zur Biographie des Dr. Johann Andreas Peter Bayer in Heilbronn', *Schwaben und Franken. Heimatgeschichtliche Beilage der Heilbronner Stimme* 33, No. 7 (1987), I–IV, No. 9 (1987), I–IV, No. 10 (1987), I–IV; Hubert Weckbach, 'Dr. med. Johann Christoph Paul Bayer. Der jüngste Sohn des Scharfrichters Johann Andreas Bayer in Heilbronn', *Schwaben und Franken. Heimatgeschichtliche Beilage der Heilbronner Stimme* 35 (1989), I–IV; Gisela Wilbertz, *Scharfrichter und Abdecker im Hochstift Osnabrück. Untersuchungen zur Sozialgeschichte zweier 'unehrlicher' Berufe im nordwestdeutschen Raum vom 16. bis zum 19. Jahrhundert* (Osnabrück 1979), 131–38, 312–16.

11 Cf. Glaser: *Curriculum Vitae* (see note 9).

register.[12] The Glaser family owned estates in and around Suhl.[13] In 1751, Johann Friedrich Glaser married Elisabeth Heintz, the daughter of an administrative clerk from Electoral Saxony, in Suhl.[14] Through this marriage he found access, if not to the upper, then to the middle strata of Saxony's territorial administration to which Suhl had belonged since 1713. Seven years after the wedding, he was appointed as town physician and health officer in waiting,[15] a position he actually assumed to three years later and held up until his death.[16]

As health officer of Suhl, Glaser joined the traditional ranks of publicly appointed and sworn practitioners. From then on, he had to make himself available in a particular building[17] at all times.[18] There, he not only saw patients, but also brewed and served beer and manufactured and dispensed medicines.[19]

---

12 Thüringisches Staatsarchiv Meiningen, Staatsministerium, Abt. D, Inneres 20. 152–75, Concession des Herzogs Moritz Wilhelm an Johann Heinrich Glaser vom 22. Januar 1710, neben der Scharfrichterei und Abdeckerei auch medizinisch tätig zu sein; Thümmler (2004), 110.

13 Thüringisches Staatsarchiv Meiningen, 499002-Zinck Mattenberg-Sammlung/Akten 1327, 9. Personengeschichtliche und genealogische Quellen, Register oder Aufzeichnungs-Büchlein auf Dreißigacker. (Tagebuch des Scharfrichters Joh. Jeremias Glaser, geb. 1653, gest. 1725), 177–78.

14 Cf. Glaser: *Curriculum Vitae* (see note 9).

15 Cf. *ibid.*

16 Cf. *ibid.*; Ferdinand Werther, *Chronik der Stadt Suhl in der gefürsteten Graffschaft Henneberg* (Suhl 1846), 262.

17 For information on Glaser's house in Suhl see Suhl Town Archives 2.1. 4772, Sammlung verschiedener ActenStücke den Suhler Brand vom 1. Maÿ 1753 [betr.], fol. 47; Johann Friedrich Glaser, *Nützliche und durch die Erfahrung bewährte Vorschläge, bey heftigen und geschwinden Feuersbrünsten Häuser und Mobilien sicher zu retten: nebst einer gründlichen Anweisung, grosse und gefährliche Feuersbrünste zu verhüten, und was man bey Auf- und Ausbauung der Häuser und Aufbehaltung der Mobilien, auf eine nutzbare Weise, zu beobachten habe* (Hildburghausen ^{3}1764), 37–48, 137–43.

18 Cf. Landeshauptarchiv Sachsen-Anhalt, MD, A 33, B IX Nr. 2, Eid des Stadtphysicus von Suhl, 1670–1690; generally for the appointment of town physicians in the territory of the German Reich cf. Manfred Stürzbecher, 'The Physici in German-Speaking Countries from the Middle Age to the Enlightenment', in Andrew W. Russell (ed.), *The Town and the State Physician in Europe from the Middle Ages to the Enlightenment* (Wolfenbüttel 1981), 123–30.

19 Both follow indirectly from entries of Glaser's predecessors: Landeshauptarchiv Sachsen-Anhalt, MD, A 33, B IX, Nr. 1, Vol. II–III, fol. 2 and fol. 56: Brief von Zapf und Streitel an Moritz Wilhelm, Herzog von Sachsen, Jülich, Cleve und Berg, und Engern und Westphalen, postulirter Administrator des Stifts Naumburg, 23 September 1696 und Brief des Medicus Johann Wolfgang Siegmund Eckart an den Statthalter wegen der ihm verliehenen Biergerechtigkeit, 31 January 1703.

In a letter, Glaser described the large jugs standing in front of his house that he would fill with home-brewed beer.[20] In his correspondence and publications we find indications that he manufactured medicines and carried out chemical experiments in his house. In an essay on his invention of a fire-proof coating for wood, he referred to a medicine chest that every medical doctor should own. He also described how he carried out experiments for the manufacture of a fire-proof alkaline solution.[21] It is likely that Glaser's house had two cellars: one for the storage of the barrels he needed for beer-brewing and another one that served as his laboratory. The medicines he produced provided the major part of his income. Glaser used his practice records to list the costs of individual powders, draughts, elixirs and so on, adding information on whether his patients were still well-stocked or whether he had to send new supplies.[22] This component of his duties was officially regulated because health officers in Suhl were obliged to manufacture and dispense medicines. The implementation of the Medicinal Act of 1768 in the Electorate of Saxony placed Glaser, however, in a difficult situation, since it demanded the strict separation of pharmaceutical and medical activities. In a letter he wrote in 1770 to Ferdinand Jakob Baier (1707–1788), president of the Imperial Academy of Sciences (Leopoldina), Glaser protested that the new regulations meant he would face considerable financial losses.[23]

But these events did not affect the period documented in the practice diary. The people who went to see Glaser often came for the powders, balms and elixirs he prescribed – indeed, he made house calls only in very exceptional cases.[24] This was due both to the fact that he had his residence in town, and to his pharmaceutical activities. Practicing in his own house and being focused on medicinal therapy, was another advantage when he assumed his office as town physician. This was an office which, in Suhl, was associated with particular spatial and visual connotations that qualified the physician's status in society. The residence in the centre of Suhl, probably a sizeable property,[25] will have

20 Letter from Johann Friedrich Glaser to Christoph Jacob Trew of 17 November 1753, Briefsammlung Trew, consulted on www.trew-letters.com; cf. also Glaser, 1764, 48.

21 Glaser, 1764, 137–38, 339–40.

22 Cf. *Medicinisches Register*; in January 1753 for instance in five of 51 consultations.

23 Leopoldina, Archiv, Matrikelmappe Sign. 630, Letter to Ferdinand Jakob Baier of 9 July 1770.

24 Cf. *Medicinisches Register*, in 1753 Glaser recorded twelve house calls out of 514 consultations in 1763 there was one house call among a total of 1179 consultations.

25 Stadtarchiv Suhl 2.1. 4772, Sammlung verschiedener ActenStücke den Suhler Brand vom 1. Maÿ 1753 [betr.], fol. 47: Glaser was allocated two instead of one plots in the town centre which suggests a sizeable property.

reflected the incumbent's dignity and his claim to equality with the members of the city council and the electoral administration.[26] The privileges of the learned town physicians[27] included a special pew in the church[28] as well as ceremonial honours during processions and municipal celebrations.[29] All this marked Glaser as a distinguished member of society, a position that his visitors would be aware of when they entered his house to make their requests – and that would induce them to forget his origins.

The Agents: The Physician and His Clients

On both his mother's and father's side, Johann Friedrich Glaser descended from families that, for several generations, had been executioners in the regions of Saxe-Meiningen, Saxe-Hildburghausen and in the Landgraviate of Hesse.[30] We have evidence that Johann Friedrich Glaser's uncle practised medicine with a special license from the Duke of Saxe-Hildburghausen, despite resistance from the local barber guild. Since this uncle also lived in Suhl, it can be assumed that Glaser's relocation there from Walsungen in 1738, was inspired by the hope that his uncle, with advancing years, would reduce

26 Glaser's successful connecting with the elite of the town is apparent, for instance, in the preface to one of his brochures: Johann Friedrich Glaser, *Physicalische Bewegungsgründe, die es wahrscheinlich und glaublich machen, daß bey der Hennebergischen Handelsstadt Suhla ein unterirdisch großer Schatz von Steinsalz verborgen liege, samt unmasgeblicher Vorschlägen, wie solcher Schatz schicklich aufzusuchen, und der gefundene sodann wohl genutzt werden kann* (Leipzig 1776).

27 On academic studies as a controversial sign of rank in Electoral Saxony cf. Thomas Weller, *Theatrum Praecedentiae. Zeremonieller Rang und gesellschaftliche Ordnung in der frühneuzeitlichen Stadt. Leipzig 1500–1800* (Darmstadt 2006), 264–96.

28 Stadtarchiv Suhl, Kirchliche Angelegenheiten, 2.1/3899, Acta den Verkauf und die Verloosung der Kirchen Stühle in der, nach dem brande 1753. neuzuerbauenden Haupt-Kirche beÿ der Stadt Suhl. und das daran fallende Accidenz des zuschreibens betreffend anno 1760, 1762.

29 Cf. the entry on Johann Friedrich Glaser's funeral: Stadtarchiv Suhl, Kirchliche Angelegenheiten, 2.1. 9263, Geborene und Begrabene 1776–1807, Eintrag December 1789.

30 Cf. the documents concerning the Glaser family in the Thüringische Staatsarchiv Meiningen, Staatsministerium, Abt. D Inneres 20.152–20.175. Files for the second half of the seventeenth and first half of the eighteenth century; Kreisarchiv Hildburghausen, Stadt Hildburghausen, 3260, Kart. No. 0237 a, Hiesige Meisterei-Schafrichter Glaser betr. Feldmeisterei und Abdeckerei 1705–1716; Rolf Wahl investigated the genealogy of the mother's family (Wahl) and made his findings available to the author; Pechaček, 2003, 72, 95, 172, 195–201, 228, 257.

his medical activity and pass on clients and prescriptions to his nephew.[31] It is also evident that Johann Friedrich Glaser's father, executioner in the service of the Duke of Saxe-Meiningen, carried out medical procedures: in addition to detailed bills relating to his tasks as executioner, he listed in his journal the cures he applied.[32]

That Johann Friedrich Glaser's family decided that he should have a higher education is an expression of the social ambitions they harboured for their son. His father went to considerable lengths to gain admission for his son to the grammar school (*Gymnasium Gleichense*) in Ohrdruf.[33] This was only possible because Johann Friedrich benefited, as a late arrival in the family, from new legislation that abolished the dishonourable status of descendants of executioners.[34] For his older brothers, academic studies had been out of the question and they had to carry on the trade of their father. Glaser wrote of one of his brothers that he failed to establish himself permanently in an alternative trade.[35]

There was a problem, arising from Glaser's family background, regarding his interaction with the people who visited his townhouse in Suhl: while he met

---

31 Cf. the letter of 4 April 1713 from Duke Ernest of Saxe-Hildburghausen to the City Council of Hildburghausen concerning the endowment of Johann Heinrich Glaser, in Kreisarchiv Hildburghausen, Stadt Hildburghausen, 3260, Kart. No. 0237 a, Hiesige Meisterei-Schafrichter Glaser betr. Feldmeisterei und Abdeckerei 1705–171; Thüringisches Staatsarchiv Meiningen, Staatsministerium, Abt. D Inneres, 20. 152–75, Concession of Duke Moritz Wilhelm to Johann Heinrich Glaser of 22 January 1710, granting him permission to practise medicine in addition to his duties as executioner and knacker; his medical practice in 1730 is confirmed by an entry in the Römhild church register (cf. Thümmler, 2004, 110).

32 Johann Jeremias Glaser: *Register*, 302–04 (see note 13).

33 Johann Jeremias Glaser: *Register*, 150–03 (see note 13); on the *Gymnasium Gleichense* in the eigthteenth century, cf. Hans Jürgen Hinrichs, *Geschichte des Gymnasium Gleichense in Ohrdruf 1564–2001* (Gotha 2001), 29–49.

34 Article 4 of the Reich Trade Act of 1731 provided that the children of executioners and knackers were to be seen as honest and allowed to follow any trade if their fathers had not pursued their ancestral occupation for thirty years, cf. Pechaček, 2003, 306. In the case of Johann Friedrich Glaser, it is not altogether clear how strictly this application was adhered to when he enrolled at Erfurt in 1725; his father died in the same year at the age of seventy-two. He recorded in his register that he had let out his business in 1694 (Johann Jeremias Glaser: *Register*, 102–06, see note 13), but at a later time there are recordings of tortures and executions up until 1715: *ibid.*, 291–301; there is a possibility, however, that they were carried out by his first-born son who bore the same name.

35 Johann Friedrich Glaser, 'Beytrag zur Lebensgeschichte des vor kurzem verstorbenen Arztes D. Johann Friedrich Glaser (Aus einem eigenhändigen Brief desselben)', *Journal von und für Franken* 6, no. 1 (1793), 411–23: 422–23.

them as a distinguished and honoured office-holder and *homo academicus*, it can be assumed that many of them were aware of his descent from a local family of long-standing executioners and surgeons. Did this ambivalence find reflection in the occupational and social configuration of his clientele? His clientele was in fact very diverse, as we can see from the fourty to sixty different occupations named in each of the evaluated annual volumes of Johann Friedrich Glaser's practice journal.[36] This diversity needs to be seen in association with the highly differentiated topographical, economic and social environment of the Ernestine duchies and the territories of the former Duchy of Henneberg. Natural geographical borders prevented large-scale farming, while special laws of succession resulted in very small compartmentalized economies in rural as well as urban districts, a development that further enhanced the political and territorial complexity arising from hereditary conflicts. The individual lords of extremely small territories tried to strengthen their economic autarchy through cameralism.[37] The town of Suhl held a special position: unlike Meiningen and Hildburghauen it had, since 1713, belonged to the much larger and politically important Electorate of Saxony and was at the receiving end of the electoral endeavours to establish a uniform territorial administration. On the surface, we might have the impression that Glaser's practice journal could constitute a rich source for a comprehensive social and economic history of the town of Suhl and its surroundings in the mid-eighteenth century, but a closer look uncovers certain social limitations.

While Glaser did not record any cures or medicines he gave out free of charge, it is possible – due to his obligation as town physician to issue such prescriptions – that he simply refrained from including them in his register, but documented them separately so as to justify his actions before the city council. The *Medical Register* only listed paying patients. Apart from maids and farmhands, the group of herders and shepherds offers itself particularly to closer scrutiny into the 'lower social strata' of his clientele.

The herders and shepherds in Glaser's practice reflect the agricultural character of the Saxon duchies.[38] We notice that, for this group, Glaser wrote down places of origin which – unlike those of the hunters for instance – do not relate

36 *Medicinisches Register*. The register years 1750, 1753, 1760 and 1763 were evaluated, transcribed and filed in a Filemaker database.

37 Cf. Günther Wölfing, *Geschichte des Henneberger Landes zwischen Grabfeld, Rennsteig und Rhön. Ein Überblick* (Hildburghausen 1992), 77–119.

38 Cf. Volker Wahl, 'Die Mehliser Hirtenzeche. Über den historischen Ursprung des Hirtenfestes in Zella-Mehlis', in Bezirkskabinett für Kulturarbeit Suhl (ed.), *Zum Hirtenwesen im Thüringer Wald. Beiträge* (Suhl 1988), 46–52.

only to the names of fields and mountains. The reason for this is that they were officially allocated to particular manors or municipalities. They represent a group that moved in the zone between town and country.[39]

There is nothing special in the way this group of patients paid their medical bills. Glaser did not grant them deferral of payment more often than usual nor did the members of this group pay particularly late or exceedingly high or low amounts.[40] Like other patients, the herders and shepherds consulted Glaser primarily to obtain medicines or a uroscopy. There is no evidence of a prevalent prescription of external treatments or folk medicine, although the group of herders and shepherds might seem to be the most likely subjects for such methods.[41] It was, rather, the opposite: this group probably accepted Glaser as an academically trained practitioner who offered services that they could not obtain from executioners, surgeons, shepherds and itinerant drug sellers. The fact that the number of herders and shepherds among Glaser's clients was quite large and constant compared to other professions,[42] suggests that Glaser had a kind of monopoly as a learned physician, although there is ultimately no reliable evidence for this because there are not sufficient sources on the clientele of the physicians in Schleusingen, Meiningen and Hildburghausen.[43]

---

39 On the geographical distribution of the patients and visits cf. Ruth Schilling, 'Raum und medizinischer Markt. Die Praxis des Stadt- und Amtsarztes Johann Friedrich Glaser (1707–1789) in Suhl', in Lina Gafner, Iris Ritzmann and Katharina Weikl (eds), *Penning Patient's Histories – Doctors' Records and the Medical Market in the 18th and 19th Century* [special issue: *Gesnerus* 69, no. 1 (2012)], 36–53.

40 In 1750 we find twelve consultations of shepherds as opposed to four in 1753, 22 in 1760 and 16 in 1763. All of them were compared in terms of diagnosis, therapy, form of payment and place of residence.

41 On the magical and folk-medical notions associated with shepherds cf. entry 'Hirte', in Hanns Bächtold-Stäubli and Eduard Hoffmann-Krayer (eds), *Handwörterbuch des deutschen Aberglaubens*, vol. 4 (Berlin ^{3}2000), 124–31.

42 In 1763 they constituted with sixteen consultations the most frequently-named occupational group next to sheriffs (seventeen times) and weavers (sixteen times) (1179 entries).

43 The medical market in the eighteenth century in these three towns must be indirectly inferred from the files. Scrutiny of the archives did not yield any sources that were comparable to the *Medical Register*. Information on the physicians who practised in Meiningen can be found in: Stadtarchiv Meiningen, Altes Archiv No. 699–257, Lists of names in the Meiningen city register from 1570; entries on the eighteenth century passim; for Schleusingen cf. Landeshauptarchiv Sachsen-Anhalt, MD, D Schleusingen, No. 24, Acta die durch das wegen Errichtung eines Sanitates Collegii zur Verbesserung des Medicinal Wesens ergangene gnädigste Mandat beschehene Verpflicht und Unterrichtung derer Hebammen in Amte Schleusingen; there are no sources known for Hildburghausen.

As we can see from their places of origin, most herders and shepherds who attended Glaser's practice were employed in the service of the aristocracy. Take the shepherds of Marisfeld, for instance: Marisfeld was a hamlet consisting of a manor, a church, a vicarage, a public house and a handful of other houses. We can assume that the herders were in the service of the Marschall of Ostheim family who owned the manor and the adjacent fields.[44] Apart from the herders, we also find the names of cooks and that of the wife of the Marisfeld steward in Glaser's practice journal.[45] For the latter he prescribed, on 14 September 1763, 'an effective laxative', because she suffered from 'severe burning in the pit of the stomach' and 'nausea'.

None of Glaser's Marisfeld patients belonged to the landowning family, however.[46] In their social range, Glaser's Marisfeld visitors represent a microcosm of his clientele in general: he treated the major domo of Marisfeld once in 1750.[47] In 1750 and 1760 he had the privilege of treating 'the old schoolmistress'.[48] The other occupations recorded for Marisfeld relate mostly to middle-class tradespeople.[49] Social mobility within the Marisfeld manorial system was probably rather limited and, as a result, the social barrier could not be removed that existed between Glaser, the academic physician who prescribed for the maids, cooks and shepherds of Marisfield, and Glaser, the executioner's son who was not socially acceptable to the residents of the manor.

Nonetheless, Glaser was not physician to the 'ordinary people' only, as is apparent if one looks at the group of sheriffs among his clients. In contrast to the group of herders and shepherds we see a marked change with this group. While Glaser treated six herders and shepherds in 1750 (total number of entries: 819) and ten in 1763 (total number of entries: 1179), the group of sheriffs grew much more significantly. Only one sheriff consulted Glaser in 1750, but there were 12 in 1753 and 17 in 1763.[50] Like the shepherds, the members of this group

44 Cf. Paul Lehfeldt and Georg Voss, *Bau- und Kunstdenkmäler Thüringens. Vol. 2: Herzogtum Sachsen-Meiningen. Kreis Hildburghausen. Amtsgerichtsbezirke Hildburghausen, Eisfeld, Themar, Heldburg und Römhild* (Jena 1904), 196–203.

45 *Medicinisches Register*, 3 March 1750, 18 March 1750, 5 May 1750, 28 August 1750, 3 September 1750; 20 February 1753, 3 March 1763, 20 June 1763, 14 September 1763.

46 Cf. *Ibid., passim.*

47 Cf. *Ibid.*, 30 December 1753.

48 Cf. *Ibid.*, 8 February 1750, 15 July 1760.

49 Cf. *Ibid.*, 12 February 1750, 27 May 1760, 6 June 1760, 30 December 1760, 14 May 1763, 11 December 1763.

50 Cf. *Medicinisches Register*, database evaluation, searching for 'Schultze' in the years 1750, 1753, 1760, 1763.

tended to settle their accounts promptly and no specific diagnoses and therapies are apparent.[51]

If we compare the catchment areas which include the places where the herders came from with those where Glaser recruited the sheriffs, we notice that they are largely similar. The catchment area of the sheriffs is also similar to those of the other occupational groups: it covers the area to the south and southwest of Suhl: the duchies Saxe-Meiningen and Saxe-Hildburghausen. A change in Glaser's clientele is, however, suggested by the fact that the herders often came from places where Glaser had fifteen or more patients, which is to say that they followed patterns established by other patients.[52] In the following maps, the black footprints mark places where more than fifteen patients dwelled, the outlined footprints places, which less than fifteen visitors named as their home. The names of places, where Glaser's relatives resided, are put into bold letters:

The sheriffs, on the other hand, came from villages that were not the same as those where Glaser's main clientele originated.[53] There are two possible explanations for this difference: the first is the different ranks of the patients. Herders and shepherds tended to belong to the bigger villages which had a feudal system, while the sheriffs represented the smaller independent farmers' communities.[54] The second reason might be that we are dealing with two very different groups of clients. The herders came from villages within Glaser's regular catchment area. We can assume that he was able to recruit them when he began practicing in Suhl as a result of former contacts, maybe through his uncle who had also practised medicine there. It was a different situation in the case of the sheriffs, who belonged to the local elite but were not of noble origin, a fact that made the contact with Glaser easier. They came to see Glaser with increasing regularity, a fact that clearly reflects his continual rise on the social scale.

A glance at the two most important ducal towns in Glaser's catchment area, Meiningen and Hildburghausen, yields similar results. Because of his father's occupation, Glaser's family ties with Meiningen were particularly close. In the *curriculum vitae* that he submitted to *Academia Naturae Curiosorum* (*Leopoldina*), he wrote that his father had been offered the city rights in Meiningen, but had declined.[55] He also mentioned several godfathers of

51 Cf. *Ibid.*

52 Cf. *Ibid.*

53 Cf. *Ibid.*

54 On the special agricultural development in Thuringia and the adjacent Franconian and Hessian territories cf. Friedrich Lütge, *Geschichte der deutschen Agrarverfassung vom frühen Mittelalter bis zum 19. Jahrhundert* (Stuttgart 1963), 61–66.

55 Cf. *Curriculum Vitae* (see note 9).

members of his family who were citizens of Meiningen.[56] Pursuing the thesis that Glaser's recruiting of patients was dependent on family connections, one might derive that, from the beginning of his medical career, he was well positioned in this ducal town. But this does not seem to have been the case. The dependence of his patient recruitment on family ties was primarily due to the medical knowledge of his executioner relatives, not to their attempts to establish connections, based on equality, with the Meiningen bourgeoisie. For the years 1750 and 1753, for instance, we find only two patients from this town: one of them was a lower-rank ducal official (*Kammerbote*), the other the landlord of the Red Bull Inn (*Roter Ochse*).[57] There was no significant increase in the number of Glaser's Meiningen patients until 1760 and 1763. An even clearer picture arises if we look at Hildburghausen, the second largest ducal town in the region. Neither in 1750 nor three years later did a single patient find his way from Hildburghausen to Glaser's practice; in 1760 and 1763 there were only eight at first and in the end as many as 56 (of 1157 in total in 1760 and 1179 in 1763). In addition to Glaser's typical clientele, that is, middle class craftspeople (passement makers, cloth-makers, potters, cooks),[58] there were representatives of the town administration, such as the wife of the court scribe[59] and a town lieutenant[60] or members of the court personnel such as a *premier musicus*[61] and the court drummer of the Hildburghausen *Schlosskapelle*:[62]

> Spielner, first musicus of Hildburghausen (1) laxative draught 1 ½ K: W:, (2) *cap(sulae) purg(antes) Mal(um) hipoch(ondriacum)*, thoughts often pass, spasms, urine was green dark. Luminescent thick full of dust, somewhat bright circle. ~~paid~~[63]

56 Cf. *Ibid.*

57 The *Rote Ochse* was a public house in the centre of Meiningen up to the end of the eighteenth century, cf. Eduard Doebner, *Bausteine zu einer Geschichte der Stadt Meiningen* (Meiningen 1902), 61.

58 Cf. *Medicinisches Register*, database evaluation of the years 1760 and 1763, with search word residence Meiningen.

59 Cf. *Ibid.*, 14 May 1763.

60 Cf. *Ibid.*, 25 April 1763.

61 Cf. *Ibid.*, 8 August 1763.

62 *Ibid.*, 13 December 1763: '*(1763, 8 August) H(err)n: Spielner premier Musicus von hildburghaußen (1) den lax(ier) tranck 1 1/2 K: W:,* (2) *cap(sulae) purg(antes) Mal(um) hipoch(ondriacum) offt vergehen die gedanken, spasmos, der urin war grün dunckel. leuchtent dick voll staub, etwas heller circkel.~~del~~*.

63 Double crossed *deletur* indicated patients who had settled their accounts.

Aristocratic persons are not mentioned unless as employers of a patient, such as a wet nurse who was in the service of a 'Frau von Thiemann' in Meiningen.[64] As in the case of the village sheriffs (and unlike that of the herders or shepherds), Glaser first had to gain a foothold in the two ducal towns as an academic doctor and assert himself over the resident physicians there who were closer at hand.[65] His Meiningen and Hildburghausen clientele could potentially open doors to new patients: through the musicians, for instance, new contacts might be established with other salaried employees of the court. But the social boundaries are obvious there, too: Glaser treated the musicians in the service of the Duke of Saxe-Hildburghausen, but not their patrons or employers. He treated the wife of the court scribe, but not the wives of the Hildburghausen or Meiningen aldermen. On the whole, however, the growing number of patients, especially from the two ducal towns, suggests that Glaser found increasing social acceptance.

On the basis of individual occupations and locations, we have analysed the link between the social profile of Glaser's patients and his own social status. The individual examples form a coherent picture if we look at all the places where he had more than fifteen patients a year. If we compare the relevant statistics for 1753 and 1763, we realize that he became increasingly established on a medical market that went far beyond the town of Suhl. In 1753, the number of Glaser's patients still decreased with the distance to Suhl: this makes sense if we consider the distance most people had to travel to consult him. By 1763, the group of patients who came to see him from within a radius of twenty to thirty kilometres had grown so much that it was larger than the group of patients who lived within a radius of one to twenty kilometres of Suhl. The reason for this was the growing number of patients who came to Glaser from the ducal towns and their surrounding villages. While Glaser did not rise to be the kind of physician who treated the nobility in his region, his catchment area kept growing steadily in a social as well as geographical sense.

The World of Executioner-medicine and the Res Publica Litterarum

There is one fact that is very obvious when one evaluates the therapies in Glaser's practice journal: he would, with almost slavish precision, adhere to the learned physician's code of practice. There are very few cases – and those only in the 1750s, not in the 1760s – where he recorded having carried out surgery

64 Cf. *Medicinisches Register*, 18 July 1760.

65 On the largely unexplored medical market in both ducal towns cf. note 43.

such as pulling a tooth.[66] Interestingly, this behaviour is not consistent with Glaser's familiarity with questions of surgery. In a publication on the cattle plague that arose from his function as veterinary health officer,[67] he displayed great interest in the dissection of beef cattle.[68] The same kind of balancing act between scientific discourse and medical practice is discernible in his approach to venesection: in his practice journal this intervention played a merely marginal part as a means of diagnosis or as a therapeutic measure. It was always performed in the patient's house and never by Glaser himself.[69] But in his essay on blood-letting devices, Glaser demonstrated his detailed knowledge of the practical requirements and problems of blood-letting from the foot.[70] Often the blood was caught in a water basin so that it was not possible to establish the exact amount of blood extracted. Glaser's invention provided a solution to this problem (see fig. 8.1). It was obviously reserved to learned physicians to monitor the activities of surgeons and barber surgeons.[71]

In the practice journal, we find surgical interventions included under domestic medicine or wound care.[72] This leads us to conclude that Glaser also adhered strictly to the rules in his relationship with other groups of healthcare providers. Although it is no longer possible to trace the exact sequences of treatments, it is highly likely that the Suhl town physician was only one of many medical practitioners consulted by the patients. We find no evidence of the derogatory view of other healers, displayed in his practice journal,[73] or in his general attitude towards non-academic healers in the district he was officially responsible for. As town physician and health officer, it was his task to

66 For the year 1750 cf. Thümmler, 2004, 92; *Medicinisches Register*, 20 March and 3 June 1753.

67 Cf. Johann Friedrich Glaser, *Auf richtige Erfahrung gegründete Abhandlung von der tödtlichen Knoten-Krankheit unterm Rindviehe und dem Rothwildprete in den Wäldern: wie solche abzuhalten und die angefallene gewiß zu curiren ist, und von der nicht tödtlichen Maul- und Fußrankheit unterm Rindviehe, wie solche leicht geheilet werden kann* (Leipzig 1780).

68 Cf. *Ibid.*, 76.

69 *Medicinisches Register*, 13 June 1753, entries in 1760: 30 March, 6 May, 1 August, 14 August, in 1763: 15 June and 25 September.

70 Johann Friedrich Glaser, *Beschreibung einer neuerfundenen Blutwaage und Blutmeßgeschirrs: womit man beym Hand- und Fußaderlassen das Blut, ob es schon unter das dabey gebräuchlichen Wasser läuft, dennoch sobald Loth zu Loth richtig und sehr nützlich wägen und messen, auch sonst durch einen angewiesenen Vortheil unter Wasser abgesondert allein auffangen kann* (Hildburghausen 1758), 79–82.

71 Cf. Annemarie Kinzelbach, Susanne Grosser, Kay Peter Jankrift and Marion Ruisinger: '*Observationes et Curationes Nurimbergenses*'. The Medical Practice of Johann Christoph Götz (1688–1733)' (in this volume).

72 Cf. *Medicinisches Register*, 21 March 1753 and 21 July 1753.

73 Cf. *Ibid.*

Johann Friedrich Glasers,
Med. Doct. und adjungirten Stadt- und
Amtsphysici zu Suhl,
Beschreibung
seiner neuerfundenen
Blutwaage
und
Blutmeßgeschirrs:
womit man
beym Hand- und Fußaderlassen das Blut,
ob es schon unter das dabey gebräuchliche Wasser
läuft, dennoch sobald von Loth zu Loth richtig
und sehr nützlich wägen und messen, auch sonst
durch einen angewiesenen Vortheil unter
dem Wasser abgesondert allein auf-
fangen kann.

Hildburghausen,
verlegts Johann Gottfried Hanisch,
Herzogl. Sächßl. privilegirter Hof- Buchhändler.
1758.

FIGURE 8.1 *Title page of Glaser's* Beschreibung einer neuerfundenen Blutwaage und Blutmeßgeschirrs, Hildburghausen 1758.

document and monitor activities of non-academic healers. He had to examine them and he issued liwcenses for barber-shops.[74] There is no evidence, not even of a single case, where Glaser ever spoke up against a non-academic practitioner. On the contrary: his conciliatory behaviour even earned him the disapproval of a superior colleague in Schleusingen.[75] In his scientific treatises, he

74 Landeshauptarchiv Sachsen-Anhalt, MD A 33, B XXV, Nr. 2, Renthereÿ-Acta, Die von Johann Caspar Bachmann gesuchte Concession zur Anlegung einer Barbierstube zu Heinrichs betr.; Nr. 4, Renthereÿ-Acta, Die Erhebung einer Badstube zu Goldlauter betr.; Letter to Ferdinand J. Baier, 1770 (cf. note 23).

75 Landeshauptarchiv Sachsen-Anhalt, MD A 33, B XXV, Nr. 4, Renthereÿ-Acta, fol. 19–20; Stadtarchiv Suhl, II. 29, Tabellarische pflichtmäßige Anzeige zu den Churfürstl. Sächsischen hennebergischen Landes Physicate der Stadt und des Landes Schleusingen Kühndroff und Benßhausen befindlichen Medicinae Practicis, Chirurgis, Barbieren, Badern, Apothekern, Hebammen und Medicastris, Schleusingen October 1782.

also appeared as willing to compromise with non-academic healers. In his dealings with them he spoke as a medical practitioner, not as a publically appointed official. In his view, a medical practitioner needed to know all areas of medicine. In his publication on blood-letting devices he wrote:

> Even if later on they [i.e. the physicians, R.S.] will not carry out these procedures themselves, they must be able – just like good architects – to help apothecaries, surgeons, oculists or eye doctors, stone and hernia cutters as well as midwives in word and deed.[76]

Glaser was also willing to include the non-academically trained groups in any necessary disease control measures. In a publication on typhus, he no longer dismissed their work but warned them to apply hygienic standards in their use of equipment:

> But they know, without my having to remind them, that they need to be cautious and circumspect; and they will doubtlessly instruct the surgeons and midwives under their supervision to be equally circumspect so that they do not unwittingly and unwarily spread the epidemic further: I will therefore not say more about this here than kindly remind the surgeons: that, with the patients infected with this epidemic, they must clean their instruments well – their venesection fleams, lancets, cups, syringes, shearing knives, clyster syringes and similar tools – before using them on other patients who are not afflicted with that disease.[77]

Glaser was the son of an executioner who also practised medicine and he was related to numerous active executioners and surgeons. He knew their curative methods. He knew about their social marginalization which was enhanced with the introduction and tightening of the relevant medicinal acts of the eighteenth century.[78] In the surgeons who were placed under his supervision as health officer he met peers of his family, a circumstance that certainly made

---

76 Glaser, *Beschreibung*, 1758, 19–20.

77 Johann Friedrich Glaser, *Nützliches Verhalten bey der jetzo in Deutschland regierenden Fleckfieberseuche: um sie zu verhüten und wieder zu dämpfen* (Hildburghausen 1758), 92.

78 Cf. Jutta Nowosadtko, 'Rationale Heilbehandlung oder abergläubische Pfuscherei? Die medizinische Kompetenz von Scharfrichtern und ihre Ausgrenzung aus heilenden Tätigkeiten im 18. Jahrhundert', in Bettina Wahrig and Werner Sohn (eds), *Zwischen Aufklärung, Policey und Verwaltung. Zur Genese des Medizinalwesens, 1750–1850* (Wiesbaden 2003), 109–30.

him more inclined to be more sympathetic towards them than other learned physicians.

In the field of diagnostics and therapeutics which, as a whole, is difficult to evaluate, Glaser also communicated with his patients at a level they were familiar with from their domestic situation and their dealings with other groups of healers.[79] In addition to popular internal remedies such as *Hirschhorngeist* (spirit of hartshorn),[80] Glaser prescribed for his patients therapies from sympathetic medicine, for instance, the drilling he prescribed for broken limbs. We find no specific information on this in his practice records. If we follow contemporary sources that were critical of sympathetic medicine, this involved nailing together pieces of wood, thus healing the broken limb by analogy.[81] These examples are particularly instructive since through them Glaser revealed his knowledge of traditional surgery. They pose the question as to why Glaser's patients turned to him with their various afflictions. Cures of this kind would probably have been considerably cheaper from a local executioner or surgeon. It is possible that such applications, when prescribed by the learned town physician of Suhl, were more highly esteemed and believed to be more effective than similar cures applied by a barber. But it is also possible that it was just this combination of familiar practices and internal medicine that attracted patients to consult Glaser. There were other 'magical' practices Glaser used as well as sympathetic applications, such as the prescription of amulets.[82] In his treatise on typhus, he justified the use of amulets by pointing out that the patients who used them would be more likely to accept other cures:

---

79 The term 'folk medicine' is not used here since it would suggest a clear dichotomy between academic and 'folk medicine' that does not apply in the case of Glaser; cf. Michael Stolberg, 'Probleme und Perspektiven einer Geschichte der "Volksmedizin"', in Thomas Schnalke and Claudia Wiesemann (eds), *Die Grenzen des Anderen. Medizingeschichte aus postmoderner Perspektive* (Köln 1998), 49–73; Eberhard Wolff, '"Volksmedizin". Abschied auf Raten. Vom definitorischen zum heuristischen Begriffsverständnis', *Zeitschrift für Volkskunde* 94 (1998), 233–57.

80 Cf. Johann Georg Krünitz, 'Hirschhorngeist', *Oekonomische Enzyklopädie, oder allgemeines System der Staats- Staat- Haus- u. Landwirtschaft* 23 (1781), 739–40 [online: http://www.kruenitz1.uni-trier.de].

81 Cf. *Medicinisches Register*, 27 May 1753, 8 May 1760, 18 June 1763, 5 July 1763, 5 November 1763. The entries alternate between '*soll es ... verbohren*' (18 June 1763) or '*heute hab ichs verbohrt*' (5 July 1763); Valentin Kräutermann, *Der Thüringische Theophrastus Paracelsus, Wunder- und Kräuterdoctor oder Zauber-Arzt* (Arnstadt/Leipzig 1730), 106–07; Erzstein, *Ertappter Briefwechsel von der Zauberey, Schröpfers Künsten, Nativitätstellen, Sympathie, Gespenstern u.d.g* (Leipzig 1777), 238–39.

82 Thümmler, 2004, 102–04; *Medicinisches Register*, 19 April 1760, 11 May 1760.

> Patients who trust in amulets or pendants and are willing to use them externally as protection against epidemics: I will not advise them against this as long as these measures are not harmful or superstitious. People assign great power to the colchicum root, if dug out in the autumn and sown into a clean cloth, then covered in red taffeta and suspended from the neck by a silk band or thread so it comes to hang in the region of the heart.[83]

The successful drug seller Glaser also had to take the needs of his clients seriously. In this capacity he relied, on the one hand, on local traditional recipes that he may have inherited from his uncle or father[84] and, on the other hand, on his ability to manufacture and sell, in Suhl, his own medicines of which the local healers were probably ignorant. For his patients this meant that he combined two advantages in his person: he could give them cures they were familiar with and, in addition, administer academically legitimate drugs. The successful combination of the two provided him with a strong base of regular customers. While he strove for scientific recognition as regards his knowledge of venesection, he had no such aspirations concerning his secret recipes. On the contrary, we see a remarkable discrepancy between the diagnoses and therapies described in the practice journal and Glaser's publications. This again has to do with his ambivalent position between two different medical and social worlds. With the exception of his treatise on the blood-letting equipment and an *observatio*[85] that was only published when he was very old, Glaser focused in his publications on topics that were not so much related to his medical practice as to his activity as health officer, such as disease control in human and veterinary medicine.[86]

Glaser's path into academia was prepared in an entirely different field and arose from his concern with fire protection. Having been severely affected personally by a devastating fire that raged in the town of Suhl in 1753, he had devoted himself to the topic since the mid-1750s with great perseverance, producing numerous writings on the theme.[87] Since this was a field that was of

83 Glaser, *Verhalten*, 1758, 117.

84 The relevant family sources are unfortunately missing. On collections of recipes from the field of executioner medicine cf. Pechaček, 2003, 227–28.

85 Johann Friedrich Glaser, 'Observatio XLIX: De praeternaturali cranii depressione absque sugillationibus & fissuris per sectionem legalem detecta', *Nova Acta Physico-Medica Academiae Caesareae Leopoldino-Carolinae* (1783), 230–32.

86 Glaser, *Verhalten*, 1758; *idem*, 1780.

87 Glaser, 1764; apart from this publication he addressed the topic in nine separate monographs.; cf. Thüringisches Staatsarchiv Gotha; Geheimes Archiv XX IV Nr. 19; Acta des

great importance to the regional and municipal authorities, Glaser's 'fire-proof wood coating' was received with interest, first by the Suhl town council and then by the Göttingen Academy of the Sciences.[88] It was this invention, rather than his blood scales, which made Glaser famous and led to the proposition, in 1759, of his admittance to the Leopoldina, an honour that would lead to membership in other academies.[89] Glaser used this development to his financial advantage and continued to issue new editions of his fire protection publication until the end of his life.

Did Glaser seek success as an expert in fire prevention? Or did he just make use of a welcome opportunity to climb up the social ladder to a level that he had no access to previously because of his origins? Several *observationes* submitted for publication in the Leopoldina's *Acta*, all of which were rejected, suggest that Glaser's efforts were coincidental rather than conscious attempts at building a scientific career for himself.[90] He used a door to the academic world that had opened for him, a door that had not been open for the physician who moved between academic and folk medicine. Glaser can stand as a case study for the possibilities and limitations of the combination of medical practice, scientific theorizing and social mobility in the Early Modern period – possibilities and limitations that we would hardly know about without insight into his 'Medical Register'. But as a case study, he is also of limited relevance since, so far, we know of no other detailed examples of physicians who stood between surgery and medicine, although there must have been many of them in the eighteenth century because of the strong trend towards professional differentiation and the breaking up of class distinctions. If we had more such case studies, they could add to the picture we have gained so far of the link between social and political developments and medical standards at the end of the early modern period, another facet of the interrelation between different social and medical worlds – a lacuna that Franziska Loetz was justified in pointing out in the context of the

Bergraths und Stadt-Physici D. Glaser zu Suhla gesuchte Einführung seines Tractats von Verhütung der Feuersbrünste in den Schulen hiesiger Lande betr. 1783.

88 Cf. Friedrich August Weiz, *Das gelehrte Sachsen oder Verzeichniß derer in den Churfürstl. Sächs. und incorporirten Ländern jetztlebenden Schriftsteller und ihrer Schriften* (Leipzig 1780), 76–77.

89 In 1759 admission to the Leopoldina, in 1774 to the Bavarian Academy of Sciences and one year later to the Mainz Academy of Sciences (located in Erfurt): *ibid.*

90 Cf. Marion Mücke and Thomas Schnalke, *Briefnetz Leopoldina. Die Korrespondenz der Deutschen Akademie der Naturforscher um 1750* (Berlin 2009), 527–28. With thanks to Thomas Schnalke for the suggestion.

medicalization debate.[91] If we see Glaser's case as a piece in this jigsaw, we can only hope that further comparable cases of medical practice in the eighteenth century will surface and reveal the uniqueness and logic of the behaviour of the executioner's son and town physician Glaser, in Suhl in the mid-eighteenth century.

---

91 Francisca Loetz, *Vom Kranken zum Patienten. 'Medikalisierung' und medizinische. Vergesellschaftung am Beispiel Badens 1750–1850* (Stuttgart 1993), 320–21.

CHAPTER 9

Medical Bedside Training and Healthcare for the Poor in the Würzburg and Göttingen Policlinics in the First Half of the Nineteenth Century

Stephanie Neuner and Karen Nolte

Christoph Wilhelm Hufeland (1762–1836), in his programmatic writings, defined the policlinic as a place where young physicians acquired practical experience and where medical care was offered to the city's poor.[1] Hufeland, a professor who also gained prominence as Goethe's private physician, described the 'clinical institution,' as he called it still in 1799, as follows:

> Help for the poorer foreign or destitute patient – perfecting medicine through more accurate observation and possible experimentation – and preparing young physicians for the practical business of healing. In the first instance, such institutes are an important part of state administration, policing and poor relief; secondly they are an essential instrument of medical culture and enlightenment, and thirdly an indispensable component of the medical school.[2]

Historical research has so far focused primarily on the university clinics, which emerged from the late eighteenth century. The out-patient policlinics, as 'decentralized' institutions of medical care, that were established at around the same time have been largely ignored by the in recent medico-historical and socio-historical writing.[3] Closing this research gap seems the more interesting

1 Cf. Christoph Wilhelm Hufeland, *Einrichtung und Gesetze der Herzoglich medicinisch-chirurgischen Krankenanstalt zu Jena* (Jena 1799); *idem*, 'Ankündigung des Königl. Poliklinischen Instituts auf der Universität zu Berlin nebst den Gesetzen desselben,' *Journal der practischen Arzneykunde u. Wundarzneykunst* 31, no. 10 (1810), 3–56; *idem*, 'Erster Jahresbericht des Königl. Poliklinischen Instituts der Universität zu Berlin vom Jahre 1810,' *Journal der practischen Heilkunde* 32, no. 6 (1811), 1–96.

2 Cf. Hufeland, 1799, 5.

3 The foundation years of policlinics in the German-speaking world: Göttingen 1773, Erlangen 1779, Hamburg 1779, Jena 1783, Kiel 1785, Marburg 1786, Bamberg 1789, Greifswald 1794, Tübingen 1792, Breslau 1793, Heidelberg 1805, Würzburg 1807, Giessen 1809, Berlin 1809, Leipzig 1812, Freiburg 1818, Rostock 1825, Zurich 1835 and Munich 1843. These foundations are largely included in the publications of Bettina Nawrotzki and Edith Heischkel. Further

 | DOI 10.1163/9789004303324_011

as the 'policlinics,'[4] which served the explicit purpose of providing healthcare to all citizens, were not only crucial for the provision of health-care to the poor but also for the practical training of medical students.

These new institutions, which offered out-patient care to poor patients who could not afford to pay for their treatment, operated very cost-effectively since they employed medical students and junior doctors to treat patients free of charge as part of their training.[5] The policlinics needed neither special buildings nor ward personnel. The only costs to be covered were those of the medication, which was paid for by funds from the universities, the *Dienstboten- und*

---

research is needed into whether all of them were genuine policlinics, cf. Edith Heischkel, 'Die Poliklinik des 18. Jahrhunderts in Deutschland,' *Deutsches Medizinisches Journal* 5, no. 1/2 (1954), 223–25 and Bettina Nawrotzki, 'Zur Geschichte der Polikliniken im 19./20. Jahrhundert. Unter besonderer Berücksichtigung der Entwicklung in Hamburg' (Med. diss., University of Aachen, 1987), 18–9, 28. Not much research literature is available yet on the other German nineteenth century policlinics. With the exception of Göttingen and Würzburg, hardly any research has so far been conducted in the everyday teaching routine and medical provision for the poor. On Göttingen cf. Renate Kumsteller, 'Die Anfänge der medizinischen Poliklinik zu Göttingen. Eine medizin- und kulturhistorische Studie zur zweiten Hälfte des 18. Jahrhunderts, gewonnen aus dem Archivmaterial der Universität' (Med. diss., University of Göttingen, 1955); Isabelle von Bueltzingsloewen, *Machines à instruire, machines à guérir. Les hôpitaux universitaires et la médicalisation de la société allemande 1730–1850* (Lyon 1997) and on Würzburg cf. Hans Franke and Joachim Schröder, *Die Würzburger Medizinische Universitäts-Poliklinik 1807–1957* (Stuttgart 1957). In their publication on the history of the policlinic at the medical school of Heidelberg University, Bauer and Ho devote only three pages to the 'Stadtpraxis,' as the policlinic was called in the early nineteenth century. The clinic, which was opened in 1801, closed again in 1815. Cf. Axel W. Bauer and Anthony D. Ho, *Nicht blos künstlich in einem Spitale. Zweihundert Jahre Medizinische Universitäts-Poliklinik Heidelberg und ihr Weg von der Stadtpraxis bis zur Blutstammzelltransplantation* (Heidelberg 2005), 14–6. The fourty-four-page publication on the Medical Policlinic in Marburg only devotes three pages to the beginnings of this institution, describing in only very general terms the tasks of a policlinic at the time, cf. Hans H. Lauer, *Die Medizinische Poliklinik in Marburg und die Anfänge ihrer Selbständigkeit* (Marburg 1994), 5–7.

4 The term 'policlinic' derives from the Greek word for city (=*polis*) and couch or bed (=*kline*). A contemporary medical encyclopaedia defines 'policlinic' as an institution for 'practical teaching on patients' who 'are not admitted to public hospitals… – one of the most important facilities for medical training that must not be neglected by any physician'. Dietrich Wilhelm Heinrich Busch, Johann Friedrich Dieffenbach *et al.* (eds), *Encyclopädisches Wörterbuch der medicinischen Wissenschaften* (Berlin 1842), vol. 24, 14–6. See also Hufeland, 1810, 12–3; Heischkel, 1954; Nawrotzki, 1987, 4–7. Policlinics were also referred to as '*Klinische Institute*,' '*Ambulatorische Kliniken*' and '*Krankenbesuchsanstalten*'.

5 Cf. Karen Nolte, '"…wohltätige Hülfe für die notleidenden Armen und Bildung junger Ärzte". Polikliniken im frühen 19. Jahrhundert,' *Historia Hospitalium* 27 (2011), 79–86.

Geselleninstitute – a kind of health insurance for servants and journeymen and the city welfare office.

The out-patient treatment of poor families served primarily to introduce trainee physicians to the diagnosis and treatment of diseases. They not only learnt how to examine patients and prescribe medicines. They could also practice communicating and interacting with patients. Dealing with poor patients in their homes was a new experience for medical students and young doctors who, for the most part, had a middle-class background.

The concept of the policlinic was by no means restricted to the European continent. In Great Britain, 'dispensaries' had emerged since the late eighteenth century, serving equally the dual purpose of allowing for the acquisition of medical expertise and of providing healthcare for the poor. Unlike the German policlinics they were, however, largely funded through private donations.[6]

The following study will investigate the policlinics in Würzburg and Göttingen, both of which were under the direction of Conrad Heinrich Fuchs (1803–1855). Fuchs was a pupil of Johann Lukas Schönlein (1793–1864), one of the founders of the so-called *Naturhistorische Schule* (natural history school of medicine). He had studied medicine in Würzburg where he went on to work as a registrar at the *Julius- Hospital* and gained his post-doctoral qualification. In 1836 he was appointed professor of pathology and director of the policlinic at Würzburg University.[7] Two years later the reformed University of Göttingen in the Kingdom of Hanover which particularly valued his teaching skills, offered him a chair as the successor of Karl Gustav Himly (1772–1837).[8] Fuchs' main publications were not released until a few years after he took the chair in Göttingen. In Würzburg he had already begun to compile 'case histories' – written by selected medical

6 Cf. Ulrich Tröhler, 'The *Doctor* as *Naturalist*. The Idea and Practice of Clinical Teaching and Research in British Policlinics 1770–1850,' in Harm Beukert (ed.), *Clinical Teaching, Past and Present* (Amsterdam 1989), 21–34; Joan Lane, *A Social History of Medicine. Health, Healing and Disease in England, 1750–1950* (London 2001), 89–95; Bronwyn Croxson, 'The Public and Private Faces of Eighteenth-Century London Dispensary Charity,' *Medical History* 41 (1997), 127–49; Donald M. Thomson, 'General Practice and the Edinburgh Medical School. 200 Years of Teaching, Care and Research,' *Journal of the Royal College of General Practitioners* 34 (1984), 9–12.

7 On the biography of C.H. Fuchs cf. Hans-Albrecht Rainer Seybold, Conrad Heinrich Fuchs (1803–1855). Ein Lebens- und Zeitbild (Göttingen 1963); Theodor Husemann, 'Fuchs, Konrad Heinrich,' *Allgemeine Deutsche Biographie* 8 (1878), 168–69; Johanna Bleker, 'Fuchs, Konrad Heinrich,' in Wolfgang U. Eckart and Christoph Gradmann (eds), *Ärztelexikon* (München 1995), 146–47 (= Heidelberg ^{3}2006, 130).

8 Cf. Universitätsarchiv Göttingen (hereinafter: UAG), Kur. 4978, personnel file Conrad Heinrich Fuchs.

students, his medical assistants or by himself – in twelve bound volumes, on the basis of observations and prescriptions which had presumably been noted down at the sickbed.[9] He took this collection with him when he moved to Göttingen in 1838, where he continued to add to it. As a result we now have access to 732 'case histories' as well as several volumes of notes about post-mortems that relate to the case histories. These case histories covered only a fraction of the treatments carried out in the policlinics. While the documents can therefore only serve as examples, they allow for vivid insights into the medical practice in the years 1832 to 1842. Fuchs' collection of 'case histories,' in conjunction with the administrative files of both policlinics, constitutes the source on which this paper is mainly based.

We will start by explaining the circumstances and aims leading to the foundation of the policlinics in Würzburg and Göttingen and the important role these academic institutions played in the provision of healthcare for the indigent. In a second step we will examine the actual everyday medical practice of caring for poor patients. What was the social structure of the clientele? How were the conditions at the sickbed in the patients' homes described? What can be said about the relationship between the trainee doctors and the poor patients? Based on one of the case histories, we will describe how the treatment of poor patients was documented and what conclusions can be drawn from this source with regard to the physician-patient relationship and the everyday diagnostic and therapeutic routine of the policlinic.

Organisation and Purpose of the Policlinics

While out-patient medical care was offered to Würzburg's poor population from 1795,[10] the institution, which would later be managed by Conrad H. Fuchs, was not formally founded until 1807, by town physician and medical professor Philipp Joseph Horsch (1772–1820).[11] Horsch, who was director of the institution until 1820, had medical responsibility for the city's poor, who made up the main clientele of the policlinics, as well as for the inmates of gaols and workhouses. After his death, a fierce controversy sprung up between the university

9 Staats- und Universitätsbibliothek Göttingen (hereinafter: SUB), Conrad Heinrich Fuchs (1803–1855), Krankheitsgeschichten, Sign.: HSD 80 Cod. MS. hist. nat. 61: I–VIII, X–XII.

10 Founded by Joseph Dömling (1771–1803) and directed, after Dömling's death, by Joseph Nikolaus Thomann (1764–1805), cf. Franke/Schröder, 1957, 6–8.

11 Karenberg also drew on Franke/Schröder, 1957, for his brief descriptions of the Würzburg policlinic; cf. Axel Karenberg, *Lernen am Bett der Kranken* (Stuttgart 1997), 78–79.

administration, the medical faculty and the city authorities, which were in charge of poor relief, as to whether or not the 'ambulatory clinic'[12] in Würzburg should continue to exist.

In principle, the conditions for establishing an out-patient clinic in Würzburg were much more problematic than in Göttingen, since Würzburg already had a comparatively well-functioning medical infrastructure that also covered the city's poor population.[13] Among the main institutions were two hospitals, the *Juliusspital* and the *Bürgerspital*, the latter holding a particularly important position among the numerous civic foundations in Würzburg. Both institutions were flagships of the town's prosperity and its catholic poor-relief tradition.[14] Würzburg also had early forms of social insurance for certain professional groups that offered help in case of illness, thus making medical care affordable. The '*Kranke-Gesellen-Institut*' (for sick journeymen) was founded as early as 1786, followed in 1801 by the '*Kranke-Dienstboten-Institut*' (for sick servants).[15]

That healthcare was generally well established is also reflected in the number of treatments given in the Würzburg policlinic under the direction of Conrad H. Fuchs: unlike in Göttingen, where access to in-patient care was much more limited, the Würzburg policlinic, in 1838, provided care for only

12 Bayerisches Staatsarchiv Würzburg (hereinafter: Bay StA Wü), Reg. Ufr. 6468, Die Königlich Bayerische Regierung des Unter-Main-Kreises, Kammer des Innern an den königlichen Staatsrat Freyherrn von Asbeck die ambulante Klinik betr. 12 May 1820.

13 On the contemporary medical infrastructure cf. Philipp Joseph Horsch, *Über die Bildung des Arztes als Klinikers und als Staatsdieners. Ein Programm, womit seine Vorlesungen über Heilmittellehre, psychische Medizin und Klinik ankündigt Philipp Joseph Horsch* (Würzburg 1807), 266, 271–301.

14 Cf. Johanna Bleker, Eva Brinkschulte and Pascal Grosse (eds), *Kranke und Krankheiten im Juliusspital zu Würzburg 1819–1829. Zur frühen Geschichte des Allgemeinen Krankenhauses in Deutschland* (Husum 1995), 24; on the history of the Juliusspital and the Bürgerspital see also Renate Schindler, 'Das Bürgerspital zum Heiligen Geist,' in Ulrich Wagner (ed.), *Geschichte der Stadt Würzburg, vol. 3/1. Vom Übergang an Bayern bis zum 21. Jahrhundert* (Stuttgart 2007), 780–85.

15 Cf. Eva Brinkschulte, 'Die Institutionalisierung des modernen Krankenhauses im Rahmen aufgeklärter Sozialpolitik. Die Beispiele Würzburg und Bamberg,' in Alfons Labisch and Reinhard Spree (eds), *'Einem jeden Kranken in einem Hospitale sein eigenes Bett'. Zur Sozialgeschichte des Allgemeinen Krankenhauses in Deutschland im 19. Jahrhundert* (Frankfurt am Main 1996), 189–207. On the general significance of the early servants' and craftsmen's health insurance cf. Reinhard Spree, 'Krankenhausentwicklung und Sozialpolitik in Deutschland während des 19. Jahrhunderts,' *Historische Zeitschrift* 260 (1995), 75–105.

about 3.5 per cent of the city's population.[16] For Göttingen, the figure was around 19 per cent in 1840.[17] While the Göttingen policlinic closed a major gap in healthcare provision this was not the case in Würzburg.

In arguing against the policlinic, the academic senate of Würzburg University referred to the existing and well-functioning infrastructure – students had sufficient opportunity for observation in the *Bürgerspital* and *Juliusspital*[18] – and dismissed the sick in the workhouses and gaols as mostly scientifically uninteresting 'patient material'.[19]

The opponents of the policlinic basically questioned whether students could be expertly and adequately taught in the confined dwellings of the poor.[20] Proponents such as Horsch, on the other hand, pointed out that, especially in the domestic situation, students could experience the impact of social circumstances on illness and healing.[21] The charge that the quality of medical provision in the 'ambulatory clinics' was not adequate was a serious one. Members of the medical faculty and the city council complained that treatments carried out by students were not, as had been agreed, by licensed physicians. The sick were 'exposed to arbitrary behaviour, if not blatant ignorance'.[22] This accusation was supported by the lawsuit of one Marianne Schenk that has been preserved in the Würzburg public record office. Mrs Schenk accused the policlinic, and particularly the town physician Horsch, of having caused the death of her 11-year-old son by allowing students to treat him without Horsch's supervision.[23]

The city authorities, who paid for medicines, put forward financial, social and moral arguments against the policlinic. Supported by the testimonials of

16 Cf. Franz von Rinecker, *Medicinische Statistik der poliklinischen Anstalt an der Julius-Maximilians-Universität zu Würzburg in ihrem vierten Decenium 1837–1847* (Würzburg 1848), 15, 19.

17 Cf. Conrad Heinrich Fuchs, *Bericht über die Vorgänge an der Poliklinik zu Göttingen im Jahre 1838 bis 39* (Hannover 1840), 3; UAG Kur 5489, letter from Conrad H. Fuchs to Universitätskuratorium, 5 September 1841.

18 The only concession they made was that one rarely came across sick children or elderly people there.

19 Bay StA Wü, Reg. Ufr. 6468, Votum des Med. Rathes Sorg über den Fortbestand einer ambulanten Clinik dahier, 19 May 1820.

20 Cf. Bay StA Wü, Reg. Ufr. 6468, Unterthäniger Bericht des akademischen Senats die ambulante Klinik [...] betr, 4 April 1820. Cf. also Joseph Nikolaus Thomann, *Über die klinische Anstalt an dem Julius-Spitale zu Würzburg* (Würzburg 1798), 4–6.

21 Cf. Horsch, 1807, 47; cf. also Hufeland, 1799, 11–12.

22 Bay StA Wü, Reg. Ufr. 6468, Bericht des Stadt-Magistrates dahier die Benutzung der Stadt-Armen zum Unterricht der medizinischen Candidaten an hiesiger Universität, 13 April 1820.

23 Bay StA Wü, Reg. Ufr. 6468, testimony by Marianna Schenk, 28 April 1820.

the city's apothecaries, the city council and the mayor criticized the wasteful use of medicines by the students who, out of ignorance, prescribed the wrong medication, corrected their prescriptions and ordered new medicaments from the apothecary which they then did not even collect.[24]

The hordes of students, their argument went on, who invaded the dwellings of the poor violated their modesty and upset the intimate, trustful physician-patient relationship.[25] Lastly, there were moral reasons why it was inadmissible to carry out physical examinations in the presence of the students. The house calls brought the students in contact with the daughters of the house leading often enough to their succumbing 'to the temptations of the senses'.[26]

Despite this overwhelming opposition, the Würzburg policlinic was maintained following a decree of King Maximilian I of Bavaria.[27] Apart from serving as a place for practical training, the policlinic was ultimately instrumental in preventing poor patients from being left sick and unable to work for long periods of time and from being a burden on the city treasury.[28]

In 1820 regulations were issued concerning the organization of the policlinic that picked up on the criticisms expressed during the debate about the continued existence of the policlinic. As a result students were only permitted to attend to the sick under strictest supervision. Regarding the high costs of remedies, it was pointed out that the 'poor people's pharmacopoeia' – a list of relatively inexpensive drugs for the treatment of the poor – was to be adhered to but that rule would not always be obeyed.[29] The regulations also reveal the kind of caregiving that was envisaged as ideal: physicians and medical students were to meet at particular times and proceed to the poor-quarters.

---

24 Bay StA Wü, Reg. Ufr. 6468, Bericht des Stadt-Magistrates dahier die Benutzung der Stadt-Armen zum Unterricht der medizinischen Candidaten an hiesiger Universität, 13 April 1820.

25 Bay StA Wü, Reg. Ufr. 6468, Votum des Med. Rathes Sorg über den Fortbestand einer ambulanten Clinik dahier, 19 May 1820.

26 Bay StA Wü, Reg. Ufr. 6468, Bericht des Stadt-Magistrates dahier die Benutzung der Stadt-Armen zum Unterricht der medizinischen Candidaten an hiesiger Universität, 13 April 1820. There seem to have been similar reservations regarding the policlinics in other places too, as one can see from a comment in the *Encyclopädische Wörterbuch* under 'policlinic' that refers to the 'unfavourable, ambiguous position' the trainees had with regard to the patients, cf. Busch *et al.*, 1842, 16.

27 Cf. Bay StA Wü, Reg. Ufr. 6468, decree of King Maximilian I of Bavaria, 25 September 1820.

28 Cf. Brinkschulte, 1996, 190.

29 Anonymous, *Die ärztliche Besuchs-Anstalt der erkrankten Armen und die ambulante Klinik in Würzburg* (Würzburg 1820). Cf. the patient history of Georg Grosch below, and Kay Peter Jankrift and Ruth Schilling, 'Medical Practice in Context. Religion, Family, Politics and Scientific Networks' (in this volume).

Each student/physician was allocated a particular district where they would treat 'their' poor. Examinations were carried out in the homes of the patients where the case histories were also taken (initially in tabular form). The standard tables specified spaces for entering the patient's name, age, place of residence, 'weekly alms,' and 'rank and occupation,' the 'name of the disease or condition,' 'beginning and end of treatment,' 'comments'.[30] The pre-printed forms were presumably filled in shortly after the end of the treatment or after a patient's death and later written out as fair copies.

Unlike Würzburg, Göttingen did not yet have a hospital in the early eighteenth century where medical students could receive practical instruction. Individual professors had poor patients brought to their homes so they could present their cases to the students. They also allowed medical students to go to the houses of the poor in order to treat patients there and to write down their histories. But these individual attempts at teaching students at the bedside '*in privatissime*' could hardly meet the wide-spread demand for practical tuition.[31] As early as 1773, Ernst Gottfried Baldinger (1738–1804) established a *Collegium clinicum* in Göttingen for the practical instruction of students. Medical students who treated poor patients had to first pay for the medicines out of their own pocket. As this way of funding the cost of treatment was not viable, not least due to the rising number of patients, Baldinger was finally granted a budget by the ministry of 200 *Florin* per year for his *Collegium*. As well as receiving practical instruction, the students were also prepared in the theory of dealing with poor patients in lectures entitled '*General instruction on medical practice with particular application for the morbos pauperum Gottingensium*'.[32] Around the year 1800 several such 'clinical colleges' seem to have existed in parallel with each other, but their responsibilities were not clearly regulated. When Friedrich Benjamin Osiander (1759–1822) was appointed professor of gynaecology and obstetrics in 1792, he did not teach his medical students in the newly-built maternity hospital, but in a general-medical 'clinical collegium' that provided healthcare for patients in their homes. In doing so the gynaecologist antagonized his colleagues at the Medical Clinic, as Jürgen Schlumbohm has shown in great detail, based on the spectacular case of the 'insect patient'.[33]

---

30 Reproduced for instance in: Besuchs-Anstalt, 1820, appendix.

31 Cf. Kumsteller, 1955, 11–20.

32 Cf. *ibid.*, 21–25. According to Kumsteller's calculations Baldinger treated 617 patients per year in his *Collegium clinicum*.

33 Cf. Jürgen Schlumbohm, *Lebendige Phantome. Ein Entbindungshospital und seine Patientinnen 1751–1830* (Göttingen 2012), 72–90. The patient was a woman with a life-threatening illness in whose vomit diverse physicians found insects such as woodlice,

It was not until 1781 that a small twelve-bed 'academic hospital' was established that offered medical students the opportunity to have some kind of practical tuition.[34] This practical training was seen as an important attraction and the medical school needed it to be competitive.[35] For the University of Göttingen, which was founded in 1737, the policlinic model was the best solution if they wanted to have a large enough number of patients for practical training without the need for a big in-patient hospital. Policlinics consequently served two purposes: poor patients, who could not afford medical treatment, received low-cost healthcare and the medical professors had enough patients to use for practical tuition. Supporters of out-patient training institutes also pointed out that there was wide-spread 'dislike of hospitals' among the 'lower ranks'.[36] As early as 1838, Conrad Heinrich Fuchs published a comprehensive report about his first year as director of the Göttingen policlinic that featured the impressive numbers of poor patients treated: all in all, 2,343 patients were treated, 1,898 of whom were seen in their homes in the town by the medical students and assistant physicians, and 445 in the surrounding villages.[37]

---

spiders etc., and even a lizard. She had initially been treated by the Göttingen town physician and university professor Johann Friedrich Stromeyer (1750–1830), before Friedrich Benjamin Osiander took on her case. Apart from Stromeyer and Osiander the patient also consulted a non-academically trained surgeon, whom Osiander, naturally, dismissed as a 'barber surgeon'. Osiander had the woman taken to his maternity hospital where she was safe from interfering rivals and had her closely observed by a midwife. Since the patient's affliction did not fall within the scope of his particular specialization and teaching remit, this encroachment drew a rebuke from the police authorities. The police were, however, unable to do anything about it since Osiander was subject to the university's jurisdiction. And the university humoured professor Osiander because of his popularity with medical students and his international renown.

34 Cf. Isabelle von Bueltzingsloewen, 'Die Entstehung des klinischen Unterrichts an den deutschen Universitäten des 18. Jahrhunderts und das Göttinger Accouchierhaus,' in Jürgen Schlumbohm and Claudia Wiesemann (eds), *Die Entstehung der Geburtsklinik in Deutschland 1751–1850. Göttingen, Kassel, Braunschweig* (Göttingen 2004), 15–30; Heike Winkelmann, *Das academische Hospital in Göttingen von 1781 bis 1850. Zur Geschichte der Göttinger Universitätsklinik* (Göttingen 1981), 62.

35 On the importance of practical instruction for the reputation of the medical school of Göttingen University see also Schlumbohm, 2012, 159–96.

36 Cf. Philipp Gabriel Hensler, 'Ueber KrankenBesuchs-Anstalten,' *Schlözers Stats-Anzeigen* 17 (1785), 273–96.

37 In 1811 Christoph Wilhelm Hufeland reported that 942 patients were treated in the founding year of the Royal Policlinical Institute in Berlin. Each of the trained physicians treated four, at most six, poor patients. Patients who were particularly 'instructive' were presented to students. Hufeland himself as well as two assistant doctors supervised the practical training at the patients' domestic bedside, cf. Hufeland, 1811, 4–5.

Thirty of the city's poor were inmates of the almshouse. The age range was as follows: 901 of the patients were children (age 0–15), 872 were aged 15–40, 428 were between forty and sixty years old and 142 were older than 60.[38] Furthermore, Fuchs listed patients according to the diagnoses which corresponded to the nosology he had developed. He gave detailed accounts of selected case histories, relating the symptoms to seasonal climatic conditions. Successful treatments at the 'policlinical institute' were listed according to gender: of 1,986 cured patients 851 were male and 1,135 female; of 120 improved patients 64 were male and 56 female; of 56 diseased patients 22 were male and 34 female. Unlike the communal hospitals, the Göttingen policlinic clearly treated more women than men. The ratio of female patients was 70 per cent in both policlinics.[39] The high proportion of women can be explained by the high number of women among the city's poor. In Göttingen, indigent patients received medical treatment in the policlinics whether or not they possessed a certificate of poverty. The communal hospitals, by contrast, began in the nineteenth century to give preference to patients whose treatment was paid for by the *Dienstboten- und Geselleninstitute* or, from the late nineteenth century onwards, by statutory health insurance.[40] The proportion of men among those who had insurance exceeded that of women so that the proportion of men in hospitals was between 60 and 70 per cent.[41] Fuchs furthermore quoted an

---

38 Cf. Fuchs, *Bericht*, 1840, 3.

39 Cf. Bueltzingsloewen, 1997, 260. In the Würzburg policlinic the percentage of women was also 70 per cent.

40 Cf. Brinkschulte, 1996 and Eva Berger, 'Frühmodernes Krankenhaus und "Pflegeversicherung" in der ersten Hälfte des 19. Jahrhunderts,' in Alfons Labisch and Reinhard Spree (eds), *'Einem jeden Kranken in einem Hospitale sein eigenes Bett'. Zur Sozialgeschichte des Allgemeinen Krankenhauses in Deutschland im 19. Jahrhundert* (Frankfurt am Main 1996), 208–35. In Göttingen, apprentices' and servants' institutes were not established until 1850 when the newly-extended Ernst August Hospital was opened at the university, while Würzburg had such institutes as early as the late eighteenth century.

41 Cf. Reinhard Spree, 'Quantitative Aspekte der Entwicklung des Krankenhauses im 19. und 20. Jahrhundert. Ein Bild innerer und äußerer Verhältnisse,' in Alfons Labisch and Reinhard Spree (eds), *'Einem jeden Kranken in einem Hospitale sein eigenes Bett'. Zur Sozialgeschichte des Allgemeinen Krankenhauses in Deutschland im 19. Jahrhundert* (Frankfurt am Main 1996), 51–90. On developments in the late nineteenth century see also Jörg Vögele, Wolfgang Woelk and Bärbel Schürmann, 'Städtisches Armenwesen, Krankenkassen und Krankenhauspatienten während des späten 19. und frühen 20. Jahrhunderts in Düsseldorf,' in Alfons Labisch and Reinhard Spree (eds), *Krankenhaus-Report 19. Jahrhundert. Krankenhausträger, Krankenhausfinanzierung, Krankenhauspatienten* (Frankfurt am Main 2001), 405–26.

overall number of 72,762 treatment days for the Göttingen policlinic with an average length of treatment of thirty-three days.[42] The high rate of cured patients (84.8 per cent) is remarkable even if we assume the figure to be strategically placed in order to document the success of the academic institution. The low mortality rate of 2.4 per cent certainly corresponds to the general mortality rate among the population.[43] Considering that the overall population was 9,000 (without the students), the 2,343 – and in later years more than 3,000 – patients treated at the 'Policlinical Institute' must have represented the majority of the ailing poor.[44] In the same year, only forty-six patients could be treated in the municipal infirmary for the poor[45] and the few beds in the 'Academic Hospital' allowed only for a small fraction of the poor patients to be treated.[46] Referring to the achievements of the 'Policlinical Institute' in this respect, Fuchs confidently demanded stronger support from the city's social welfare administration. First he tried to exploit the debate about the relocation of the poor-infirmary to the city's Albani Gate in order to request a new kind of hospital for the poor of which he would be the director.[47] While Fuchs' suggestion found the support of the Minstry of Education,[48] the City of Göttingen refused to appoint the professor as director of the poor-hospital, not least since the town physician Dr Adolph W.C. Ruhstrat performed this duty more than satisfactorily. But Fuchs did

42 Cf. Fuchs, *Bericht*, 1840, 7. Fuchs published a brief report in the *Göttingischen Gelehrten Anzeigen*, cf. Fuchs, *Bericht*, 1840 and Conrad Heinrich Fuchs, 'Göttingen. Policlinic daselbst 1838/39,' *Göttingische gelehrte Anzeigen* 1 (1840), 81–85.

43 General figures on the Göttingen population in the first half of the nineteenth century can be found in Karl Friedrich Marx, *Goettingen in medicinischer, physischer und historischer Hinsicht* (Göttingen 1824), 248–331.

44 On the history of the Göttingen Policlinical Institute see also Bueltzingsloewen, 1997, 241–68. Bueltzingsloewen's considerations are based exclusively on the annual reports published by Fuchs. The collections of the university archives were evaluated by Bueltzingsloewen for the time before Fuchs' nomination.

45 Cf. Stadtarchiv Göttingen (hereinafter: StadtAGö), AA, Bd. 4: Wohlfahrt, Sign.: 293: Verzeichnisse der im Hospital am Albanitor befindlichen Kranken, 1838.

46 In contrast to that, the maternity hospital, which re-opened in a new building in 1792, was comparatively well equipped. A minimum of 16 women could be accommodated at a time in the good-sized bright and airy rooms The maternity hospital received general funding from the university, cf. Schlumbohm, 2012.

47 Cf. UAG, Kur 5490, report (*Gutachten*) by Conrad Heinrich Fuchs, 25 December 1838.

48 Cf. StadtAGö, AA, Bd. 1: Gesundheitswesen, Schreiben des Ministeriums des Königlich Hannoversches Ministerium der geistlichen und Unterrichts-Angelegenheiten. Abteilung der Universitäts-Sachen an den Magistrat der Stadt Göttingen, 26 October 1839.

not accept the refusal or the fact that there would be no hospital patients available for the instruction of his students. In his report to the ministry he emphasized '...how unpleasant it is here for the physician and preceptor, if he has to refer interesting patients, whom he has treated for weeks or months on end because they found no care at home, to other hands'.[49] Fuchs also deplored the fact that, in the houses of the poor, only 5–6 trainee doctors could be instructed at the bedside. These trainees (*Practicanten*) were advanced medical students who treated patients in the poorer districts of Göttingen by themselves, but supervised by an assistant physician. They were trained in 'examination, exploration, diagnosis and prescription' and had accompanied older *Practicanten* or assistant physicians on their rounds. *Auscultanten* was the term used for beginners who received their first theory lessons from Fuchs. They first had to 'learn to look at and observe patients before taking active steps'.[50] Finally in 1839, after Fuchs had been offered the post as successor to Schönlein in Zurich, he managed to negotiate, as part of his conditions for staying in Göttingen, the allocation of eight beds for clinical instruction in a small separate infirmary.[51]

The tremendously high costs of drugs presented a great challenge for the in-patient infirmary.[52] Fuchs had to regularly apply for an increase in funds. In 1840/41 for instance, the cost of drugs was 1,455 *Reichsthaler* – plus a fee of 39 reichsthaler for the 'surgeon Vogel' for cupping and leeching. This sum constituted the major share of the expenses of the infirmary in the *Heynische Haus*, while the general operating costs amounted to only 250 *Reichsthaler*. Feeding the poor cost 69 *Reichsthaler* in that year. Apart from the actual medicines,

49 UAG Kur 5489, Conrad Heinrich Fuchs: Bericht zur Jahresabrechnung 1838/39, 8.

50 Conrad Heinrich Fuchs, *Bericht über die medicinische Klinik zu Göttingen im Jahre 1853/54* (Göttingen 1855), 5. For the difference between *Auscultanten* and *Practicanten* see also Hufeland, 1799, 19–23.

51 Cf. UAG, Kur 5489, letter from Conrad H. Fuchs to Geheimer Kabinetsrath, 2 August 1842, 3. In his letter Fuchs points out that Göttingen 'of all hospitals had fallen behind another German university in medical terms, while outdoing it in others, and it might be time now more than ever to do something in this respect since Berlin had greatly enhanced its reputation by appointing Schönlein'. – Fuchs was probably referring indirectly to the offer he had received from Zurich.

52 The problem obviously also arose in the work of other policlinics: the town practice in Heidelberg for instance was closed down due to the high costs of drugs, cf. Bauer *et al.*, 2005, 15–16. Hufeland put together a pharmacopoeia for the poor to keep down the cost of treating poor patients. *Armenpharmakopöe*, cf. Christoph Wilhelm Hufeland, *Armen-Pharmakopöe. Zugleich eine Auswahl bewährter Arzneimittel und Arzneiformeln, Fünfte vermehrte Auflage* (Berlin 1828).

the undernourished poor patients needed to be prescribed meat broth, meat and other restorative foods.[53] The professor also quoted the number of poor people treated which had risen steadily since 1838. And he complained about the healthcare for the poor in and around Göttingen having to be paid mostly out of the policlinic's own resources. The 150 *Reichsthaler* granted by the city covered only a very small portion of the overall expenses.[54] The professor also bemoaned that, unlike in Würzburg, there were no plans in Göttingen to set up 'servants' and apprentices' institutes,' so that the employers as well as the servants, journeymen and apprentices who lived on the breadline even though they had work, could contribute to the costs of medical healthcare. This kind of facility that would relieve the university and city coffers, did not materialize in Göttingen until 1851 when the academic hospital was extended and became the *Ernst-August-Hospital*.[55] The city of Göttingen had its own reasons for not wanting to pay more than a small contribution towards the costs of the policlinic: since the foundation of the 'Policlinical Institute' the town physician, who was by contract appointed to care for the poor, had to leave all 'instructive patients' to the university. In return, the university had pledged to take on the costs for the treatment of these poor patients.[56] This proved that

53 This was true particularly for the group of patients who suffered from consumption or cancer, cf. SUB Göttingen, Conrad Heinrich Fuchs (1803–1855), Krankheitsgeschichten, Sign.: HSD 80 Cod. MS. hist. nat. 61: I–VIII, X–XI. On the group of consumptive patients cf. in particular Katharina Schilling, 'Ach gieb mir doch nur etwas Luft, Du hast der Luft so viel!' Palliativmedizin im frühen 19. Jahrhundert (Duisburg 2011), on cancer patients cf. Karen Nolte, 'Ärztliche Praxis am Sterbebett in der ersten Hälfte des 19. Jahrhunderts,' in Walter Bruchhausen and Hans-Georg Hofer (eds), *Ärztliches Ethos im Kontext. Historische, phänomenologische und didaktische Analysen* (Göttingen 2010), 39–58 and Karen Nolte, 'Vom Umgang mit unheilbar Kranken und Sterbenden in "Kranken-Besuchs-Anstalten" zu Beginn des 19. Jahrhunderts,' *Würzburger medizinhistorische Mitteilungen* 26 (2007), 28–52.

54 Cf. UAG, Kur. 5489, Bemerkungen zur Rechnung über Einnahme und Ausgabe des K. Poliklinikums zu Göttingen für das Rechnungsjahr vom 1. Julius 1840–1841, 37.

55 Cf. Ferdinand Oesterley, *Bemerkungen über die Bekanntmachung vom 23. December d.J. die ärztliche Behandlung und Aufnahme hiesiger kranker Gesellen. Lehrlinge und Dienstboten in das neue academische Hospital betreffend* (Göttingen 1851). On the foundation of the Ernst-August-Hospital cf. Gunda Jentzsch, 'Zur Geschichte der Klinischen Medizin in Göttingen. Das Ernst-August-Hospital 1850–1890' (Med. diss., University of Göttingen, 1988).

56 Cf. StadtAGö, AA, Vol. 4: Wohlfahrt, 1. Armensachen und Stiftungen, Armen-Hospital-Sachen, No. 290: Überweisung der armen Kranken zur ärztlichen Behandlung an das klinische Institut von Professor Conradi: 1818–1875; letter by Königliches Universitätskuratorium Hannover, 22 November 1839.

the out-patient clinic was first and foremost an institution devoted to the training of physicians and to medical research.[57]

Poor Patients and Their Doctors

Almost all patients of the Würzburg and Göttingen policlinics belonged to the lower classes. Two thirds were manual labourers, most of them women, who hired themselves out as seamstresses or washerwomen. The men worked as servants or messengers. Around a third of the patients in the policlinics were craftspeople and worked as tailors, cobblers or butchers. The sources do not always reveal whether or not manual labourers or servants belonged to a guild. Detailed information on a person's professional grade (journeyman, apprentice or master) was usually not included either.[58]

Since they were mostly penniless labourers, the patients of the policlinics were usually financially insecure and this situation was aggravated when their work was disrupted by illness and other family members were unable to compensate for the loss of income. Unmarried or widowed workers were mostly affected, as were even the widows of middle-class municipal or public administrative clerks.

The individual patient histories of the policlinics that were investigated accentuate the link between poverty, poor health, inability to work and lack of financial security, especially in old age.[59] The (temporary) poverty revealed by the assessment of healthcare provision through the policlinics related to cases that were, according to contemporary criteria, deemed worthy of support.[60]

57 Poor women who were pregnant or about to give birth, mostly maids and servant girls, could from 1751 on have themselves admitted to the university's *Accouchierhaus* where they could give birth free of charge. In return for the free obstetric help and accommodation before and after the delivery they were expected to make themselves available for research and practical instruction, cf. Schlumbohm, 2012.

58 Evaluation of the case histories of C.H. Fuchs on the basis of occupations. 185 of the overall 710 patient histories mention the patients' occupation reliably; in sixty-three cases the father's or husband's occupation is entered.

59 On the concept of poverty in more recent historiographical research cf. Martin Krieger, *Arme und Ärzte, Kranke und Kassen. Ländliche Gesundheitsversorgung und kranke Arme in der südlichen Rheinprovinz (1869 bis 1930)* (Stuttgart 2008), 18–20. On the history of healthcare for the poor see also Gerd Göckenjan, *Kurieren und Staat machen. Gesundheit und Medizin in der bürgerlichen Welt* (Frankfurt am Main 1985), 286–304.

60 Cf. Krieger, 2008, 21; Christoph Sachße and Florian Tennstedt, *Geschichte der Armenfürsorge in Deutschland, vol. 1: Vom Spätmittelalter bis zum ersten Weltkrieg* (Stuttgart 1980), 107; Brinkschulte, 1996, 189.

The patients in question were assigned 'poor' status because they had become unable to work through illness or they had become poor through hardship, but they were seen as members of the municipal community and therefore entitled to its support.[61]

The physicians of the policlinic who called on their poor patients found themselves in situations that were foreign to them and that they had to adapt to, because they came from a different social background. They were expected to empathize as much as possible with the patients' milieu so they would be able to explore the medical case in all its layers,[62] since the social context was seen as a crucial factor in the treatment and cure of diseases.[63] The place where physician and patient met, the patient's home, moved into the physician's focus. Contemporary medical journalism usually referred to this place as a 'harmful atmosphere' (*schädlicher Dunstkreis*),[64] characterizing it as dirty, pokey, dark and damp.[65] According to the Würzburg medical professor Franz von Rinecker (1811–1883) one could, even in Würzburg, a wealthy bourgeois city with a comparatively small poor population,[66] 'behold the whole misery of proletarian living' in the poor districts.[67] The smell of sickness added olfactory irritation to the impressions that the physicians and medical students of the policlinics came away with from their house-calls.[68] The following vivid description of just such an experience is by the contemporary Göttingen town physician Ruhstrat: 'The patient's consumption had reached the fifth degree and such a malodorous stench prevailed in his chamber that I could not stomach it and had to step outside the door while conversing with him'.[69]

61 Katrin Marx-Jaskulski, *Armut und Fürsorge auf dem Land. Vom Ende des 19. Jahrhunderts bis 1933* (Göttingen 2007), 23.

62 Cf. Karen Nolte, 'Die medizinische Versorgung der Göttinger Stadtarmen durch das Poliklinische Institut der Universität,' *Würzburger medizinhistorische Mitteilungen* 29 (2010), 131–57: 136–37.

63 Cf. Hufeland, 1810; Rinecker, 1848.

64 Hensler, 1785, 276.

65 Occasionally, negative moral connotations were included in the physicians' assessment of dirt and disorder. Cf. Ludwig Gerhard Wagemann, *Nachricht über die Versorgung der Armen in Göttingen vom Jahr 1796* (Göttingen 1797), 9–10.

66 Bay StA Wü, Reg. Ufr. 6475: Medizinische Topographien, XIII (VII): Wohlstand, ca. 1850.

67 Rinecker, 1848, 3.

68 Cf. also Göckenjan, 1985, 233. The opponents of the policlinics considered the 'disgusting staunch' in the houses of the poor not only to be harmful for the patients, but to present an unacceptable health risk for the treating physicians.

69 StadtAGö, AA, Bd. 1: Gesundheitswesen, Sign. 1902, Stadtphysicus Dr. med. Adolph W.C. Ruhstrat, Krankengeschichte Nr. 44.

The idea was that the medical students of the policlinics should get used to the reality of private practice by visiting the houses of the poor as early as possible in their training. Dealing with the 'stubbornness'[70] of patients and their relatives, whose behaviour is described in contemporary medical guidelines as ranging from indifferent to hostile and prejudiced, was another aspect of this acclimatization.[71] Patients might insist on doing what they thought adequate and ignore doctor's orders.[72] The physicians were also explicitly warned against the 'workshy' poor who were a burden to the social welfare fund.[73] At the same time the concept of the policlinic included compassion as a physician's central quality.[74] Apart from the 'moral education'[75] of young physicians, the primary task when calling on the sick was to study the symptoms they displayed. In 1837 the town physician of Stettin elaborated on the particular value, from the scientific point of view, of seeing poor patients in their own domestic environment:

> ...here the physician finds what he looks for in vain in the houses of the wealthy, and what he longs for most: true crisis situations; it is here that he is confronted with the plain, unadulterated symptoms; that he learns about the certain regular course of acute illnesses; here he can finally put to the test and experience the fruitful, harmful or disturbing influence of his medical actions.[76]

Such views about medical provision for the poor make one wonder whether Hufeland's programmatic appeal that patients must not be seen as 'mere objects of his art' by the physician,[77] was able to withstand the scientific impetus of the policlinics. The poor patient was very much the object of the 'physician's focus' which, in the environment of the poor, could easily assume a dimension of control and discipline. The physicians and medical students of the policlinics invaded the privacy of their patients when calling on them. House calls opened up insights not only into the patients' state of health, but also regarding their

70 Hufeland, 1810, 14.

71 Cf. W. Schmidt, 'Vorschläge zur Verminderung der Arzneikosten,' *C.W. Hufeland's Journal der practischen Heil-Kunde, fortgesetzt von Dr. E. Osann* 84 (1837), 24–48: 32.

72 SUB Göttingen, Conrad Heinrich Fuchs (1803–1855), Krankheitsgeschichten, Sign.: HSD 8 COD MS H NAT 61: Heft VII, 1836, Nr. 22: Siegmund Kraus; Heft I, 1833/34, Nr. 11: Conrad Kuhn.

73 Cf. Besuchs-Anstalt, 1820, 23.

74 Cf. Nolte, 2007, 32–33.

75 Hufeland, 1810, 13.

76 Schmidt, 1837, 27.

77 Cf. Hufeland, 1810, 12–13.

way of life and moral conduct. Physicians like Horsch and Fuchs used such information statistically, to establish 'medical topographies' for instance. Apart from that, the physicians' observations were also of interest for the municipal welfare offices in their assessment of a family's entitlement to poor relief.[78] The medical staff of the policlinics were, therefore, like agents of a 'medical policy' (*Medizinische Polizey*). Horsch, the Würzburg town physician and first director of the city's policlinic, summarized the tasks of a physician in the words: as an administrator of health in the service of the common good, the physician had to ensure through their actions that the state would not lose its subjects to illness.[79]

The Case of Georg Grosch

The case histories compiled by Conrad H. Fuchs give us an idea of how physicians perceived the patients, their state of health and life situation, but they also present the patients as active agents and reflect their experiences of illness. One of these case histories is that of the 20-year-old hairdresser Georg Grosch, who received his first treatment in the Würzburg policlinic in January and then again in July and August 1836.[80] After the first seemingly unproblematic and promising treatment for 'pulmonary phthisis' he came back to the policlinic some time later. His main complaints – haemoptysis, pains in the chest and larynx – that initially seemed to have improved, had deteriorated and weakened him at a rate that led to his death within two weeks after the second treatment was started.

78 Philipp Joseph Horsch, *Versuch einer Topographie der Stadt Würzburg in Beziehung auf den allgemeinen Gesundheitszustand und die dahin zielenden Anstalten* (Arnstadt/Rudolstadt 1805); Conrad Heinrich Fuchs, 'Über die Sterblichkeit der Stadt Würzburg vom 1ten Juli 1819 bis zum 30sten Juli 1829,' *Zeitschrift für Staatsarzneikunde* 25 (1833), 368–403; *idem*, 'Ueber den Einfluß der verschiedenen Gewerbe auf den Gesundheitszustand und die Mortalität der Künstler und Handwerker in den Blüthejahren nach den Tabellen des Instituts für kranke Gesellen zu Würzburg 1786–1834. Ein Beitrag zur medicinischen Statistik,' *Wissenschaftliche Annalen der gesammten Heilkunde* 32 (1835), 385–419; *idem*, Bericht über die Vorgänge an der Poliklinik zu Würzburg im Jahre 1835. Mit einigen Bemerkungen über die Krankheits-Constitution dieses Jahres. Würzburg 1835. On the medical historical context of medical topographies cf. Jan Brügelmann, 'Der Blick des Arztes auf die Krankheit im Alltag 1779–1850. Medizinische Topographien als Quelle für die Sozialgeschichte des Gesundheitswesens' (PhD diss., Free University of Berlin, 1982).

79 Horsch, 1805, 258. See also Göckenjan, 1985, 286–305.

80 Cf. SUB Göttingen, Conrad Heinrich Fuchs (1803–1855), Krankheitsgeschichten, Sign.: HSD 8 COD MS H NAT 61: Heft VII, 1835/36, Nr. 14, Nr. 61: Georg Grosch.

The case history illustrates that the patient hoped that he was not seriously ill and would not have to die. He asked to discontinue his previous treatment, playing down the symptoms. On 8 July the treating physician, Dr Ratke, noted:

> Patient insists that cough and sputum are abating. I called on him in the evening and found clear signs of the hectic fever that was not present when I saw him first and that the patient professes not to have noticed, even though he admitted of feeling a slight shudder now and again.

The young hairdresser tried to find ways of curing himself, by resorting to 'popular' healing methods such as drinking water from a well said to have 'miraculous healing powers'.[81] Dr Ratke did not oppose the patient's wishes for medication as long as he did not consider it harmful. He accepted that Georg rubbed in dog fat, because the patient had 'great confidence' in this method. Georg also discussed the cause of his affliction with the physician. As we can see from the physician's note, the patient thought that

> his illness was related to haemorrhoids and asked me for the appropriate treatment. I did not argue with that, but tried to convince him that his chest problem needed attention before we would treat his haemorrhoids.

Even when Georg's condition became increasingly aggravated, the physician reassured him that recovery was possible:

> I did nevertheless encourage him in his hope for a recovery, and dissuaded him from the belief that he was suffering from tracheal consumption, by assuring him that the affection of the larynx was due to a cold alone, and that the continually changing climate was to blame for the persistence of this affection.[82]

A few days later, Dr Ratke wrote down that the patient 'who had always been full of hope,' had lost his 'usual serenity' and no longer believed that he would get better.[83] Georg Grosch passed away in the evening of 16 August 1836.

---

81 SUB Göttingen, Conrad Heinrich Fuchs (1803–1855), Krankheitsgeschichten, Sign.: HSD 8 COD MS H NAT 61: Heft VII, 1835/36, Nr. 61: Georg Grosch.

82 *Ibid.*

83 *Ibid.*

From this case history we gain a rare glance into the relationship between a physician and a poor patient and find a precise account of treatment, diagnosis and therapy. In keeping with the *Naturhistorische Schule*, the physicians of the policlinic noted down detailed descriptions of the symptoms they perceived and the therapeutic steps they took. Especially during the second, terminal, phase of his illness, Georg Grosch was closely observed. We find records of 15 consultations. Especially when the patient's state deteriorated rapidly, he was seen almost daily and all observations were written down.

Two different physicians treated Georg Grosch. While Dr Münchmeyer, who saw him first, did not investigate the hereditary background of Grosch's 'pulmonary phthisis,' Dr Ratke, who was the second physician to treat him, focused on that aspect right from the beginning. He took an in-depth family history and finally diagnosed *phthisis pulmonalis tuberculosa hearedetaria c. tracheae exulceratione*, while the diagnosis established after the first examination had been 'simple' tubercular *pneumophthisis*.[84] Dr Ratke recognized Grosch's *habitus phtisicus* as an expression of his hereditary condition and noted down the following characteristic signs: 'thin emaciated body, long neck, sagging shoulders, narrow but slightly arched chest, hollow cheeks, *rubor genarum circumscriptus* [author's note: circumscribed red cheeks], dull, sunken eyes with blue shadows'.[85]

The case history supplies information on other causal aspects of the disease: the patient's attitude and personal circumstances – the drinking of cold well water which, according to popular belief, had miraculous healing powers and his frequent contact with clients – had triggered and aggravated his pulmonary phthisis. Both explanations rely on the contemporary belief that tuberculosis could be caused by infection.[86]

The diagnostic process that can be derived from the case history of Georg Grosch was based on physical examinations, such as percussion and auscultation, for which the corresponding instruments – plessimeter[87] and stethoscope[88] – were used. Sounds and noises as well as the diverse manifestations of perspiration, stool and pulse were written down meticulously.[89]

84 On hereditary aspects of consumption cf. Karen Nolte, 'Schwindsucht – Krankheit, Gesundheit und Moral im frühen 19. Jahrhundert,' *Medizin, Gesellschaft und Geschichte* 29 (2010), 47–70.

85 *Ibid.*

86 Cf. Nolte, 'Praxis,' 2010, 47–8, 52; Schilling, 2011, 40.

87 A flat plate made from ivory or silver which enhanced the sound in percussive examinations.

88 UAG, Kur. 5489, inventory list of the policlinic Göttingen 1841/42, 54–64.

89 Cf. Kay Peter Jankrift *et al.* (in this volume).

The case histories not only reveal which medicines were prescribed by the trainee physicians on the basis of which symptoms and complaints, they also show that the patient rejected marshmallow tea because he did not like the taste and was given an alternative remedy instead. On another occasion, syrup was added to his medicine to improve the taste. Consideration for the patients' wishes and the addition of (costly) sweeteners to their medicines was common practice in the policlinics and is proof of the students' and physicians' reliance on the compliance of poor patients whose illnesses they were keen to study.

In the case of Georg Grosch the derivative treatments common at the time were prescribed, which included leeches, 'drawing' plasters and inunctions. Active agents such as salep root or hemlock were used to break open pulmonary ulcers. The patient was also given mixtures containing henbane and wild lettuce to alleviate his pain and soothe his cough. Marshmallow appears in almost all recipes because of its assumed analgesic and ulcer-inhibiting effect. Grosch's physician also administered Iceland moss and verbascum to promote expectoration. The most expensive part of the therapy consisted in a 'non-warming' diet of lentil soup, veal, meat broth and pigeon.[90] Many of the substances Georg Grosch was given as part of his therapy were narcotics or analgesics that would have been poisonous in higher dosage. Towards the end of his suffering he received the 'classic' palliative pain relief: opium mixed with cherry-laurel water.[91]

The body of Georg Grosch was dissected one day after his death.[92] Autopsies served to verify or correct any diagnosis or assumption made when the patient was still alive. The knowledge of pathological anatomy was described in the contemporary technical literature as equally essential to 'professional' medical action as the knowledge of physiology and the application of physical examination techniques.[93] Because of their importance for research and training, autopsies were carried out frequently in the policlinics.[94]

90 Cf. Schilling, 2011, 65–73.

91 Cf. Nolte, 'Praxis,' 2010.

92 SUB Göttingen, Sectionsprotocolle, Sign.: HSD: 8 COD MS H NAT 60/II, Nr. 196: Georg Grosch. The post-mortem showed that the larynx was also affected and C. H. Fuchs accordingly added "with an ulceration of the trachea" to the headline of the Grosch's case history.

93 Cf. Carl Ernst Bock, *Lehrbuch der pathologischen Anatomie und Diagnostik*, 2 vols (Leipzig 1853), vol. 2 (*Lehrbuch der Diagnostik*), 7; Ferdinand Weber, *Kurze Bemerkungen über die Section der Leiche zu pathologischen Zwecken: ein Gruss an seine Herren Collegen in den Herzogthumern Schleswig-Holstein und Lauenburg, von Dr. F. Weber, Prosector an der Universitat in Kiel* (Kiel 1847), 7–9.

94 Fuchs' case histories included five volumes of 'autopsy reports'. Cf. SUB Göttingen, Sectionsprotokolle, Sign.: HSD: 8 COD MS H NAT 60/I–V.

Conclusion

The out-patient policlinics in Würzburg and Göttingen were places of 'decentralized' healthcare. They were institutions of research and training on the one hand, while being instrumental in supplying healthcare to the poor on the other. This dual purpose affected not only their status within the healthcare system and their relationship with the universities and municipal authorities, but also the quality of services provided.

Conditions in the cities of Würzburg and Göttingen differed widely when the policlinics were founded, especially with regard to the existing medical infrastructure. Both institutions nevertheless faced similar problems and conflicts with medical colleagues at the hospitals, with the authorities and especially with the social welfare office on questions of funding. Both policlinics owed their existence primarily to the decision of the city and university administrations to augment the attractiveness of the medical schools by optimizing the training facilities. For the social welfare offices, the contact that the students and junior physicians had with the poor population meant that they could obtain information about poor patients and their families and, above all, about whether or not they were capable of working.

The multi-layered interest of the students and physicians at the policlinics in medical training and scientific research as such constituted a challenge for the doctor-patient relationship in the everyday healthcare situation. In addition to that there was the treatment situation, which was unfamiliar to both physicians and patients, and the ambivalent role of the physicians as persons of trust for the patients on the one hand and as representatives of the social services on the other. That the physicians' authority and ability to discipline the poor was limited is apparent from contemporary healthcare guidelines that, at length, bemoaned the 'stubbornness' of the poor. Sources relating to poor-relief healthcare in Hamburg likewise confirm that measures introduced to discipline the poor were far from successful.[95]

As a consequence, university-trained physicians who wanted to win the trust of the poor they called on, had to humour their patients' interpretations of their illnesses and their wishes for particular therapies. As we can see from the case history of the hairdresser Georg Grosch, the assistant doctor at the Würzburg policlinic did exactly that in order to keep his patient, after the

95 Cf. Göckenjan, 1985, 286–304; Fischer Christian Ernst, *Versuch einer Anleitung zur medizinischen Armenpraxis* (Göttingen 1799). On the problems of healthcare for the poor in Hamburg cf. Johann Jacob Rambach, *Versuch einer physisch-medizinischen Beschreibung von Hamburg* (Hamburg 1801).

consumptive barber had already once discontinued his first course of treatment, making himself unavailable as an object of training and research. Self-medication and alternative treatment suggestions were accepted by the physician, who, on his part, applied scientifically proven examination and treatment methods as well. This approach is documented in many other of Fuchs' 'case histories'.[96] While the proponents of the policlinics emphasized, in their theoretical writings, the advantages of domestic bedside teaching, Conrad Heinrich Fuchs was critical of the approach. He complained about the inadequacy of the homes of the poor for training purposes. His unceasing efforts to be allocated as many in-patient beds as possible for research and teaching are a further indication that Fuchs found the hospital situation preferable. In the hospitals, there was no need for physicians to heed the wishes and 'idiosyncrasies' of poor patients and their relatives[97] or to apply for permission to conduct post-mortems from bereaved relatives who, in the case of the Göttingen policlinic, often withheld that permission from Fuchs.[98]

Case histories such as the one of Georg Grosch illustrate how Fuchs' collection was intended for teaching purposes and also needed to be accessible as teaching material by later generations of physicians. His fastidious recording of case details was part of his medical scientific impulse. The director of the policlinics aspired to establish a differentiated clinical methodology and a reliable nosology based on the experience gained from treatments and observations. The compilation of case histories by Fuchs as a representative of the Natural History School of Medicine must be seen against this background. He incorporated his research results, which included, and primarily relied on, the experience gained in the policlinics in a comprehensive medical textbook.[99]

---

96 Cf. Nolte, 'Praxis,' 2010 and Nolte, 2007.

97 Lachmund and Stollberg underline that physicians, around 1800, not only dealt with the patient, but with his or her entire social environment, referred to as 'bystanders,' cf. Jens Lachmund and Gunnar Stollberg, *Patientenwelten. Krankheit und Medizin vom späten 18. Jahrhundert bis zum frühen 20. Jahrhundert im Spiegel von Autobiographien* (Opladen 1995), 124. On dealing with 'obstinate' relatives in domestic healthcare cf. Martin Dinges, 'Medizinische Aufklärung bei Johann Georg Zimmermann. Zum Verhältnis von Macht und Wissen bei einem Arzt der Aufklärung,' in Martin Fontius and Helmut Holzhey (eds), *Schweizer im Berlin des 18. Jahrhunderts* (Berlin 1996), 137–50.

98 Cf. UAG, Kur 5489, Conrad Heinrich Fuchs: Bericht zur Jahresabrechnung 1838/39, p.11, cf. also Nolte, 'Versorgung,' 2010, 146–47.

99 Conrad Heinrich Fuchs: *Lehrbuch der speciellen Nosologie und Therapie* (Göttingen 1845), vol. 1 (*Klassen und Familien*); *id.*, *Lehrbuch der speciellen Nosologie und Therapie* (Göttingen 1846), vol. 2,1 (*Gattungen und Arten. Erste Abtheilung. Hämatonosen. Erste bis dritte Ordnung*); *id.*, Lehrbuch der speciellen Nosologie und Therapie (Göttingen 1847), vol. 2,2

The medical practice of the policlinics as reflected in the case histories, which Fuchs adapted and edited, shows that the practice of medicine and the accumulation of new stores of knowledge were mutually enhancing processes: the diagnostic and therapeutic insights gained in practice, once they had been confirmed in autopsies, edited, categorized and systemized, were fed back to the medical practice via the training of medical students.

(*Gattungen und Arten. Zweite Abtheilung: Hämatonosen. Vierte bis sechste Ordnung*); *id.*, *Lehrbuch der speciellen Nosologie und Therapie* (Göttingen 1848), vol. 2,3 (*Gattungen und Arten. Dritte Abtheilung: Neurosen, Morphonosen und Register*).

CHAPTER 10

Unlicensed Practice: A Lay Healer in Rural Switzerland

Alois Unterkircher and Iris Ritzmann

> ...a woman aged forty-five has venous ulcers on her feet; she has thin blood; she sought treatment from quacks and contracted severe gangrene and pain; now she asks me to advise her and accept her as a patient...[1]

This comment on the alleged malpractice of medical laypersons is consistent with a number of similar complaints about 'quacks', 'dilettantes' or 'dabblers'. The main criticism, in each case, was that such persons with their doubtful therapies aggravated rather than improved the patients' state of health. As if that was not reprehensible enough, they took away customers from the 'true' medical experts (physicians, surgeons, apothecaries). What is surprising is not the fact that such criticism is expressed in a practice journal, but who expressed it: it was penned by a certain Gottfried Wachter from Hugelshofen, a village in the northeast of Switzerland that, around 1830, counted 800 inhabitants. In the first half of the nineteenth century, Wachter was active as a lay healer in this remote community, away from busy trade routes or flourishing commercial centres. It is not a *doctor medicinae*, that is, a practitioner ranking highly in the medical hierarchy, who is complaining here about 'quacks', but a healer who had no diploma or certificate and who stood on the lowest rung of that hierarchy. Wachter's only credentials consisted of a permit issued by the Thurgau health authorities allowing him to carry out a restricted range of medical treatments. It is remarkable that this healer, who had no academic education or practical training, expressed such disdain for the medical activities of persons to whom he was much closer from a training point of view than he was to physicians.

---

* The research results presented here form part of a project on 'Rural Healers in the First Half of the 19th Century' and 'A Phase-Out Model: The Medical Practice and Life Situation of a Lay Healer in the First Half of the 19th Century', supported by the Swiss National Science Foundation.

1 '*...eine Frau von 45 Jahren hat offne // Füß Hat ein flüßig geblüt hat sich von // Quaksalbern Behandlen Laßen bekam // ersäglichen Brand und schmerzen so Bitt // sie Mich doch um einen rath sie doch in // Cur zu Nehmen...*'. Archiv für Medizingeschichte, Universität Zürich (AfM), MsH 17.4, fol. 90.

 | DOI 10.1163/9789004303324_012

It is possible that Gottfried Wachter chose the term 'quacks' deliberately in order to demonstrate to potential official inspectors that the treatment of the patient had been 'botched' before he was consulted himself. The use of such a medical 'combat term'[2] might be expression, too, of a new self-image that Wachter gained as a result of the fact that his therapeutic competence could no longer be questioned after he had been granted limited permission for medical practice in 1805. Having secured this official document made it easier for him to stand his ground in the face of the authorities. At the same time, this letter, with its official seal, created a distance between him and those healers who, while they drew their medical knowledge from similar sources, were banned by the Thurgau authorities from carrying out any healing activities whatsoever.

The kind of self-styling we find with Gottfried Wachter has, so far, not been investigated in the heterogeneous group of lay healers. For educated physicians, on the other hand, there are numerous examples which document this kind of demarcation behaviour. These examples show that diatribes against non-academic healers have, since early modern times, been the traditional means of expressing the most profound contempt.[3] Most academic physicians applied certain distinctions, however: right into the nineteenth century they assigned to the practically trained surgeons the more or less clearly defined responsibility for external, especially surgical, afflictions and as a result of this, conflicts were largely avoided.[4] It was different with lay healers, who had no

2 On the change in the meaning of concepts cf. Robert Jütte, *Geschichte der Alternativen Medizin. Von der Volksmedizin zu den unkonventionellen Therapien von heute* (München 1996), 18–23.

3 Philip Rieder, 'The Physician Louis Odier and the Medical Market in Geneva (1774–1817)', *Gesnerus* 69 (2012), 54–75; Elisabeth Dietrich-Daum, 'Die Klage der Ärzte. Marktprobleme und Professionalisierung in der zweiten Hälfte des 19. Jahrhunderts in Österreich', in Alois Unterkircher, with Reinhard Bodner and Kathrin Sohm (eds), *Medikale Kulturen* [special issue: *Bricolage* 5 (2008)], 202–21; Alois Unterkircher, '"…für die Zukunft auf selbe ein wachsames Auge zu halten." Die Ausgrenzung medikaler Laienbehandler aus dem Netz medizinischer Versorgung zu Beginn des 19. Jahrhunderts anhand Quellen aus dem Tiroler Raum' (diploma thesis, University of Innsbruck, 1999); Iris Ritzmann, 'Der Verhaltenskodex des "Savoir faire" als Deckmantel ärztlicher Hilflosigkeit? Ein Beitrag zur Arzt-Patient-Beziehung im 18. Jahrhundert', *Gesnerus* 56, no. 3/4 (1999), 197–219; Gabriele Danninger, *"…daß sie auch vor den Krancken=Betten müsten das Maul halten…". Frauen zwischen "traditioneller Heiltätigkeit" und "gelehrter Medizin" um 1800 anhand Salzburger Quellen* (Wien 1998); Robert Jütte, *Ärzte, Heiler und Patienten. Medizinischer Alltag in der frühen Neuzeit* (München 1991).

4 Sabine Sander, *Handwerkschirurgen. Sozialgeschichte einer verdrängten Berufsgruppe* (Göttingen 1989); Michael Stolberg, *Experiencing Illness and the Sick Body in Early Modern Europe* (Basingstoke 2011).

officially recognized training and who often practised without a licence. These lay healers were frequently the target of aggressive attempts at ostracism that manifested in accusations of incompetence, of material interests and a failure to subject to the leading medical authority.[5] With the growing academization and specialization of medicine in the nineteenth century the lay healers became increasingly marginalized or forced to specialize in treatments that the learned physicians did not offer.[6]

Medical history has, in recent decades, focused on the period of time when the lay healers were ousted from the health market where they had previously played an important part.[7] The main questions are: who was responsible for what kind of medical provision? How did the diverse professional groups differ from one another? And how did the learned physicians, as one group among many, ultimately succeed in gaining supremacy in the course of the 19th century?[8]

By adopting the questions and methods known from social history,[9] medical historians arrived at a new evaluation of those groups of healers that offered medical services without licence. Especially since the 1980s, scholars abandoned the view that the medicine practised by lay healers was by definition of

5 Cf. Michael Stolberg, 'Heilkunde zwischen Staat und Bevölkerung. Angebot und Annahme medizinischer Versorgung in Oberfranken im frühen 19. Jahrhundert' (Med. diss., Techn. Universität München, 1986); Oliver Stenzel, *Medikale Differenzierung. Der Konflikt zwischen akademischer Medizin und Laienheilkunde im 18. Jahrhundert* (Heidelberg 2005), 85–112. For Tyrol as a regional example cf. Unterkircher, 1999.

6 Cf. the overview in Jütte, *Geschichte*, 1996, 115–261. For case studies relating to the German Empire cf. Thomas Faltin, *Heil und Heilung. Geschichte der Laienheilkunde und Struktur antimodernistischer Weltanschauungen in Kaiserreich und Weimarer Republik am Beispiel von Eugen Wenz (1856–1945)* (Stuttgart 2000), 224–39; for the example of a German homeopath cf. Marion Baschin, *Wer lässt sich von einem Homöopathen behandeln? Die Patienten des Clemens Maria Franz von Bönninghausen, 1785–1864* (Stuttgart 2010).

7 That this marginalization has failed so far is apparent from the research of the sociologist Andreas Obrecht into modern spirit healers in Austria, cf. Andreas Obrecht, *Die Welt der Geistheiler. Zur Renaissance magischer Weltbilder* (Wien 1999).

8 Cf. Claudia Huerkamp, 'Ärzte und Patienten. Zum strukturellen Wandel der Arzt-Patient-Beziehung vom ausgehenden 18. bis zum frühen 20. Jahrhundert', in Reinhard Spree and Alfons Labisch (eds), *Medizinische Deutungsmacht im sozialen Wandel des 19. und 20. Jahrhunderts* (Bonn 1989), 57–73; Ute Frevert, *Krankheit als politisches Problem 1770–1880. Soziale Unterschichten in Preußen zwischen medizinischer Polizei und staatlicher Sozialversicherung* (Göttingen 1984); Stenzel, 2005. For a summary cf. Michael Stolberg, 'Heilkundige. Professionalisierung und Medikalisierung', in Norbert Paul and Thomas Schlich (eds), *Medizingeschichte. Aufgaben, Probleme, Perspektiven* (Frankfurt am Main 1998), 69–86.

9 Cf. the overview in Robert Jütte, 'Sozialgeschichte der Medizin. Inhalte – Methoden – Ziele', *Medizin, Gesellschaft und* Geschichte 9 (1991), 149–64.

lower quality than academic medicine and that its widespread disappearance from the medical market was due entirely to the improvement of medical provision. New questions moved to the centre of scientific inquiry, asking how academic physicians succeeded in rising in society and in gaining the sole power of definition in medical health care.[10]

There is a problem with using the term 'lay healers' since it has connotations that are informed by the views of physicians of that time. In order to define the profession in a historically valid way, the present contribution uses the term 'lay healer' as meaning providers on the medical market who were not in possession of the officially required diploma.

This contribution is part of the more recent medical history research into medical markets,[11] 'folk medicine'[12] and 'alternative' healing methods[13] as well as their agents; its central focus is the marginalization of unlicensed persons with healing knowledge and its main question is how one can trace – on the basis of the practice journals of a Swiss lay healer – the changes and consistencies in medical lay-practice against the background of growing medical academization.

Gottfried Wachter and His Practice Journal

As there are not many sources offering information on the practices of lay healers, the practice journal of the Swiss healer Gottfried Wachter (1776–1861)

10 Philip Rieder, *La figure du patient au XVIIIe siècle* (Genève 2010); Otto Ulbricht, 'Der Kampf des "Kurpfuschers". Friedrich Frantz Heinitz, 1786–1814', in Otto Ulbricht (ed.), *Mikrogeschichte. Menschen und Konflikte in der frühen Neuzeit* (Frankfurt am Main 2009), 298–336; Faltin, 2000; Christian Probst, *Fahrende Heiler und Heilmittelhändler. Medizin von Marktplatz und Landstraße* (Rosenheim 1992); Jütte, *Ärzte*, 1991; Reinhard Spree, 'Kurpfuscherei-Bekämpfung und ihre sozialen Funktionen während des 19. und zu Beginn des 20. Jahrhunderts', in Alfons Labisch and Reinhard Spree (eds), *Medizinische Deutungsmacht im sozialen Wandel des 19. und frühen 20. Jahrhunderts* (Bonn 1989), 103–21.

11 Cf. the individual contributions on the topic in Lina Gafner, Iris Ritzmann and Katharina Weikl (eds), *Penning Patient's Histories – Doctors' Records and the Medical Market in the 18th and 19th Century* [special issue: *Gesnerus* 69, no. 1 (2012)].

12 On the criticism of the term 'folk medicine' cf. Michael Stolberg, 'Probleme und Perspektiven einer Geschichte der "Volksmedizin"', in Thomas Schnalke and Claudia Wiesemann (eds), *Die Grenzen des Anderen. Medizingeschichte aus postmoderner Perspektive* (Köln 1998), 49–73; Eberhard Wolff, '"Volksmedizin". Abschied auf Raten. Vom definitorischen zum heuristischen Begriffsverständnis', in *Zeitschrift für Volkskunde* 94 (1998), 233–57.

13 Cf. Jütte, *Geschichte*, 1996.

from Thurgau is a rarity. Twelve volumes have been preserved which are kept in the Archives of Medical History at the University of Zurich. These volumes form the basis of a research project on rural healers that began at the University of Zurich in 2009.[14]

The records consist of a total of 643 pages, most of them closely filled with writing in ink. They cover the remarkably long period from 1803 to 1843. The first journal, from 1803, consists mostly of a list of foreign words. At this point in time we know as little about the purpose of this list as we know about its origin. The medical terms are mostly Latin designations of parts of the human skeleton, to which more foreign words and the German translation of each term are added. The remaining space is filled with the records of individual consultations written in Gottfried Wachter's hand. He also wrote the title of the journal and the date – 1803 – which presumably refers to the parts that constitute a practice journal. The remaining eleven journals contain records of his practice and notes on treatments, also written by Wachter. To two journals which originally only contained the records of his practice in 1806 and 1808 additional records were added later, in 1819 and 1818, probably as a means of saving paper: Wachter wrote into the spaces between individual entries and in the margins, adding the respective year to these new entries. Enclosed in the journals are, moreover, ten folded sheets of paper with lists of accounting details and prescriptions. The journals are set up like diaries, but bear no precise dates. Wachter entered the gender, place of residence and illnesses of his patients with some regularity and added the therapies. The approximately 2000 cases recorded document the everyday practice of the healer, the order of recording suggesting that in each period all treatments were entered.

Who was this lay healer? Gottfried Wachter was born in Hugelshofen in 1776. He was the son of Hans (Johann) Kaspar Wachter (1728–1800), a farmer who practised medicine on the side. The authorities in Thurgau described his father as 'a chimera of a quack of Hugelschoffen who cannot write or read but has the cheek to treat people'.[15] Gottfried's brother and his senior by twenty years, Hans Rudolf (1754–1805), was apparently also active as a 'doctor' until his premature death. He practised further to the west, in the village of Emmishofen, while Gottfried Wachter took over the practice of their father in Hugelshofen.

14 AfM MsH 17.1–14. We thank Katharina Weikl for the first transcriptions of the practice journals of 1805 (AfM MsH 17.2), 1808/1818 (AfM MsH 17.3) and 1806/1819 (AfM MsH 17.4).

15 Staatsarchiv Thurgau (StATG) 0'02'21, 26.09.1791. Quoted in Alfons Bieger, *Schröpfende Heiler – schwitzende Kranke. Das Thurgauer Medizinalwesen im 18. und frühen 19. Jahrhundert* (Frauenfeld 2004), 53.

Hans Rudolf as well as Gottfried gave the district physician cause to complain when, after an outbreak of dysentery, they 'continued to practise in spite of the interdiction they had received'.[16] Wachter's son-in-law, Hans Ulrich Keller (born 1804), who was also a lay healer, had to contend with similar difficulties.[17] Only Gottfried Wachter's son Jakob made the academic leap by going to medical school. He opened a practice in 1831 in Hessenreuti, a community situated around ten kilometres from Hugelshofen.[18] Gottfried Wachter himself practised in rural Thurgau, in and around Hugelshofen, from 1801 and probably until shortly before his death.

Because Gottfried Wachter was taught by his father, who was not an officially trained surgeon, he received no official certificate or diploma. Up until the early twentieth century it was customary for sons to learn the trades of their fathers and take on the family business, a tradition that also applied to health-related occupations. In the official lists one finds, however, apart from the physicians' dynasties, mostly the families of practically trained and certified surgeons and barber-surgeons.[19] Sabine Sander established that for surgeons in Württemberg the rates of self-recruitment in the eighteenth century ranged between 50 and 78 per cent.[20] The Prussian statistics on the social background of students at all Prussian universities and their choice of subjects shows for the late nineteenth century that, while only slightly over seven per cent of all medical students were sons of physicians, the proportion of sons whose fathers had also enrolled in medical schools was 55.5 per cent (these figures relate to the academic years 1887/1888–1891).[21] This shows that, certainly in late-nineteenth century Prussia, many physicians' sons tended to choose the occupation of their fathers. From among the medical practitioners investigated by the Research Network, only Friedrich von Bönninghausen (1820–1910) followed in his father's footsteps when he took over the latter's successful homeopathic practice in Münster,[22] but other physicians of that group had

16 Alfons Bieger, *Prüfung der Ärzte und Wundärzte im Thurgau (1798–1867)* (Dietikon 1988), 13.

17 StATG 0'02'21, 26.09.1791. Quoted in Bieger, 2004, 53.

18 Jakob Wachter passed the board of health's admission examination in a second attempt on 15 Feb. 1831, cf. StATG Sig. 4'870'2, Sanitätsrats Protokolle, vol. 7, fol. 155.

19 On the situation in Zurich in the eighteenth century cf. Sebastian Brändli, *"Die Retter der leidenden Menschheit". Sozialgeschichte der Chirurgen und Ärzte auf der Zürcher Landschaft (1700–1850)* (Zürich 1990).

20 Cf. Sander, 1989, 140.

21 Cf. Claudia Huerkamp, *Der Aufstieg der Ärzte im 19. Jahrhundert. Vom gelehrten Stand zum professionellen Experten. Das Beispiel Preussens* (Göttingen 1985), 77, Table 9.

22 Cf. Marion Baschin, 'Warum besucht man einen Homöopathen? Eine Praxis in Münster im 19. Jahrhundert', in Lina Gafner *et al.*, 2012, 126–40: 127.

other family members who practised medicine. Johann Friedrich Glaser (1707–1798), for instance, had two uncles who were officially appointed town or country physicians.[23]

As with many healer families[24] the practice of medicine was in Wachter's case also a tradition that stretched over several generations. In most cases the sons, after receiving their practical instruction, would be given a certificate of apprenticeship by their fathers that would allow them to apply as journeymen to other master healers. It was the lack of such a document that caused ongoing problems for Wachter.

Wachter's family tradition can be seen as a mirror of the changing health system. In the eighteenth century, Gottfried Wachter's father, Hans Kaspar, was still able to practise as a 'peasant doctor' without having any restrictions imposed on him by the health authorities. In introducing his son to his own field of work he followed the tradition of generations. By the time Gottfried Wachter took over from his father, the health authorities in the Thurgau region of Switzerland were beginning to implement regulations. Over the years that followed, the local health board would put increasing pressure on Gottfried Wachter, by pestering him with examinations, temporarily banning him from practising and putting restrictions on the range of treatments he was allowed to offer. While Gottfried Wachter resisted exclusion and continued to carry out his profession until his death, he made sure to send his son Jakob to medical school, thus concluding the metamorphosis of a family tradition from peasant doctor to academic physician that mirrored the general historical trend towards professionalization.[25]

Not Certified but Tolerated: Wachter and the Authorities

In the early nineteenth century the Thurgau medical system underwent fundamental modernization. The region's status as 'subject territory', which had

23 Cf. Ruth Schilling, Sabine Schlegelmilch and Susan Splinter, 'Stadtarzt oder Arzt in der Stadt? Drei Ärzte der Frühen Neuzeit und ihr Verständnis des städtischen Amtes', *Medizinhistorisches Journal* 46 (2011), 99–133: 121.

24 According to Danninger's regional historical study most women-practitioners received their knowledge from their mothers, occasionally also from their fathers. Cf. Danninger, 1998, 132–33. For the passing on of verses within families of modern conjurors cf. Jütte, *Geschichte*, 1996, 99.

25 For examples of surgeon families in Wurttemberg cf. Dominik Gross, *Die Aufhebung des Wundarztberufs – Ursachen, Begleitumstände und Auswirkungen am Beispiel des Königreichs Württemberg (1806–1918)* (Stuttgart 1999), 237–53.

applied up until 1798, had not supported a centralized health system, but rendered the formulation of separate medical regulations at a cantonal level redundant. Due to the lack of academic facilities any aspiring medical students went to universities elsewhere. The possibilities for learning surgery were also very limited because there were no guilds in the area that is now Thurgau.[26] This decentralized distribution of health authorities as well as the lack of clearly defined areas of competence in the territories without uniform administration facilitated the emergence of a wide range of self-taught medical practitioners who – like the Wachter family in tranquil Hugelshofen – were able to pursue their medical activities without much official interference or the need for professionalization.

Gottfried Wachter was unfortunate in that the reconstitution of the Thurgau region began just as he started to set up his practice as a healer.[27] As part of this political reorientation the health system was restructured when the region received the status of an independent canton in 1803.[28] The health commission that was appointed to develop a legal medical framework began to control many areas of health care.[29] As in other countries, the academic physicians succeeded, as part of this process, in claiming central positions in the medical administration of Thurgau and in gaining power over other groups of healers.[30] In parallel with this administrative reorganization, the medical landscape became increasingly academized, a development that, regionally, was strongly driven by the foundation, in 1833, of Zurich University with its medical school.

Even before the institution of a board of health, Wachter had to undergo his first medical examination on 30 July 1801 as part of a newly introduced obligation to obtain a licence. Wachter was the only candidate failed by the examining physicians who declared that he 'had presented the most obvious proof of

26 The only surgeons' guild whose certificates and diplomas were recognized was founded in Diessenhofen in 1735. On this association cf. Bieger, 2004, 171–210.

27 First under Napoleonic rule; later an independent state; from 1848 a canton within a federal state. On the political history of Thurgau in the early nineteenth century cf. Albert Schoop, *Geschichte des Kantons Thurgau, vol. 1: Chronologischer Teil* (Frauenfeld 1987), 37–80.

28 On the institution of the new cantons from 1803 cf. Daniel Frei, 'Mediation', in Hanno Helbling (ed.), *Handbuch der Schweizer Geschichte* (Zürich 1977), vol. 2, 841–69: 849–52.

29 Cf. Bieger, 2004, 54–62.

30 Generally cf. Stolberg, 'Heilkundige', 1998, 69–78; on Wurttemberg cf. Sabine Sander, 'Die Bürokratisierung des Gesundheitswesens. Zur Problematik der "Modernisierung"', in Werner Kümmel (ed.), *Jahrbuch des Instituts für Geschichte der Medizin der Robert Bosch Stiftung* 6 (1987), 185–218.

utter ignorance in medical and surgical matters'.[31] It was apparently not only the diploma that he was lacking: unlike other non-academic healers he did not possess the theoretical and practical knowledge tested in examinations. The fact that Wachter was not granted the desired licence did not, however, keep him from continuing his practice of healing.

The bureaucratization of the medical system moved to the next level when all healers active in the canton were registered on the basis of reports supplied by the eight newly appointed district physicians. This list of registered healers was submitted to the board of health in the summer of 1805.[32] Inclusion in this register implied a licence to practise, but required the necessary diplomas and certificates. If the board of health deemed that these documents were not sufficient, the respective healer had to demonstrate his competence before a board of examiners. While physicians and surgeons rarely found it difficult to secure a licence once they had submitted their qualifying certificates, practitioners without a recognized, formal training were faced with a growing number of obstacles. Many of the 'plaster-appliers', cuppers, bloodletters and other experts were not granted a licence to practise after undergoing thorough examination. In the end, the register included 74 persons divided into four groups.[33] It was the first time that all medical practitioners in the still young canton of Thurgau were systematically registered, their skills examined and evaluated on the basis of a uniform set of criteria.[34]

The first category included, in addition to the academic *doctores* and surgeons, the practically trained barber-surgeons ('operators'). The second category comprised the 'practitioners for less serious cases'; these were healers who were licensed to treat patients, but had to call in other practitioners in more serious cases. The third category was that of 'bloodletters' and 'cuppers'. The fourth and lowest category included only a few 'plasterers', some of whom were only licensed to apply one single kind of plaster against one specific affliction.[35]

Although Gottfried Wachter had, years before, failed the examination when he first attempted to gain a licence to practise, he was now admitted as

31 Quoted in Bieger, 1988, 22.

32 Cf. *ibid.*, 1988, 28–30.

33 Cf. *ibid.*, 1988, 40, for this list.

34 Cf. *ibid.*, 2004, 57–62.

35 Dominik Gross presents an analogous categorization of surgeons into four classes of competence for Württemberg; Dominik Gross, 'Zur Rekonstruktion der Wundarztdichte im Königreich Württemberg. Ein Beitrag zur Sozialgeschichte der Handwerkschirurgen im 19. Jahrhundert', *Medizinhistorisches Journal* 32, no. 1 (1997), 57–81: 65–67.

a 'practitioner for less serious cases'. Of another practitioner of the same category one reads in this context that he had to keep an 'orderly diary'[36] which the officially appointed district physician was entitled to inspect at any time. It can therefore be assumed that Wachter also had to keep records of his patients, their illnesses and the therapies he applied. This method of control, imposed on him from above, seems to have been the main reason why he kept a practice record and it might explain the unusually frequent allusions to successful therapies in the earlier volumes.[37]

In 1805 the new board of health in Thurgau introduced, as mentioned earlier, an examination for all persons offering medical services. The corresponding documents have been meticulously studied by Alfons Bieger.[38] They form the basis for Fig. 10.1, which shows the distribution of healers according to category.[39] The map shows that in Wachter's area there were numerous practicioners and, among them, several other healers with restricted licences (cf. Fig. 10.1).

Based on the distribution of 'practitioners for less serious cases', Bieger suggested that the board of health only tolerated these healers because they 'filled the gaps or margins in the network of communities offering health care provision within a one-hour radius'.[40] The distribution of healers around Hugelshofen

36 Cf. Bieger, 1988, 39–40.

37 In the 1805 journal, approx. 149 successful treatments were recorded and 17 deaths. Cf. Katharina Weikl and Iris Ritzmann, 'Ein Auslaufmodell. Spuren einer ländlichen Heilerpraxis im Zeitalter der Akademisierung', in Lina Gafner *et al.*, 2012, 76–94: 86.

38 For the medical situation in the Canton of Thurgau the authors consulted mainly the following works: the two-volume diary of the physician Elias Haffter that mentions the healer Gottfried Wachter (D Med. Elias Haffter, Bezirksarzt und Sängervater, 1803–1861. Tagebuch 1844–1853, ed. by Prof. Dr med. Carl Haffter and Hermann Lei senior, vol.1: 1844–1848, vol. 2: 1849–1853 (Frauenfeld 1985) and, in addition, Alice Denzler, *Medizingeschichtliches aus der 1. Hälfte des 19. Jahrhunderts. Aus den Berichten der thurgauischen Bezirksärzte 1839–1862* (Basel 1938)), the two fundamental works on the Thurgau medical system by Alfons Bieger (Bieger, 2004; *idem*, 1988), Markus Oettli's dissertation (Markus Oettli, 'Das Amt des Bezirksarztes im Rahmen der Thurgauischen Sanitätsorganisation 1803–1869' (Med. diss., University of Zürich, 1982)) and the publication of Johann Adam Pupikofer (Johann Adam Pupikofer, *Der Kanton Thurgau historisch, geographisch, statistisch geschildert* (St. Gallen 1837)) which partly counts as a source.

39 The chart is based on information by Bieger, 2004, 58, cf. Fig. 10.1. As a result of new insights Bieger, 2004 changed his categorization of healers slightly compared to that contained in his 1988 dissertation, cf. Bieger, 1988, 39–40.

40 Bieger, 2004, 119. For Baden cf. Francisca Loetz, *Vom Kranken zum Patienten. "Medikalisierung" und medizinische Vergesellschaftung am Beispiel Badens 1750–1850* (Stuttgart 1993), 199–201.

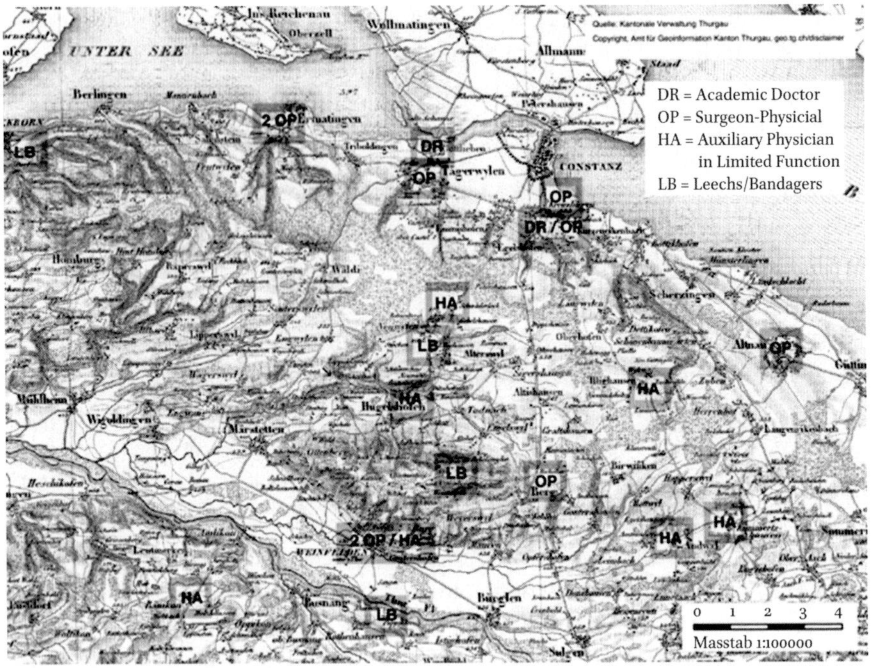

FIGURE 10.1 *Distribution of healers around 1805, based on the official classification in four categories.*

corroborates this thesis. Only the bigger communities and towns had more than one healer, while across the smaller villages and hamlets the web of medical provision proved to be rather loose.

The Range of Wachter's Activities

What kind of services did Gottfried Wachter offer? How did the treatments administered by this – uneducated and untrained – 'quack'[41] compare with those of licensed physicians or surgeons to account for their perceiving him as a threat? These questions are best approached by looking at a particular case history. In the third journal, which contains several patient histories from the years 1808 and 1819, there is an entry on page 8 concerning a man from Hugelshofen. The man had accidentally hit his index finger with a hammer and contacted the healer who lived in the same village. After inspecting the injured finger, Wachter described the wound as follows: 'flesh and tendons

41 On the history of this term cf. Jütte, *Geschichte*, 1996, 18–23.

shattered, severe inflammation, swollen'.[42] As regards medication, Wachter recorded that *Unguentum lithargyrum* was to be applied every two hours. *Lithargyrum* or litharge is a by-product in the smelting of certain mineral ores (*Fahlerze*) from which, under addition of lead, the silver or copper content is isolated.[43] Medical practitioners appreciated such lead preparations primarily for their cooling, dehydrating and mildly irritating effect on the skin tissue.[44] Prepared as salves or plasters they remained a popular topical medicine right into the nineteenth century.[45] In using this lead-based ointment, Wachter probably took the right decision since the entry in question concludes with the word 'cured'.

The treatment of this male patient served precisely the purpose for which the Thurgau board of health had, in 1805, granted Wachter limited permission to practise. It was at least an hour's walk from Hugelshofen to the next physician and, this being an occupational accident, it counted as an emergency that needed urgent attention. This kind of injury would, moreover, have constituted a 'less serious case' which Wachter was able to treat adequately and without difficulties, using the remedies he kept in his medicine chest.

Treating a bruised finger was a procedure 'where healing is mainly induced by manual intervention',[46] as one author wrote in a treatise of 1785, and therefore, strictly speaking, the task of a surgeon. While there are by now regional case studies that illustrate in what kind of situations surgeons encroached on the province of physicians or apothecaries, ignoring all kinds of medical rules and regulations in the process,[47] there is as yet little empirical documentation on the extent to which lay healers without recognized training infringed on areas officially reserved to surgeons.

---

42 *'Fleisch und Senen zerschlagen be komt starken Brand ge schwulln'.* AfM MsH 17.3, fol. 8.

43 Cf. Johann Hermann Pfingsten, *Deutsches Dispensatorium oder allgemeines Apothekerbuch nach den neuesten und besten lateinischen Dispensatorien und Pharmacopoen zum gemeinnützlichen Gebrauch nach alphabetischer Ordnung eingerichtet* (Frankfurt/Leipzig 1795), 191–92; Ludwig August Kraus (ed.), *Preussische Pharmakopöe. Nach der neuesten Ausgabe aus dem Lateinischen übertragen und mit vielen Zusätzen versehen* (Braunschweig 1805), 142.

44 Cf. Ludwig August Kraus, *Dr Justus Arnemann's chirurgische Arzneimittellehre* (Göttingen 1818), 226.

45 Cf. Ulrike Zeber, *Die Geschichte des Pflasters. Von der traditionellen Arzneiform Pflaster zum Heftpflaster* (Stuttgart 2001), 113–15.

46 Cf. Thomas Kirkland, *Untersuchung über den gegenwärtigen Zustand der Medizinal-Chirurgie* (Leipzig 1785), 13.

47 On Württemberg in the eighteenth and nineteenth century cf. Sander, 1989; Gross, 1999.

The fact that Wachter's medical activities were restricted to 'less serious or less critical cases'[48] suggests that he offered mainly services pertaining to the lower category of surgery. If that was the case he would, however, have been more frequently in conflict with the surgeons than with academic physicians. A number of Wachter's journal entries reveal that he observed the official regulation to call in a physician in more serious cases. When a woman from Engwilen, who had consulted Wachter repeatedly, suddenly became delirious (*Entzükung*) and suffered convulsions (*Gichter*), he correctly referred her to the superior physician, in this case a Dr Johann Melchior Äppli (1744–1813) in Gottlieben.[49] The question is whether cases such as the treatment of a bruised limb constituted routine situations or exceptions in Wachter's practice.

At this stage of our research it is not yet possible to perform a systematic search the database for particular complaints or reasons for consultation. But it is possible to establish how often Wachter prescribed an internal or external treatment when he was consulted, whether he used both forms of application simultaneously or whether he even carried out minor surgery. Such investigation, based on the practice records of 1808 and 1835, provides first insights into this lay healer's actual sphere of activity. Both records contain approximately the same number of patients – 92 for 1808 and 82 for 1835 – and are therefore suitable for quantitative comparison (cf. Fig. 10.2). The number of individual consultations was, however, considerably lower in the first than the second of the years investigated (155 individual visits with 188 therapeutic interventions in 1808 as compared to 246 visits and 362 interventions in 1835). These differences are due to the more exact and detailed journal-keeping of the older Wachter, with more extensive (cross-)references, that resulted in the greater complexity of the later journals. The figures presented below were established on the basis of individual consultations.

This shows that, at around seventy per cent, the dispensation of internal medicines made up the majority of services delivered by Wachter in 1808. This means that, as a healer, he appears to have mostly prescribed internal medication, a service that falls within the ambit of licensed physicians or apothecaries. But the high percentage of internal medicines can be explained by Wachter's habit of administering laxatives in the form of pills or teas before starting the treatment proper.[50] In the early years of his practice, Wachter used

48 Cf. Bieger, 1988, 39.

49 AfM MsH 17.4, fol. 19.

50 In 1808 Wachter noted for instance: '*Monshaus ein Man ist auf seiten gefallen leg Pfst* [= plaster, author's note] *auf Nerv Geist zum befeuchten gab ihm ein Laxant Specii gesund*'. AfM MsH 17.3, fol. 14.

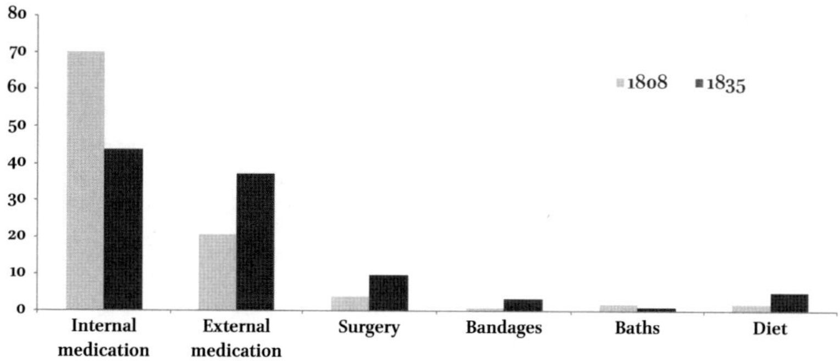

FIGURE 10.2 *Percentage of different kinds of therapeutic interventions. Wachter performed in two selected years (n = 188 (1808) and 362 (1835)).*
ARCHIV FÜR MEDIZINGESCHICHTE, UNIVERSITÄT ZÜRICH, MS H 17.3 AND MS 17.13

his home-made 'herbal wine' as a kind of panacea for congestions of diverse body juices or misdirected fluxes. In the journal of 1806 we read, for instance: 'Engwillen a man full of flux = gave a laxative herbal wine, recovered' (*Engwillen ein man fol fluß = Gab ein Lagxant Crt win geßund*).[51] The conclusions regarding internal medicines are therefore not as reliable as they appear to be at first glance.[52]

Wachter's dispensation of external medicines is more easily evaluated: almost a fifth of all services rendered involved the application of ointments or cataplasms, plasters, poultices or compresses and diverse alcohol- or oil-based inunctions. At 3.7 per cent the percentage of minor surgical interventions, such as venesection or the setting of limbs, was very low. In his early practice years Wachter hardly ever recommended foot baths or special diets for his patients. There are only a few such references in the journals, as for instance in 1835, when he wrote 'let her eat buttermilk' for a patient who suffered from the aftereffects of a 'bilious fever'.[53]

By 1835 the relationship between the various treatments had shifted significantly. Internal medication was no longer predominant because the administration of topical medicines, which had risen to 36.7 per cent, was almost as high as that of pills, powders, mixtures and juices. The proportion of 'surgical interventions' had also risen compared to 1808 and now constituted a tenth of

51 AfM MsH 17.4, fol. 33.

52 Changes in medication in the course of the nineteenth century were analysed by John H. Warner, *The Therapeutic Perspective. Medical Practice, Knowledge and Identity in America 1820–1885* (Cambridge 1986).

53 AfM MsH 17.13, fol. 2.

all medical treatments. The application of bandages rose in parallel with the number of treatments of dislocated limbs. Baths or diets were not part of Wachter's main repertoire. The diagram reveals that the medical treatments offered by Wachter were progressively more those of a 'lower category surgeon', because they were increasingly restricted to manipulations on the surface of the skin and ever less frequently aimed at processes inside the body.

Minor Surgical Interventions

This development is also apparent from the frequency with which Wachter carried out minor surgical interventions in the two years chosen for investigation. In addition, the corresponding search provides information on the medical equipment and instruments owned by this lay healer. Sabine Sander showed in her study that eighteenth-century surgeons in remote areas of Württemberg possessed more 'professional tools' than those close to flourishing centres;[54] for there, the greater competition among healers as well as the generally higher doctor-to-patient ratios resulted in the restriction of surgical activities to clearly delineated areas that would very rarely bring a surgeon of the second or third category into contact with ear, nose or throat disorders. Vessels for the storage of remedies were also more common, with surgeons at the periphery because pharmacies were rare and usually situated at a great distance from the dwelling places of the patients. Due to the proximity of the higher health authorities and stricter inspection regimes it was very difficult for practitioners to become active in areas bordering on those of physicians, pharmacists or surgeons. Wachter's presumably modest possession of medical tools suggests that he was assigned clearly defined areas of competence which he could not disregard (cf. Fig. 10.3).

This chart shows that more minor surgeries were conducted in 1835, when Wachter had been active in Hugelshofen for close to thirty years. In 1808, only nine of the 155 (or 5.5 per cent of) consultations recorded refer to such measures. By 1835 this proportion had almost tripled to 13.4 per cent (Wachter carried out some type of minor surgery in 38 of 246 consultations in that year). In the early years of his practice Wachter apparently very rarely applied measures that involved injuring the skin of his patients.

With rates of approximately fifty-five per cent in 1808 as well as 1835, blood extractions make up about half of all recorded 'surgical interventions'. Opening a vein with a lancet was one of Wachter's preferred techniques, even though he

54 Cf. Sander, 1989, 73–80.

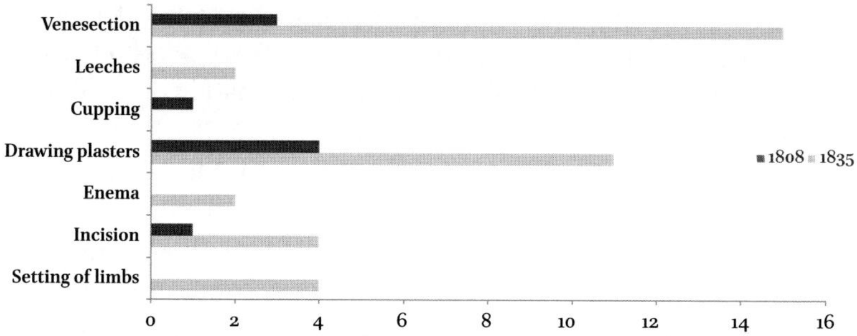

FIGURE 10.3 *References to minor surgery in Wachter's journals of 1808 and 1835.* ARCHIV FÜR MEDIZINGESCHICHTE, UNIVERSITÄT ZÜRICH, MSH 17.3 AND 17.13

also applied cupping glasses or leeches in some cases. When, in 1835, a Mr Auer could not tolerate the 'eye waters' given to him by other physicians for an 'inflammation in his left eye and pain',[55] the Hugelshofen healer used leeches (*blut Eglen*) in addition to his mandatory preparatory laxatives. The second most frequent measure used by Wachter were blistering plasters (*vesicatores*) which aimed to irritate the skin so that bad fluids could exit. The fact that he applied enemas, if only in two cases, suggests that he must have been in possession of the necessary equipment. Repeated mention of the setting of limbs indicates that he most probably owned a selection of splints and bandages. His choice of words in his records can be confusing, however: the difference between a bandage and a warm or wet compress is often not clear which means that precise percentages cannot be established.

Cases of interventions requiring incisions or piercing the skin tissue appear to have been very rare for Wachter. In 1808 he used a knife or lancet on one occasion only,[56] in 1835 four times (but the overall number of consultations was higher in that year). In two cases Wachter merely opened the blisters drawn out by an irritant plaster to allow the evil humours to be discharged. On another occasion he used a knife to remove the scab on a severe head injury (*Schneit über wunden den schlanz weg*[57]).

55 *'wegen Sein Linken Aug die Entzündung mit schmerzen'*. AfM MsH 17.13, fol. 36.

56 A boy had developed a 'flux on the left side of his buttocks towards the inside, against his genitals' (*ein Fluß am Linken gesäß ihn werts gegen gemacht*), which Wachter cut open and dressed with a camphor plaster.

57 AfM MsH 17.13, fol. 59.

The treatment of a young woman who presented with an open and 'flowing ulcer' on her lower leg[58] proved more difficult. She consulted Wachter when the flow ceased, causing congestion that resulted in the formation of an 'ulcer... with severe inflammation'[59] below the hollow of the knee. Wachter's initial attempt to treat the induration with salves, a cataplasm and little bags filled with softening herbs remained unsuccessful and the application of more salves and plasters of diverse composition did not get rid of the swelling either. The 'flux' that had by then accumulated 'for almost 3 weeks'[60] began to wander around the body, manifesting also in the face and mouth. When the young woman experienced difficulties swallowing due to the swellings in her throat and all other attempts by Wachter to dissolve the ulcer below the knee with plasters, salves and the use of silver nitrate (*Lapis infernalis*) failed, he saw no alternative to surgical intervention. 'So I opened the flux under the hollow of the knee with the lancet & and a brick-red substance came out'[61] – he noted in his journal.

The chart also reveals the modesty of Wachter's 'professional tools'. He apparently possessed only a few cupping glasses, an enema and a lancet. In the two years investigated he used the lancet only for very minor skin incisions or for the opening of blisters. References to more expensive or mechanically more demanding equipment such as saws, forceps or instruments for tooth-pulling are missing altogether from his records. Wachter presumably kept in larger numbers only tools for the preparation and storage of plasters, salves, oils, spirits or compresses and other pharmaceutical accessories.

If one looks at the number of minor surgical interventions Wachter carried out and the instruments he owned as indicators of the services he provided, one can conclude that, even in areas with a lower surgeon- or physician-to-patient ratio, the barber-surgeons and surgically trained practitioners in Thurgau did not lose much of the 'classical' surgical work – such as complicated fractures, more serious injuries like stabbings, shot-wounds and bruises from brawls or the treatment of all kinds of malignant ulcers and growths – to unlicensed healer groups. In her comparison of eighteenth-century surgeons in administrative towns on the periphery and close to the commercial centres of Württemberg,

58 On the idea of harmful fluxes in the early eighteenth century cf. Barbara Duden, *The Woman Beneath the Skin. A Doctor's Patients in Eighteenth-Century Germany* (Cambridge 1991), 104–78.

59 '*Eine geschwulst...mit starker Entzundung*'. AfM MsH 17.13, fol. 38.

60 '*gegen 3 Wochen zusamen samlet*'. AfM MsH 17.13, fol. 45.

61 '*So hab ihr den fluß Unter der Knie beügen mit der Lancet geöfnet & komt Ziegelrothe Materie*'. AfM MsH 17.13, fol. 52.

Sabine Sander showed that the surgeons on the periphery could – and probably had to – offer a much wider range of services, because of the small number of physicians available and the less frequent official inspections.[62] But the patients in Wachter's catchment area could not go to him for the amputation of a leg, any more than they could ask him for an abdominal operation. But if they were looking for a practitioner to allay inflammation by applying a plaster or a salve, Wachter was the right person to turn to.

Both charts suggest that the way Wachter's practice developed was an expression of the growing monopolization and control of the medical market – as far as internal medicine was concerned – by the physicians in the area. The first physicians' association (the *Wertbühlia*) was, after all, founded in 1833 as a society of physicians from the northern and central parts of Thurgau.[63] The growing influence of physicians on medical training is apparent from various petitions. One petition by the medical-surgical society from 1832 demanded a stricter examination regime for young physicians, because there was concern about their deficient knowledge of Latin, German and mathematics.[64] As is also apparent from the activities of the Thurgau physicians, medical associations and the foundation of such organizations were often expression of growing professional aspirations and professional policies that aimed at joint agitation. It must have been increasingly difficult for Wachter to move between the boundaries of internal and external medicine without arousing the displeasure of the medical authorities. Compared to the time when the cantonal health system was introduced, in the first decade of the new century, the areas of competence of the various healer groups were much more strictly monitored thirty years later. A look at the kinds of therapies Wachter used in 1808 and 1835 suggests, however, that he encroached, at least in 1835, less on the domain of the physicians than on that of the (lower) surgeons and 'operators'.

Wachter's Clientele

Which references to Wachter's clientele in his records are suitable for analysis? The journals offer information on the patients' places of residence, their gender, symptoms and the treatments they received. Earlier volumes also contain frequent references to the success of treatments, such as 'is healthy', 'cured' or

62 Cf. Sander, 1989, 73–80.

63 In 1840 the physicians around Lake Constance followed suit and founded their own association. 'Münsterlingia'. Cf. Bieger, 1988, 47.

64 Cf. *idem*, 1988, 46.

'died', while the later ones mention names and occupations such as cobbler, chimney-sweep, school teacher, joiner or farm labourer, and, in a very few rare cases, office-holders or clergymen. For the most part the clientele consisted of families and persons without occupational designation who – as the village names suggest – had a peasant background.

Some entries show that Wachter tended to call on and treat his patients in their own homes: the journal of 1805 mentions a woman in Märstetten who suffered from intermittent fever. She was treated by Wachter with his mandatory herbal wine and laxatives followed by heating and cooling mixtures. The next entry begins with the words: 'In the same house a fellow with a nosebleed that almost killed him'.[65] The man, a bachelor, had tried diverse remedies. He consulted the lay healer because he was in the house anyway and Wachter bled him from the foot. Wachter's one-hour walk to Märstetten was profitable for all parties involved: he received payment for two treatments that resulted in the recovery of both patients.

Did Wachter's catchment area change over the decades? The two moments in time – when the health authorities were instituted (1805) and when the new medical legislation came into force in the mid-1830s – can be juxtaposed with Wachter's practice journals of 1805 and 1835 (cf. Fig. 10.4).

Most patients who found their way into Wachter's 1805 journal came from within a radius of five kilometres; less than 25 per cent lived further away from Hugelshofen. The range is similar for 1835, but the percentage of patients who lived more than five kilometres away had risen to thirty per cent. By then Wachter also had a greater number of patients who lived more than a day's journey away and the question arises whether he made use of public transport, such as the mail coach.

At this stage of our research we cannot establish whether any of the journal entries referred to remote consultations. The proportion of contacts via third persons seems to have been limited in Wachter's practice compared to those of the academic physicians investigated by the Research Network. In 1806 we find a reference to a messenger who collected a remedy from Wachter on behalf of a certain *Herr Kronen wurt* from Bohlingen in Baden (across the German border).[66] The messenger took the opportunity to have treatment for 'flux in the index finger of the right hand'.[67]

How many treatments Wachter carried out per period can, at this stage, only be estimated on the basis of a number of clues. Journal 11, which covers the

65 *'Am gleichen hauß ein Kärlj blut sich noch zu tode aus der Naßen'*, AfM MsH 17.2, fol. 36.

66 AfM MsH 17.4, fol. 55.

67 *'Fluß an der rechtn hand am zäg finger'*. AfM MsH 17.4, fol. 55.

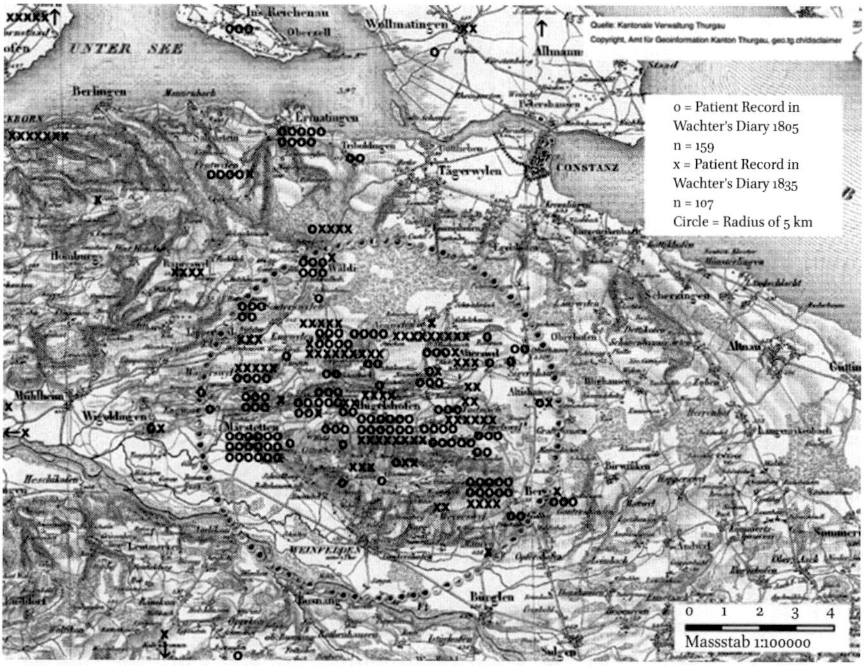

FIGURE 10.4 *Places of residence of Wachter's patients according to the practice journals of 1805 and 1835.*

thirty-five-day period from 27 November 1831 until the end of that year, contains fifty-two entries, each referring to two to three visits. The symptoms noted down by Wachter for that period of time do not suggest the presence of an acute epidemic, which would have driven up the number of consultations. They rather correspond with those in other practice journals. A clue can be found also in the 1835 journal where Wachter recorded on page 52 that a female patient was transported back home on 17 September. The journal started on 15 July. During those two months Wachter recorded more than 180 consultations.

What do these numbers imply? We can compare them with the average number of consultations Michael Stolberg established for academic rural physicians in nearby Bavaria for the 1840s. The numbers were lower in the early journals than in the later ones, a fact that is in line with the information given by the academic physicians, even though Wachter also worked as a publican at that time. The frequency of consultations as reflected by Wachter's later journals was high compared to that of academic physicians.[68]

68 Michael Stolberg, 'Patientenschaft und Krankheitsspektrum in ländlichen Arztpraxen des 19. Jahrhunderts', *Medizinhistorisches Journal* 28 (1993), 3–27: 12, also Loetz, 1993, 200.

The allocation of consultations to individual patients, which becomes possible in the later years, underlines the static localization of Wachter's practice. Several regular patients can be clearly identified for the later years. In his 1835 journal Wachter entered at least 247 individual consultations with only eighty-two patients, including four persons who used Wachter's assistance between twelve and seventeen times. One woman even saw him twenty times: the case history of this widow (Frau Sigerist) consists of four longer entries with the corresponding cross-references.[69] This regular client lived in the little village of Dattenhub, not far from Hugelshofen. In his first entry on her Wachter described her afflictions as bilious fever, breast inflammation and stitches in the side; other descriptions followed, such as 'dropsy with tight chest'. He called on her more than two dozen times in the period covered by the practice journal of 1835 and treated her with a variety of remedies.[70] Wachter obviously found it difficult to identify the condition and to judge its severity. His last entry concluded the treatment history with the words: 'After losing more and more weight and becoming increasingly confused she passed into blessed eternity'.[71]

Wachter practised until shortly before his death in 1861 despite the many admonitions, restrictions and even interdictions aimed at him by the ever more powerful Thurgau health authorities. Comparison within the range of treatments he offered shows a gradual narrowing down of services. While the earlier journals frequently mention venereal diseases, bad fluxes and cancerous afflictions, those of later years dwell more on minor surgical procedures after accidents and palliative care. Like the analysis of treatments, the evaluation of the conditions treated confirms that Wachter's activity at the beginning of his practice coincided more with that of academic physicians, while in later years it came closer to that of the surgeons and barber-surgeons.

There are signs to suggest that Wachter specialized in conditions associated with sexual organs, sexuality and reproduction. A great number of women sought assistance with menstrual problems, during pregnancy or in childbed. He was not authorized, however, to take on obstetric tasks since the official regulations specified that only midwives and specially educated obstetric doctors were allowed to assist at births.[72] In the journals examined so far there is

---

69 AfM MsH 17.13, fol. 6 (2 entries); fol. 31; fol. 46.

70 AfM MsH 17.13, fol. 6.

71 '*Da sie Nach Und Nach abnahm mit verwirrung So Starb sie in die Seligen Ewigkeit*'. AfM MsH 17.13, fol. 46.

72 Franziska Jenny, *"Sie habe 3024 Kinder gehoben". Das Thurgauische Hebammenwesen 1799–1869* (Dietikon 2005).

only a single reference to such assistance. 'A woman comes for childbirth',[73] the young Wachter wrote in 1805. The birth appeared to take a 'happy' course, 'but suddenly she swooned and went into convulsions and the after-birth stayed behind' (*überfalt sie ein Ohnmacht und gichten und bleibt die Nachgeburth hinden*). Wachter treated the woman with a mixture of cinnamon water, spirit and camphor which she had to take by the spoonful until the placenta was delivered. The woman was then given a cooling mixture, barley coffee and a light brandy. For the bilious belching that occurred later Wachter prescribed 'a gentle emetic so she vomited'. The treatment concluded with cardiotonic drops and the comment: 'and now she is well'.

Because the records have not been fully preserved it is not possible to establish if this record was entirely exceptional. Wachter mentioned miscarriages occasionally in his earlier journals, but these indications must not be interpreted as referring to definite abortions. Wachter treated women for white discharge or the absence of the 'monthly cleansing', for instance in a spinster who had 'a cold flux and whose period was not normal'.[74] In her case Wachter used his laxative herbal wine as well as 'natural drops' and bloodletting. According to his records she was cured. Wachter's male patients occasionally suffered from venereal diseases or had difficulty passing water. Although it is not possible at this stage to present a quantitative evaluation, these consultations seem to indicate that he specialized in the 'secret illnesses'. This conclusion is corroborated by the learned physicians who considered that patients preferred to call on their non-academic rivals when they suffered from conditions they felt shameful about.[75]

Unlike the range of conditions treated, the actual therapies offered by Wachter hardly changed across the years. From the first to the last preserved practice journal he mainly used remedies that he prepared himself – mostly laxative teas, pills and his herbal wine – as well as a variety of drops, pills and powders as tonics for heart, stomach or the general constitution and numerous salves, compresses and plasters. Some of his special applications are reminiscent of the practices found in 'popular medicine', a system that is no longer strictly separated from academic medicine.[76] For burns and ulcers Wachter

73 For this and the following quotations: AfM MsH 17.2, fol. 6.

74 '*Kalte Fluß bey ihr ihr reinigung nicht recht*'. AfM MsH 17.4, fol. 5.

75 Cf. Stolberg, 1993, 25.

76 Cf. Eberhard Wolff, '"Volksmedizin". Abschied auf Raten. Vom definitorischen zum heuristischen Begriffsverständnis', in *Zeitschrift für Volkskunde* 94 (1998), 233–57; Stolberg, 'Probleme', 1998.

occasionally prescribed frogspawn plasters[77] and he was familiar with hay-flower compresses.[78] In the houses of his patients he occasionally availed himself of objects of everyday use such as a flour bag to weigh down a swelling.[79] Foodstuffs such as butter or milk he would use as dietetic measures or topical remedies. But his repertoire also included the medicines from the official pharmacopoeia, among them compounds such as *Hoffmann's drops* or *Laudanum Sydenhami* as well as simple drugs such as *Digitalis purpurea* or *Opium*, which he presumably obtained from a pharmacy. It is in this area that the services offered by lay practitioners overlapped with those provided by academic physicians.

Comparison across the practice journals reveals changes in the individual therapies: Wachter used *Liquor anodinus mineralis Hoffmanni* and *Elixirium Acidum Halleri* much more frequently in later years while his use of the popular *Sydenham's Laudanum* decreased. He also increasingly abandoned coffee although it used to be one of his preferred remedies earlier in his career. These kinds of changes may reflect an on-going learning process. It is unlikely that Wachter had any official vocational training. He was probably self-taught or learned his of the local population, in choosing a practitioner, was perhaps that he could be reached within a radius of five kilometres. Many of Wachter's clients were probably culturally close to or personally connected with him, but he also had a few patients of higher rank. His frequent references to previous treatments by other practitioners or to self-medication suggest that there were patients who came to him as a last resort. Analysis of the therapies he applied reveals considerable developments in his practice over the years. Wachter's activities overlapped increasingly with those traditionally assigned to the artisanal surgeons. A certain focus on 'illnesses' relating to sexuality, sexual organs and reproduction is also discernible.

Gottfried Wachter was no longer alive when, in 1867, the Thurgau administration, in collaboration with the university cantons of Bern and Zurich and six other German-Swiss cantons, introduced a concordat which combined the mutual recognition of medical qualifications with the corresponding licence to practise and excluded practitioners who did not meet the relevant requirements.

77 AfM MsH 17.13, fol. 12, 17, 55.

78 AfM MsH 17.3, fol. 21.

79 AfM MsH 17.2, fol. 36.

CHAPTER 11

Administrative and Epistemic Aspects of Medical Practice: Caesar Adolf Bloesch (1804–1863)

Lina Gafner

The journal of the physician Caesar Adolf Bloesch, who practised in the Swiss town of Biel, is impressive in its scope and continuity. The fifty-five volumes – all written in the same hand, lined up neatly in a row, labelled and numbered – were composed over a period of thirty years, between 1832 and 1863. On the spines we read: 'Diary of medical observations', and on the covers: 'Doctor's visits and prescriptions'.[1] The former suggests an epistemic, the latter rather an administrative function. On opening the books we find vertical and horizontal lines drawn by pencil on each page. They served to guide the hand and eye of the writer and transformed the book into a register that did not need to be reinvented every day but just filled in, making writing and reading more economical.[2]

Bloesch had started his record as a medical student in Göttingen and continued to add entries 'up to the final days':[3] under the last entry in volume 57 somebody has written Bloesch's date of death in pencil. But let us first return to Göttingen where the young student Bloesch began his record in the hospitals he worked in by taking notes 'about each patient entrusted to me'.[4] From Göttingen he travelled to Berlin and Paris to visit the hospitals there. He continued to log his observations in his record and these stood often, as he later noticed, 'in curious contrast to the reports I found about the same patients in the *Journal des progrès des sciences et institutions médicales*'.[5] Disappointed that physicians apparently adjusted their observations on occasion so as to

1 Stadtarchiv Biel, 4 A 1, Nachlass C. A. Bloesch, Tagebuch ärztlicher Besuche und Verordnungen, 1832–1863, 55 vols.

2 On the genealogy of the register cf. Hagen Keller, 'Vom "heiligen Buch" zur "Buchführung". Lebensfunktionen der Schrift im Mittelalter', *Frühmittelalterliche Studien* 26 (1992), 1–31; Cornelia Vismann, *Files. Law and Media Technology* (Stanford 2008).

3 Caesar A. Bloesch, 'Allgemeine Grundsätze der theoretischen und der praktischen Medizin, ausgezogen aus meinem Tagebuch ärztlicher Besuche und Konsultationen', *Mitteilungen der medizinisch-chirurgischen Gesellschaft des Kantons Bern* 1 (1866), 1–150: 28.

4 Bloesch, 1866, 10.

5 The two volumes of the Journal that were written during Bloesch's training have unfortunately not been preserved.

 | DOI 10.1163/9789004303324_013

shine with their publications, he recorded: 'There I learned how case histories are made and used. I was not impressed with the Paris hospitals'.[6]

Bloesch's teacher and 'benefactor' in Göttingen, the famous anatomist Johann Friedrich Blumenbach, offered him a post as private physician at the Brazilian court, but Bloesch declined the offer – reputedly out of consideration for his mother – settling instead, in 1827, as medical practitioner in his hometown of Biel.[7] In the same year he married Elisa Pugnet, the daughter of the renowned Napoleonic military surgeon Jean-François-Xavier Pugnet. After Pugnet's retirement from the army he moved to Biel with his family and went on to make a name for himself as a committed physician for the poor. He would be an important influence on his son-in-law's professional life.[8] In his first year Bloesch found little work in Biel and devoted himself to the study of botany and Latin and to his dissertation (*Statistik und Topographie von Biel und der Umgegend*). Already during his sojourn in Göttingen he had ordered literature on the subject to be sent to him from Switzerland. Now that he was in Biel, research for his studies led him to the town archives where he came upon 'great disorder' and spent the next twelve years 'organizing and documenting everything'.[9] His exploration of the plentiful material available in the archives resulted in a 900-page chronicle of Biel which to this day links Bloesch's name closely with the history of the town.[10] It was this archival work, as he himself stated, that paved his way into the town administration: the school board, the poor relief office, the legislative – and eventually in 1832 – the executive authorities of Biel.[11]

Bloesch came from a long-established family in the town. His father had died as a young physician during a typhus epidemic carried to Biel by Imperial troops in 1814. The town put his mother, a mayor's daughter and now a widow with four young children, in charge of the local boarding school. Caesar Adolf was the eldest child and in choosing to become a doctor he followed in his father's footsteps. His brother Eduard embarked on a political career in the

6 Bloesch, 1866, 11.

7 Cf. *ibid.*, 8–9.

8 Bloesch translated Pugnet's '*Beobachtungen und Erfahrungen aus dem Gebiete der praktischen Heilkunst*' (Observations and experiences from the field of practical medicine) from French to German and published it in two volumes in 1837.

9 Bloesch, 1866, 13.

10 Cäsar A. Bloesch, *Geschichte der Stadt Biel und ihres Panner-Gebietes*, 3 vols (Biel 1855–1856).

11 Cf. Werner Bourquin and Marcus Bourquin, *Biel. Stadtgeschichtliches Lexikon von der Römerzeit (Petinesca) bis Ende der 1930er Jahre, historisch, biographisch, topographisch, mit Ergänzungen für den Zeitraum bis 1999* (Biel 1999), 77–78.

newly constituted Switzerland, and Friedrich, the other brother, went on to be a successful industrialist. Little is known, rather typically, of their sister Louise.[12]

After a slow start, Bloesch succeeded in building up a flourishing medical practice in Biel. The smallpox vaccination for children was an important factor in this development as it induced the families of the vaccinated children to choose him as their family doctor.[13] Bloesch treated the workers from the cotton mill as well as, due to his own social rank, members of the Biel bourgeoisie.[14] Patient numbers grew steadily (1834: 515, 1842: 755, 1852: 800), but the average number of patients treated per day better reflects Bloesch's everyday practice (1834: 8, 1842: 18, 1852: 14). He mostly called on his patients in their homes which was only feasible because the majority of his clients lived in Biel or neighbouring parishes. The average number of treatments per patient remained relatively stable throughout the entire period (1834: 6, 1842: 9, 1852: 6).[15] Long-term treatments appear to have been typical of Bloesch's practice from the beginning.

With the growing number of regular patients, the doctor found himself facing new administrative challenges. In June 1832 Bloesch began writing his journals and continued this form of record-keeping over the next thirty years. The accuracy and tenaciousness he had displayed when bringing order to the Biel archives were now applied to his own note-keeping. Nonetheless, his practice journals are more than just the work of a tidy-minded personality. They need to be seen against the background of the growing nineteenth-century trend towards historicization, administration and book-keeping – the latter in the financial sense, but also as a means of producing knowledge by induction.[16]

---

12 Cf. Arnold Bähler, *Biel vor hundert Jahren* (Biel 1916), 48–50; Wilhelm Lindt, *Zum Jubiläum der medizinisch-chirurgischen Gesellschaft des Kantons Bern* (Bern 1909), 133–38; Bourquin *et al.*, 1999, 77–79.

13 Cf. Marion Baschin, Elisabeth Dietrich-Daum and Iris Ritzmann, 'Who Consulted the Doctor? The Patients and Their Motives' (in this volume). On Bloesch's establishment on the medical market cf. Philipp Klaas, 'Cäsar Adolf Bloesch – ein Geburtshelfer? Einblicke in seine ärztliche Praxis zwischen 1832 und 1850', in Lina Gafner, Iris Ritzmann and Katharina Weikl (eds), *Penning Patient's Histories – Doctors' Records and the Medical Market in the 18th and 19th Century* [special issue: *Gesnerus* 69, no. 1 (2012)], 110–25.

14 On the social composition of the patients cf. Baschin *et al.* (in this volume).

15 Calculations from the dissertation of Philipp Klaas.

16 Cf. Gianna Pomata, 'Praxis Historialis. The Uses of Historia in Early Modern Medicine', in *eadem* and Nancy G. Siraisi (eds), *Historia. Empiricism and Erudition in Early Modern Europe* (Cambridge 2005), 105–46; Mary Poovey, *A History of the Modern Fact. Problems of Knowledge in the Sciences of Wealth and Society* (Chicago 1998), 7–16.

Journal Keeping, Diagnosis and Therapy

Bloesch's journals constituted a register and served, among other things, as the basis for his accounts. The visits of patients are listed in chronological order, with the fees for the services received added at the end. At the end of the month these fees were transferred to a separate list and an outstanding amount was recorded for each patient. After three years of practical experience with a growing number of patients, unpaid bills and journal volumes, Bloesch introduced an alphabetic index of his debtors at the end of each volume. Starting with volume 19 (1841) the indices, which had grown ever longer, disappeared from the journals. Since many patients paid their bills annually or not at all, the billing became more and more complex. It appears that Bloesch gradually transferred these entries to a separate account book which has, however, not been preserved.[17]

The fact that the journal was structured according to the date not only brought order to the entries, making them more easy to read, it also impelled the author to take notes every day – or probably, rather, every evening: 'In this way I would recall every patient, and what I had in many cases considered but fleetingly during the day would often give occasion for serious consideration in the evening'.[18] We find hardly any gaps or indications that Bloesch sometimes deferred the record-keeping to the next day. In the last consultations registered before his embarking on one of his military medical assignments or before withdrawing to a spa for the summer, we occasionally find only the patient's name and the fee he charged, but on the whole Bloesch appears to have devoted much time to his journal. The journal contains not only references to consultations, but also copies of letters he wrote and the expert reports, autopsies and certificates he issued. It constituted the administrative fulcrum of his medical practice and Bloesch administered his practice very meticulously.

The entries are structured in the same way throughout the whole period, with the exception of the first four days: from 8 to 11 June 1832, having seen his first patients, Bloesch noted down a diagnosis in Latin, followed by the therapeutic measures taken. The valuator Moser in Madretsch, for instance, was treated for 'caries under the lower lumbar vertebrae with feb[ris] lenta and insomnia, diarrhoea' and was ordered to drink rice-water, a glass of poppy-head infusion in the evening and to chew asafoetida.[19] From the 12 June onwards Bloesch no longer recorded a diagnosis but focused on the signs described by the patients that were also visible or palpable to him:

---

17 Cf. Philipp Klaas, Hubert Steinke and Alois Unterkircher, 'Daily Business: the Organization and Finances of Doctors' Practices' (in this volume).

18 Bloesch, 1866, 28.

19 StA Biel, 4 A 1, Tagebuch, vol. 3, 7.

> Montbaron, cotton spinner, pale-faced and skinny; frequent bouts of dizziness, especially when standing upright or walking; tongue clear, poor appetite; pulse small and slow: cort[icis] chinae unc. 5 infuso=decoct. with syr. g. arab unc. 1 every two hours 1 tablespoon; rice-water to drink.[20]

Bloesch's approach is an example of the good medical observation postulated by Johann Georg Zimmermann: 'Only the vain temptation to mix one's judgement into one's observations is to blame for the fact that every truth taught us by a great spirit comes with a hundred false verdicts…We must describe the afflictions of the sick in their words if they are able to express them'.[21] It is likely that Bloesch did not always find it easy to establish a diagnosis right at the beginning of a treatment and to derive the corresponding therapy. In his memoirs he admitted to having 'felt abandoned by the theory at the sickbed' in the early days of his practice. This juxtaposition of clinical experience and 'useless' theory is characteristic of his time.[22] Bloesch said that, inspired by his father-in-law, he had found his way to applying 'dietetic and symptomatic treatment', describing the method in retrospect as very successful indeed.[23] The practice journals reveal that Bloesch often had recourse to dietetic prescriptions and that with each consultation he adapted the therapeutic measures, depending on how the symptoms developed. This semiotic approach – the observing of the signs of illness – he would combine with the physical examination procedures required in the individual case. His technique marks him as one of the empirical-eclectic physicians who, in the first half of the nineteenth century, turned against theories such as Brownianism and who were, as a matter of principle, suspicious of philosophically inspired views of medicine.[24]

Towards the end of his life, Bloesch, bemoaning his initial sense of unease at the sickbed, not only called attention to an early change of direction in his medical practice: he supplied evidence of his own emancipation from university learning and asserted his own independent medical identity. While establishing a diagnosis and deriving a treatment from it may have been the hub of

20 *Ibid.*, vol. 4, 167.

21 Johann Georg Zimmermann, *Von der Erfahrung in der Arzneykunst. 1. Theil* (Zürich 1763), 214–15.

22 Bloesch, 1866, 23; cf. Michel Foucault, *The Birth of the Clinic. An Archaeology of Medical Perception* (London 2006), 131–33.

23 Bloesch, 1866, 24.

24 Cf. Urban Wiesing, *Kunst oder Wissenschaft? Konzeptionen der Medizin in der deutschen Romantik* (Stuttgart 1995); Jens Lachmund, *Der abgehorchte Körper. Zur historischen Soziologie der medizinischen Untersuchung* (Opladen 1997), 251; Jens Lachmund, 'Between Scrutiny and Treatment. Physical Diagnosis and the Restructuring of 19th Century Medical Practice', *Sociology of Health & Illness* 20 (1998), 779–801.

medical training,[25] they took a backseat, as implicit knowledge, in Bloesch's dealings with his patients.[26] In his memoirs Bloesch did not name any academic teachers as his role-models, but showed himself as a follower of Hippocrates, Asclepiades, Sydenham and his father-in-law, placing himself in a long tradition on the one hand, while advocating 'direct experience' as the most important principle of his medical practice on the other.

Even though we find the occasional diagnosis, such as ascaris, chlorosis or herpetic eczema, particularly when he was confronted with venereal diseases or epidemics, Bloesch almost always, when he met new patients, began by writing down their ages and some external features as well as the history of the present complaint, followed by information on appetite, thirst, bowel movements and sleep. Women patients would always be asked about their menstruation. Bloesch attributed numerous complaints to irregular menses and he would always try to regulate the flow of blood.[27] Over the years, with the number of clients rising and his experience widening, the entries took on a more schematic form. Many words are abbreviated; where he once wrote 'complains about headache', he now simply stated 'headache'. In later years all entries are of about equal length, with first six – and later five – lines given to each of them. In the very rare case of an unusual mental illness or other special circumstances Bloesch would write more, but he never wrote less. Over the years the diary increasingly resembled a standardized form (see Fig. 11.1).

---

25 On the significance of the diagnosis in clinical training cf. Félix Séverin Ratier: *Coup-d'oeil sur les cliniques médicales de la faculté de médecine et des hôpitaux civils de Paris* (Paris 1830); Volker Hess, *Von der semiotischen zur diagnostischen Medizin. Die Entstehung der klinischen Methode zwischen 1750 und 1850* (Husum 1993); John H. Warner, *Against the Spirit of System. The French Impulse in Nineteenth-Century American Medicine* (Princeton 1998); Othmar Keel, *L'avènement de la médecine clinique moderne en Europe 1750–1815. Politiques, institutions et saviors* (Montréal 2001).

26 Because of the official requirements attached to such reports and certificates, Bloesch was forced to include definite diagnoses and prognoses. Cf. Annemarie Kinzelbach, Stephanie Neuner and Karen Nolte, 'Medicine in Practice: Knowledge, Diagnosis and Therapy' (in this volume) and Lina Gafner, 'Ärztliche Zeugnisse auf dem medizinischen Markt. Die professionalisierte Zeugenschaft im wachsenden Verwaltungsapparat des frühen 19. Jahrhunderts, am Beispiel einer Bieler Arztpraxis', in Lina Gafner *et al.*, 2012, 95–109.

27 On the significance of this practice in the nineteenth century cf. Janet Farrell Brodie, 'Menstrual Interventions in the Nineteenth-Century United States', in Elisha P. Renne and Etienne van de Walle (eds), *Regulating Menstruation. Beliefs, Practices, Interpretations* (Chicago 2001), 39–63; Michael Stolberg, 'Menstruation and Sexual Difference in Early Modern Medicine', in Andrew Shail and Gillian Howie (eds), *Menstruation. A Cultural History* (New York 2005), 90–101.

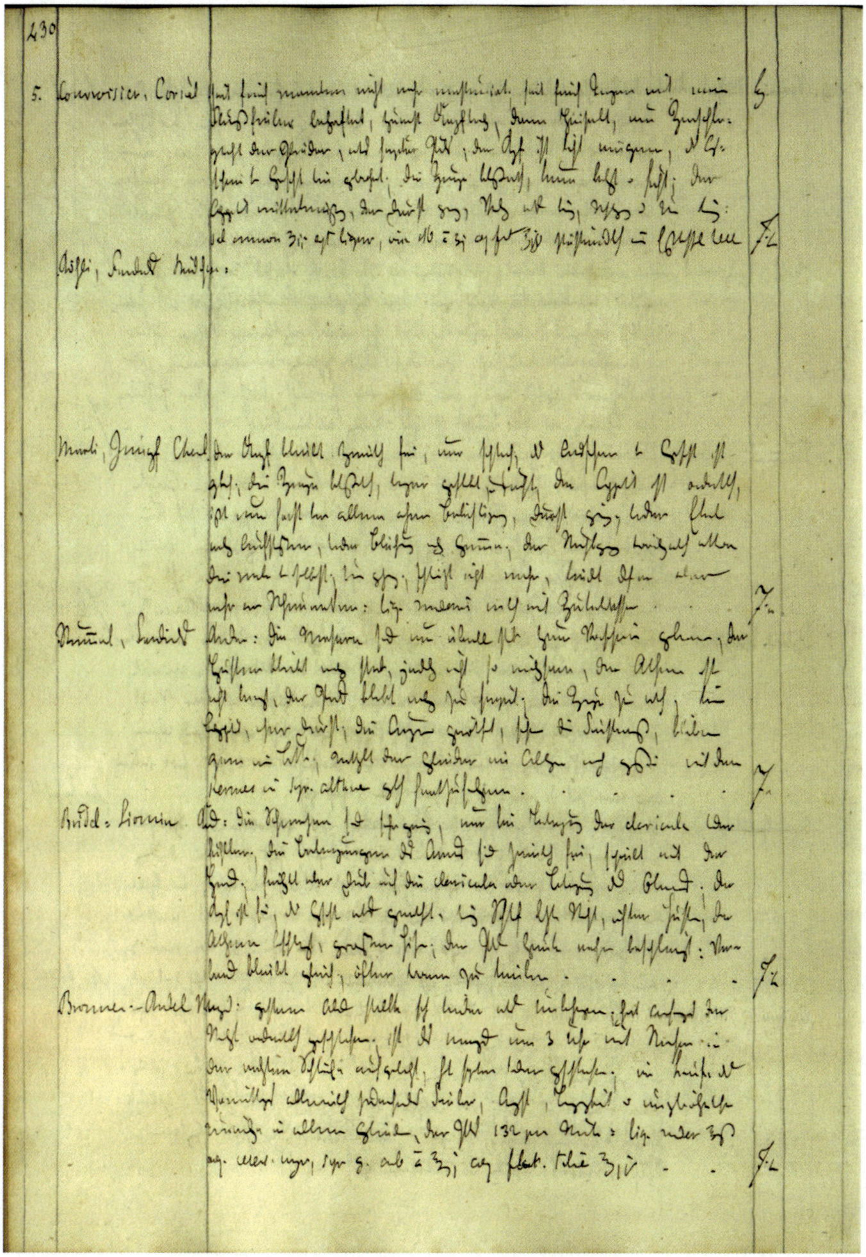

FIGURE 11.1 *Extract from C. A. Bloesch's journal for the year 1849.*
STADTARCHIV BIEL, 4 A 1, TAGEBUCH, VOL. 33, P. 430.

We observe a certain standardization and routinization in the consistent sequence of inquiry, observation, record-taking and treatment. With the growing number of patients the death rate also increased, requiring the appropriate administrative work. Bloesch listed all patients who had passed away by name in a register at the end of the year, adding the cause of death in each case. 'Apoplexy' we read, or *gastro-enteritis nervosa*. He would differentiate between 'his' deceased and cases where 'I was merely asked for advice or where no medical help was given'. There was the case of 2-year-old Maria Simmen, who had drowned. She was pulled out of the Schuess, the river that flows through Biel, and various attempts were made to resuscitate her, such as applying tobacco smoke enema, dropping sealing-wax onto her chest and holding snuff to her nose. But to no avail.[28] Similarly, Bloesch could not do anything to help a 10-year-old boy, August Frey, who lived in the barn of the textile factory. He diagnosed *febris gastrico-cerebralis*. The boy was already comatose when Bloesch saw him for the first time (he only saw him twice).[29] The primary purpose of the list of deceased patients was to provide a basis for transferring the names and causes of death to the Biel register of deaths at the end of the year.[30] Naming the cause of death would also have been important for the relatives. Diagnosis – having been removed from the treatment process – assumed new relevance in the case of death.

We rarely find reflections, doubts or more extensive entries in the journal, but here or there the routine of the increasingly uniform notes is interrupted. Unexpectedly dramatic cases which brought the physician to the limits of his knowledge would induce him to write down a more detailed report in the evening. This happened, for instance, in the case of the 7-year-old vicar's son from Nidau who was run over by a cart loaded high with wood. The wheel went straight over the boy's stomach. Bloesch administered leeches and camphor oil inunctions, which he covered with flannels and oil cloths. He prescribed lukewarm baths, a diet and enema, but the boy died a day after the accident.[31] The occurrence of death often induced Bloesch to critical introspection. In the case of the stenographer Alexander-Wilhelm Engel-Schneider he wrote as the cause of death *febris catarrh[alis] gastrica*, adding that he should not have tried to stop the liberating nosebleed as the flow of blood might have saved the patient's life.[32]

---

28 Cf. StA Biel, 4 A 1, Tagebuch, vol. 8, 38, 40.

29 Cf. *ibid.*, 13, 15.

30 Cf. StA Bern, K Biel 14, 15, Totenrodel (1816–1870).

31 Cf. StA Biel, 4 A 1, Tagebuch, vol. 11, 202.

32 Cf. *ibid.*, 253, 264.

Epidemics also tended to bring the physician to the boundaries of his abilities and strength. The first severe epidemic Bloesch was confronted with as a practitioner was the dysentery that scourged the district of Biel in the autumn of 1837. The number of Bloesch's consultations rose to an average of 25 per day in September that year, compared to the daily number of thirteen in an average year. Bloesch called on some of his patients up to three times a day. While he continued to omit the diagnosis with first consultations at the beginning of the epidemic, he abandoned this routine after some weeks, making instead entries such as 'has suffered from dysentery for three days'.[33] The fact that he now diagnosed 'dysentery' might have to do with the naming of the epidemic by the authorities. Bloesch wrote that rice-water had proved effective as a therapy. He tended to prescribe it in combination with leeches, a strict diet of gruel and, occasionally, 'opium with gum-arabic, taken internally, and with enemas of *decoct[um] sem[inis] lini*'.[34] At the end of the year five of his patients had died of dysentery. Closer investigation reveals that some other patients had also suffered from dysentery, but in Bloesch's opinion it had not been the immediate cause of their death.

The dysentery gave Bloesch access to houses that had been unknown to him beforehand and where he came across living conditions that he later declared to be in part responsible when he lost his fight against the disease. In the case of Sophie Menzel, the 5-year-old daughter of a cotton mill worker, Bloesch stated as the cause of death: *dysenteria neglecta*, adding that

> [Sophie Menzel] died from the consequences of neglected dysentery; all through the illness, which started four weeks ago, she hardly ever wore a shirt, the first fourteen days she went naked to the chamber pot and often sat entirely naked under the window, only recently she kept to her bed, still mostly without wearing a shirt.[35]

The link between social circumstances and the incidence of illnesses would continue to occupy the physician, and not only in individual cases, as will be shown below.

During the epidemic Bloesch began to settle on a largely uniform treatment for different patients, standardizing, as it were, his therapeutic approach to dysentery. This was uncharacteristic for him in that it was at variance with his conviction, which he adhered to even in epidemics, 'that each afflicted person

33 *Ibid.*, 428.

34 *Ibid.*, vol. 12, 206.

35 *Ibid.*, 31.

has something which is entirely their own'.[36] This change was manifest in Bloesch's way of recording: he no longer devoted one entry to each person treated, but subsumed entire households in one entry.[37] In this kind of epidemic the individuality of the patient was secondary while the disease affecting each of them would determine the course of medical action. Because there was not much time and many phenomena repeated themselves, the entries are noticeably shorter during the epidemic. Bloesch largely maintained this reduced style even after the dysentery was over and the number of treatments had abated.

At first glance it seems that the notation of patient examinations hardly changed over the thirty years. Closer inspection reveals alterations, however, which correspond to the changing conditions of Bloesch's medical practice. While the number of consultations rose steadily up until the 1840s, it exploded during the epidemics. The doctor's growing experience as well as the particularly work-intensive phases and the cases which brought him to the limits of his knowledge resulted in a process of routinization that kept being disrupted by those very events.

Observation and Experience

From the end of 1835 Bloesch not only listed fatal cases annually, but also began to refer to them in his reflections on the events of the previous year. He was particularly interested in frequent and deadly diseases. The impact of the dysentery epidemic of 1837 even led him to take stock every month:

> 23 people died and were buried in the parish of Biel this month. 10 of them died of dysentery: 12 children under the age of 7; 5 between 7 and 21; 5 between 21 and 60; one person of over 60; 14 died in Boezingen, 11 within the town of Biel; the average age of all fatalities is 17 years 6 months 26 days; the average age of those who died of dysentery is 13 years, 3 months, 8 days.[38]

A year later, in August 1838, when Bloesch was concerned that another dysentery epidemic was imminent, he observed a temperature drop in Biel at the end of the month which he considered to be responsible for the fact that the

36 Bloesch, 1866, 81.

37 Cf. StA Biel, 4 A 1, Tagebuch, vol. 11, 389.

38 *Ibid.*, 395.

symptoms subsided again.[39] From that moment, monthly meteorological observations and comments on the vegetation appeared regularly in his journals, usually preceded by a description of the health situation 'in general' and followed by an account of the most interesting or 'remarkable' cases of the previous month.[40]

The Canton of Bern did not carry out comprehensive surveys on the incidence of epidemic diseases in 1837. Only in 1868 did the Canton's medical association and the board of the healthcare services appeal to the physicians to submit their monthly reports on the diseases they had encountered.[41] Bloesch had clearly not kept his notes to satisfy official regulations, but for entirely different reasons. They were more like the reports about the 'weather and health situation in Berlin' (*Witterungs- und Gesundheits-Constitution von Berlin*) that used to appear in Hufeland's Journal under the heading of 'short news and extracts' (*Kurze Nachrichten und Auszüge*), or in the annual reports of hospitals and infirmaries.[42] Such lists ultimately constituted attempts to reflect the state of health of individual territories in medical topographies.[43] Bloesch's interest in this genre, which we spoke of previously in relation to his dissertation, is manifest in the format he gave his journals, a fact that leads us to assume that scientific or public health-related ambitions were part of Bloesch's professional self-image. His regular publications ensured that the scholarly world did not forget about him and, in 1845, he founded, with a colleague, a medical association for the Biel district of Seeland, with the aim of merging science, professional exchange and political aspirations. It was the place where the latest publications, discoveries and personal insights as well as problematic health policies were discussed and where Bloesch regularly presented his statistical research.

---

39 Cf. *ibid.*, vol. 14, 50.

40 This 'casebook' style differs from the otherwise register-like entries. Bloesch included cases that were especially instructive in his publications, some of them were also published as case histories, such as: Caesar A. Bloesch, 'Gutachten über einen gerichtlich-psychologischen Fall', *Schweizerische Zeitschrift für Medizin, Chirurgie und Geburtshülfe* 1 (1842), 369–80.

41 *Statistisches Jahrbuch für den Kanton Bern. Herausgegeben vom kantonalen statistischen Bureau*, 3. Jahrgang (Bern 1870), 95–96; cf. Christian Pfister, '"Uss gewüssen Ursachen". Hintergründe und Methoden statistischer Erhebungen im Kanton Bern 1528–1928', *Schweizerische Zeitschrift für Geschichte* 45 (1995), 44–47.

42 Cf. 'Witterungs- und Gesundheits-Constitution von Berlin im September', *Journal der practischen Heilkunde* 55 (1822) 6, 129–33; Emanuel Eduard Fueter: *Achter Jahresbericht der poliklinischen Anstalt in Bern für das Jahr 1842* (Bern 1843).

43 Cf., for instance, Jan Brügelmann, 'Der Blick des Arztes auf die Krankheit im Alltag 1779–1850. Medizinische Topographien als Quelle für die Sozialgeschichte des Gesundheitswesens' (PhD-thesis, Freie Universität Berlin, 1982).

In this research Bloesch did not follow any particular patterns of data collection nor was he guided by a question or a hypothesis. It was more a matter of unusual occurrences awakening his interest. He would, for instance, compare the mortality rates of men and women; or he would focus on the age of the deceased. The question as to whether meteorological conditions had an influence on health seems to have motivated him for a long time. In 1843 he published observations he had made during a typhus epidemic, concluding: 'I do not know if the weather had any kind of influence; if it did, I did not notice any'.[44] In 1844 he registered a particularly high number of fatalities and fewer births than usual, without being able to identify a predominant cause of death.[45] In 1845 the general state of health remained good despite unusual weather conditions, a strong winter and the potato blight.[46] When the incidence of births was lower than that of deaths in Biel in 1846, Bloesch concluded that the 'misgrowth' of the potato, which was the staple food of the 'large majority of our residents', was to blame.[47] He increasingly doubted that the weather made any difference at all. In 1848 he published an article, elaborating on these considerations:

> This shows clearly that the higher mortality affected especially the poorer classes who will have felt the rising food prices most painfully, and that the cause of this higher mortality in a year that generally favoured people's health is to be found in just that rise of food prices.... This phenomenon is therefore confirmation of the principle that mortality and average life expectancy in a place depend on the prosperity of the residents.[48]

In this respect, Bloesch agreed with Louis-René Villermé, who saw the mortality rate as directly related to poverty and food prices.[49] Bloesch found confirmation of this view in the high death rate among the children of factory workers in Boezingen, a small parish adjacent to Biel that accommodated the cotton mill and the factory workers with their families. In a presentation he gave to the district medical association he opined that the weather was only to

44 Cäsar A. Bloesch, 'Die Nervenfieber Epidemie von 1842 in Biel', *Schweizerische Zeitschrift für Medizin, Chirurgie und Geburtshülfe* 2 (1843), 241–52: 252.

45 Cf. StA Biel, 4 A 1, Tagebuch, vol. 25, 448.

46 Cf. *ibid.*, vol. 27, 377.

47 *Ibid.*, vol. 29, 121.

48 Cäsar A. Bloesch, 'Einige Rückerinnerungen aus dem Jahr 1846', *Schweizerische Zeitschrift für Medizin, Chirurgie und Geburtshülfe* 7 (1848), 345–56: 355.

49 Cf. Gérard Jorland, *Une société à soigner. Hygiène et salubrité publiques en France au XIXe siècle* (Paris 2010), 96–98, 11–113.

a very limited extent responsible for the state of health.[50] In his monthly reviews he nevertheless persisted with his weather observations. Documenting the weather, it seems, was part of the full assessment of the general level of health in the area – even if the results seemed to prove the opposite. In 1855 Bloesch published an article on cholera, introducing his new theory regarding the pathogenesis of this disease and blaming 'galvanic, electrical and magnetic processes' inside the earth for such 'global epidemics' – a view that not many of his colleagues in the district association were willing to subscribe to.[51]

The process of collecting and evaluating data that had become increasingly institutionalized from 1800 onwards found its way also into hospital administration and clinical observation.[52] It combined the more efficient administration of individuals with the production of a new form of knowledge that aimed at providing general outlines. As a result scholars, from around 1800, increasingly turned into bureaucrats whose primary tasks consisted of bookkeeping and data management.[53] Despite being sceptical of the medical theories of his time, Bloesch did not see himself as a mere practitioner either, but as part of a theoretically educated guild of physicians. From his hometown he tried to enter into the scientific discourse and to generalize the knowledge he had gained in his practice.[54] The basis for this production of knowledge was his patient journal which, looked at from this perspective, constitutes one single gigantic research report. In that respect the composing of the journal is a form

50 Cf. Institut für Geschichte der Medizin, Bern, Akten des medicinischen Bezirksvereins des Seelandes, Ms A122.

51 Cf. Cäsar A. Bloesch, 'Einige Bemerkungen über die Choleraepidemie in Aarau im Herbste 1854', *Schweizerische Zeitschrift für Medizin, Chirurgie und Geburtshülfe* 14 (1855), 1–16: 10; Institut für Medizingeschichte Bern, Akten des medicinischen Bezirksvereins des Seelandes, Ms A122.

52 Cf. Volker Hess, 'Formalisierte Beobachtung. Die Genese der modernen Krankenakte am Beispiel der Berliner und Pariser Medizin (1725–1830)', *Medizinhistorisches Journal* 45 (2010), 293–340; Libby Schweber, *Disciplining Statistics. Demography and Vital Statistics in France and England, 1830–1885* (Durham 2006).

53 Cf. Wolfgang Schäffner, 'Nicht-Wissen um 1800. Buchführung und Statistik', in Joseph Vogl (ed.), *Poetologien des Wissens um 1800* (München 1999), 125; Friedrich A. Kittler, 'Das Subjekt als Beamter', in Manfred Frank, Gérard Raulet and Willem van Reijen (eds), *Die Frage nach dem Subjekt* (Frankfurt am Main 1988), 401–20; Friedrich A. Kittler, *Discourse Networks 1800/1900* (Stanford 1990); Hans Erich Bödeker, 'On the Origins of the "Statistical Gaze". Modes of Perception, Forms of Knowledge and Ways of Writing in the Early Social Sciences', in Peter Becker and William Clark (eds), *Little Tools of Knowledge. Historical Essays on Academic and Bureaucratic Practices* (Ann Arbor 2001), 169–95; Vismann, 2008.

54 As early as 1832 he processed his practical experience, cf. Caesar A. Bloesch, *Beytrag zur Pathologie und Therapie der gastrischen Krankheiten* (Bern 1832).

of epistemic writing.[55] In trying to advance medical research by diligently writing down his experiences as a medical practitioner, Bloesch put into practice the ideas of Johannes Schmidtmüller, professor of 'state medicine' in Landshut, who in a publication of 1806 had emphasized the importance of the practice journal, not only as a memory aid, but as a service to medicine as a science. The note-keeping of the physicians would help medicine to attain a degree of precision that had so far been limited to the mathematical sciences. Posterity would be able to draw upon this rich store and medicine would develop into a very precise empirical science.[56] The practice journal was therefore not only a means of monitoring the patients, but also a way of transmitting knowledge to future generations. This latter function is apparent in Bloesch's work 'General principles of theoretical and practical medicine, extracted from my diary of medical visits and consultations'[57] that was published after his death: a review of his own activities and an account of the doctrines he had followed or derived from his own medical experience. This book was meant to give lasting significance to the painstaking note-keeping Bloesch had carried out through decades. Even the dedication in the book points towards the future: 'dedicated to my younger son, Julius, as a guide and to my friends as a memory'. The fifty-seven volumes assembled by Bloesch over the years were to keep alive the memory of his actions and ideas and serve posterity as a treasure trove of knowledge. The author seems to have known from the start that self-discipline and continuity would be required to make this journal a work of lasting value. But his inclusion of reflective sections such as the monthly and annual reviews is proof of his scientific aspirations too. Because of his perseverant note-keeping, his experiences ultimately became a store of expertise, a factor that helped establish the physician's own identity and legitimize his practice.

Self-Administration

With his ever increasing workload as a physician Bloesch also extended his political activities. He was a member of the social elite in Biel and co-founder,

55 Cf. Christoph Hoffmann, 'Schreiben als Verfahren der Forschung', in Michael Gamper (ed.), *Experiment und Literatur. Themen, Methoden, Theorien* (Göttingen 2010), 185; Christoph Hoffmann, 'Festhalten, bereitstellen. Verfahren der Aufzeichnung', in *idem* (ed.), *Daten sichern. Schreiben und Zeichnen als Verfahren der Aufzeichnung* (Zürich 2008), 7–20.

56 Cf. Johann A. Schmidtmüller, *Beiträge zur Vervollkommung der Staatsarzneikunde. Eine Beilage zum Handbuch der Staatsarzneikunde* (Landshut 1806), 129–30.

57 Bloesch, 1866.

in 1830, of the 'Casino Society' (an institution that played a prominent part in the societal life of the town); his political views were first radical, later liberal-conservative. In 1840 Bloesch and his wife purchased a mansion built in the classical style in the *Seevorstadt,* a noble suburb of Biel, where they moved with their two children and Bloesch's father-in-law.[58] Despite the political frictions that divided the town in those days of turmoil, in which Bloesch was also involved, his practice journals prove that he saw patients of all social ranks and political convictions, even though Biel boasted quite a wide array of physicians to choose from.[59] That Bloesch was highly regarded in the town is apparent from the fact that he issued numerous forensic expert reports and medical certificates.[60] In 1850 he was appointed chairman of the board of directors of the newly founded 'support and credit association' (*Hülfs- und Darlehenskasse*) and from 1854 to 1862 he officiated as President of the commune.

During these later years the entries in the practice journal grow more uniform. Months go by without the addition of any longer notes that would point to an extraordinary case. Bloesch's medical practice seems to have run its routine course, in parallel with his other activities. On top of his political commitments, Bloesch devoted himself to the history of the town, publishing smaller historical and philosophical texts and finally also the chronicle of the city of Biel mentioned earlier.[61] At the same time he increasingly used his journal to reflect, in shorter or longer comments or contributions, on his own position in the world of medical science. He expressed his dislike of philosophy and of 'medical journalism' pledging himself to the empiricism of Thomas Sydenham.[62] From 1860 these notes were consolidated to form the publication mentioned above, in which Bloesch reflects on his biography, his philosophical views and

58 Cf. Bourquin *et al.*, 1999, 105–06, 77–78. On Bloesch's income cf. Klaas *et al.* (in this volume), on his social standing cf. Kay Peter Jankrift and Ruth Schilling, 'Medical Practice in Context. Religion, Family, Politics and Scientific Networks' (in this volume).

59 In 1837 Biel had five physicians and one surgeon for 2796 inhabitants; cf. Daniel Alexander Mahlberg, 'Der ärztliche Bezirksverein Bern-Seeland. 1845–1895' (med. diss., Universität Bern, 2005), 51–56; Raffael von Niederhäusern, *Sozialtopografie der Stadt Biel Mitte 19. Jh.* (Bern 2011, unpublished typescript), 10. On the conflict-free everyday practice cf. Elisabeth Dietrich-Daum, Marina Hilber and Eberhard Wolff, 'An Alpine Doctor's Private Practice: Franz von Ottenthal (1847–1899)' (in this volume).

60 On the significance of medical certificates in Bloesch's practice cf. Gafner, 2012.

61 Cf. Cäsar A. Bloesch: *Bern und Biel. Eine Erinnerung an die grosse Vorzeit* (Biel 1853); *idem*, 'Einige Bemerkungen über pantheistische Weltanschauung', *Schweizerische Zeitschrift für Medizin, Chirurgie und Geburtshülfe* 9 (1850), 29–48; *idem*, 1855–1856.

62 Cf. StA Biel, 4 A 1, Tagebuch, vol. 41, 21–24.

his view of illness. The volume, which was published after his death, can also be seen as a narrative attempt to lend meaning to his life.[63]

When in 1863 his son Julius contracted *typhus cerebralis*, Bloesch's consistent record-keeping stopped and time seemed to stand still in his journal. What follows is a patient history, written down in neat handwriting, documenting the father's despair and concern as much as the physician's search for the right remedy. The precise observations Bloesch made at his son's bedside are enclosed in the journal: stained slips of paper in scrawly handwriting, that follow the days of illness from 1 to 41 rather than the calendar days of September 1863. It can be assumed that Bloesch wrote such notes for all his patients, probably holding on to them until an acute illness was cured because they enabled him, in a better way than the journal entries, to monitor the course of illness and interpret it according to the principles of medical crisis development. But only in the case of his son have these notes been preserved.

The son survived, but soon after his recovery his mother, Bloesch's wife Elisa, died, 'the victim of her unceasing motherly love!'[64] Bloesch entrusted the pain of his loss to his diary, concluding the entry that ends volume 56 with a prayer.[65] The last volume opens with a lyrical-religious self-interrogation that seems to suggest that he had a presentiment of his own death and wanted to take courage. Routine then took its course again, the time which seemed to have stopped moved on again from 12 October, with patient examinations being presented in their usual form. But not for long. A few more entries, then the journal comes to an abrupt end. Bloesch died a month after his wife, supposedly from 'typhoid house endemic'.[66] The fact that, during the last phase of his life, he continued to write down his journal in as meticulous a manner as ever seems to indicate that the journal had by that time become much more than a functional tool.

The verse on the first page of volume 57 is taken from the song 'Eximination in the evening' (*Prüfung am Abend*) by the German poet Christian Fürchtegott Gellert. The devout Bloesch only wrote down the one verse of a very long poem that referred to the practice of one's profession and seems to allude to the evening not so much of the day, but of life:

> Did I, in the vocation
> That God consigned to me
> With diligence and industry

63 Cf. Hayden White, 'The Value of Narrativity in the Representation of Reality', *Critical Inquiry* 7 (1980), 5–27.

64 StA Biel, 4 A 1, Tagebuch, vol. 56, 431.

65 Cf. *ibid.*, vol. 56, 431.

66 Lindt, 1909, 138.

praise the almighty Lord?
Did serve myself and the world,
Did all I did
For the will of God,
Not for others to see?[67]

FIGURE 11.2 *C. A. Bloesch posing with his practice journal in his library. Oil portrait by Aurèle Robert, 1865.*
KUNSTSAMMLUNG DER STADT BIEL.

67 StaA Biel, 4 A 1, Tagebuch, vol. 57, 1. For the complete poem cf. Christian Fürchtegott Gellert: *Werke* (Frankfurt am Main 1979), vol. 1, 223–25.

By the end of volume 57 administration had become self-adminstration, the practice journal a testimony to his life's work. He no longer needed to justify his actions to the authorities, the relatives of patients or his medical colleagues. But his looking back at the end of his life was also a way of preparing himself for the Last Judgement. As Cornelia Vismann so aptly wrote, keeping track of his actions was a way of gathering 'enough material to match the evidence accumulated in heaven'.[68]

Conclusion

The two titles referred to at the beginning suggest that Bloesch's journal served various purposes and that there were consequently various aspects to his activity and identity as a physician. As a collection of patient information the journal constitutes an archive of medical actions that required the writer to be reliable and neat, while it reflected these very qualities. When new problems presented themselves in Bloesch's medical practice, they would stir his scientific ambitions. The journal is therefore also the epistemic basis and evidence of his gradually evolving medical expertise. Unlike pre-printed forms, the journal's structure could accommodate differing formats such as lists or longer passages of prose. The journal impresses with its sheer scope and with Bloesch's persistent, continual note-keeping that is testament to his extraordinary self-discipline. The daily effort of writing down the entries required a combination of humbleness, diligence and scientific ambition. The records were not meant to remain invisible.[69] They were destined to confidently showcase the physician's industriousness and reliability (Fig. 11.2). Keeping the journal involved administrative as well as epistemic actions. Because it was such an essential part of Caesar Adolf Bloesch's medical practice it cannot be ignored as a heuristic key to the historical investigation of this practice.

---

68 Vismann, 2008, 112.

69 Cf. Bernhard Siegert and Joseph Vogl (eds), *Europa. Kultur der Sekretäre* (Zürich 2003).

CHAPTER 12

Franz von Ottenthal: Local Integration of an Alpine Doctor's Private Practice (1847–1899)

Elisabeth Dietrich-Daum, Marina Hilber and Eberhard Wolff

Introduction

It was by chance that the general medical practice, maintained by Franz von Ottenthal (1818–1899) between 1847 and 1899 in the municipality of Sand in Taufers in Southern Tyrol, came to be the object of medical historical research. In 1998, an archivist of the *Südtiroler Landesarchiv* (South Tyrolean State Archives) was fortunate enough to discover 244 quarto-sized books in the attic of Neumelans Castle in Sand.[1] The books provide insights into over fifty years of a country doctor's medical activities in the eventful nineteenth century.

Since this momentous discovery, the books, which constitute the main part of Franz von Ottenthal's legacy, have been deposited in the *Südtiroler Landesarchiv* where they can be accessed in their original form and on microfilm. The online database that has been set up at Innsbruck University as part of the six-year Interreg IIIA project (Austria–Italy 2002–2007) offers a more user-friendly alternative to visiting the archives in person. The full set of case histories from Ottenthal's time as private GP have been transcribed and transferred to an Oracle database.[2] Early in 2008 this database was made available

1 Cf. Christine Roilo, 'Medizin auf dem Lande. Die Historiae morborum des Franz von Ottenthal', in Elisabeth Dietrich-Daum, Werner Matt und Hanno Platzgummer (eds), *Geschichte und Medizin. Forschungsberichte – Fachgespräche. Dokumentation zur internationalen Tagung "Geschichte der Medizin", 5. Dornbirner Geschichtstage, 9. bis 12. Juni 1999* (Dornbirn 2001), 151–72.

2 For details of the project cf. Elisabeth Dietrich-Daum, '"Historiae Morborum" – Die Krankengeschichten des Franz von Ottenthal. Ein Interreg IIIA Projekt. Österreich–Italien 2002–2007', *Virus. Zeitschrift für Sozialgeschichte der Medizin* 4 (2004), 83–89; Elisabeth Dietrich-Daum and Andreas Oberhofer, 'Die "Historiae Morborum" des Franz von Ottenthal. Ein Interreg IIIA-Projekt. Österreich–Italien 2002–2007', in Rainer Alsheimer and Roland Weibezahn (eds), *Körperlichkeit und Kultur 2004. Interdisziplinäre Medikalkulturforschung. Dokumentation des 7. Arbeitstreffens des "Netzwerk Gesundheit und Kultur in der volkskundlichen Forschung", Würzburg, 31. März–2. April 2004* (Bremen 2005), 203–14; Elisabeth Dietrich-Daum and Elena Taddei, '"Historiae Morborum" – Le anamnesi del dottor Franz von

 | DOI 10.1163/9789004303324_014

to the international research community and can be used free of charge via a multilingual (German/English/Italian) online query portal.[3]

The following considerations compile the results of a number of studies previously carried out by the project group. The purpose of this paper is, in particular, to investigate, on the basis of this extensive discovery, Ottenthal's position within the medical market of a remote inner alpine region, and to describe his integration into the rural society as an aristocratic landowner, academic physician and local politician. Ottenthal's *Historiae Morborum* – 'histories of illnesses' – which document more than 87,500 consultations, represent the most comprehensive source of its kind known so far in the German-speaking world.[4] However, it is not only the scope of the documentation that makes it historically valuable, but also the fact that this nineteenth-century physician wrote his notes in Latin, the language of scholars. Franz von Ottenthal probably acquired this method of logging or recording his medical activity when training at Vienna University and he continued to apply it consistently in his private practice.[5] The notes constituted a reference system that allowed the physician an overview of his patients' pre-existing conditions and previous treatments and therefore of the individual course of their illnesses. While the consistency of record-keeping would have made these records suitable for scientific analysis, there has so far been no evidence of any scientific ambitions on Ottenthal's part. Rather than publishing the empirical insights gained in his own practice for the benefit of the general medical community, he used his notes exclusively as a reference to further develop his own medical and therapeutic accuracy.[6] The *Historiae* do not contain many diagnoses as such. The case histories are strongly symptom-oriented and are based on the complaints described by the patients as well as on symptoms elicited by the physician on questioning and examination of the patient.[7] The overall impression one gains is that the narrative style suited Ottenthal's view of illness, a view that was informed by humoral

Ottenthal. Un progetto Interreg IIIA. Austria–Italia 2002–2007', in Elisabeth Dietrich-Daum and Rodolfo Taiani (eds), *Medikalisierung auf dem Lande – Medicalizzazione in area alpine* [special issue: *Geschichte und Region / Storia e regione* 14, no.1 (2005)], 157–60.

3 Cf. the project homepage URL: http://www.uibk.ac.at/ottenthal/index.html.

4 Cf. Martin Dinges, 'Arztpraxen 1500–1900. Zum Stand der Forschung', in Elisabeth Dietrich-Daum, Martin Dinges, Robert Jütte and Christine Roilo (eds), *Arztpraxen im Vergleich. 18.–20. Jahrhundert* (Innsbruck 2008), 23–61: 44.

5 Cf. Christine Roilo, '"Historiae Morborum" des Franz v. Ottenthal – ein Zwischenbericht', *Medizin, Geschichte und Gesellschaft* 18 (1999), 57–80: 61–63.

6 *Ibid.*; Elena Taddei, *Franz von Ottenthal. Arzt und Tiroler Landtagsabgeordneter (1818–1899)* (Wien 2010), 16–18.

7 Cf. Roilo, 1999, 64.

pathology. For Ottenthal, illnesses could not be rigidly defined as entities; they developed individually. Lastly, the *Historiae* served a more mundane, economic, purpose since Ottenthal used them to write down the fees he charged.[8] As no receipts are included, the source does not constitute an account book as such and his actual income as a physician cannot, therefore, be inferred from it.[9]

The Country Doctor and His Medical Network

Franz Seraph Joseph Anton von Ottenthal was born on 23 May 1818 in Sand where he grew up, the son of an old aristocratic family of Tyrolese landowners who had served as judges for generations. After finishing grammar school in Brixen, Ottenthal went against family tradition when he chose his future profession. In 1836/37 he took up medicine at Vienna University, graduating five years later with a doctorate in medicine and surgery and a diploma in obstetrics. In 1844 the young doctor took up a salaried post as community and forensic physician in Windisch-Matrei (today: Matrei in East Tyrol). The journals he kept there between 1844 and 1845 were already called *Historiae Morborum*, with the added subtitle 'Wendo Matrii' (for Windisch-Matrei). After only two years the young physician left his post in the public health service and returned to his home town of Sand, with its 500 inhabitants, where he began to practise medicine.[10] Although Sand already had an academically trained, publicly appointed community physician, Ottenthal had no qualms about offering his services as a private practitioner. As a private practitioner who was not under any contract he had to meet the general requirements of the health authorities for medical doctors, but he did not draw a salary from the state.[11] His aristocratic background will have allowed him to embark on a career as private physician without any major financial strains because, due to his considerable private property and means, Ottenthal was not dependent on patient fees.[12] His economic and social situation, which was secure compared to that of other country physicians in Tyrol, largely determined Ottenthal's attitude and actions.

The area where Ottenthal now worked was the *Taufer Ahrntal*, an alpine valley situated in what is today South Tyrol (Italy); it extends for more than

8 Cf. *ibid.*, 63.

9 Cf. Andreas Oberhofer, 'Eine Landarztpraxis im 19. Jahrhundert am Beispiel der Ordination des Dr Franz von Ottenthal (1818–1899)', in Elisabeth Dietrich-Daum *et al.*, *Arztpraxen*, 2008, 175–76.

10 Cf. Roilo, 1999, 57–80; for more detail cf. Taddei, 2010, 27–48.

11 Cf. *ibid.*, 75.

12 Cf. Oberhofer, 2008, 188; Taddei, 2010, 60–70, 216–20.

fourty-six kilometres in a north-eastern direction from the town of Bruneck towards the Zillertal Alps. The valley is framed by high mountain ranges and, except for a few high passes, is only accessible from the South. The catchment area of Ottenthal's practice included not only the immediate surroundings of the town of Sand, but all the parishes of the Taufer Valley, including its tributary valleys, and the Ahrntal parishes situated higher up in the mountains. Politically, the region belonged to the district court of Taufers which had originally comprised nineteen very small parishes until, in the late nineteenth century, their number was reduced to fifteen through amalgamation.[13] The 1909 register of community physicians refers to the districts of *Sand-Taufers* and *Ahrntal* as particularly 'spread-out and onerous'.[14]

In total, the population of the Taufer Ahrn Valley at the time of Franz von Ottenthal was around 9,000 and declining.[15] According to Unterkircher it consisted, throughout most of the nineteenth century, of 'an upper and middle stratum of farmers, small and very small craft enterprises and a broad substratum of day labourers and servants'.[16] The traditional agricultural economy based on dairy and fat stock farming could not support the labourers for more than short periods of time and, what had been periods of seasonal migration, developed into a permanent rural exodus in the course of the nineteenth century. The valley community which relied primarily on farming was therefore one of many areas of depopulation in nineteenth century Tyrol, a trend that could not be halted by the alternative source of income provided by the mining and smelting works in Prettau, deep in the Ahrn Valley.[17]

13 Cf. Alois Unterkircher, *Jungen und Männer als Patienten bei einem Südtiroler Landarzt (1860–1900)* (Stuttgart 2014) [special issue: MedGG-Beiheft 51], 63; Elisabeth Dietrich-Daum and Andreas Oberhofer, 'Medizingeschichte als Sozialgeschichte. Die Krankengeschichten des Südtiroler Landarztes Franz von Ottenthal (1818–1899), ein Interreg III A-Projekt. Österreich–Italien (2002–2007)', in Christoph Haidacher and Richard Schober (eds), *Bericht über den 24. Österreichischen Historikertag in Innsbruck* (Innsbruck 2006), 646–55: 647–49; Taddei, 2010, 76–82.

14 Sanitätssprengel 1909, 11–13, quoted from: Alois Unterkircher, 'Ausser Konkurrenz? Strategien und Inanspruchnahme einer Südtiroler Landarztpraxis um 1890 bei Erkrankungen von Kindern', in Lina Gafner, Iris Ritzmann and Katharina Weikl (eds), *Penning Patient's Histories – Doctors' Records and the Medical Market in the 18th and 19th Century* [special issue: *Gesnerus* 69, no. 1 (2012)], 165.

15 Cf. Unterkircher, Jungen, 2014, 63–64.

16 *Ibid.*, 75.

17 Cf. Wolfgang Messner, 'Wirtschafts- und Sozialgeschichte des Tauferer Tales/Ahrntales im ausgehenden 18. und im Laufe des 19. Jahrhunderts' (Master thesis, University of Innsbruck, 2004), 21–24.

In 1847 Franz von Ottenthal returned to this environment, joining the small elite of aristocrats and civil and clerical officials as a representative of both aristocracy and educated bourgeoisie. This 'in-between' societal position was further enhanced when the 34-year-old physician married Katharina, the 27-year-old daughter of the Brixen country judge Ignaz Theodor von Preu zu Lusenegg und Koburg.[18] There are no sources to reveal to what extent Ottenthal's patients were aware of his unusual personal career and special social position or whether they accepted him as one of them. What is remarkable is how fast and smoothly the private practice of the aristocratic physician became an established part of the local medical market. As soon as word had spread that a 'new doctor' had come to the valley, Ottenthal's practice was attended by a great number of patients and was operating successfully within a relatively short period of time.[19] In his very first year in Sand, 741 patients came to see the young doctor.[20] This is not astonishing considering the scarcity of physicians on the rural periphery of the Taufer Ahrn Valley. In 1847, there was, apart from Ottenthal, only the community physician Dr Josef Daimer, who also practised in Sand, and a surgeon in Steinhaus, further up the valley. Even though the inhabitants of the valley could, in theory, consult the physicians and surgeons in the town of Bruneck, this was not a realistic option as the remote situation of some valley hamlets made the journey too long and difficult.[21] Statistics from 1880 show the entire region to have been medically undersupplied.[22] While a physician in Tyrol would look after 1633 patients on average, the number of patients per doctor in Bruneck, the district that Ottenthal's catchment area belonged to, was 2,352. Only in the Trentino, a region generally disadvantaged due to its lack of infrastructure, had a less favourable doctor–patient ratio.[23] The absence of academic physicians[24] had a

18 Cf. Taddei, 2010, 49.

19 Cf. Oberhofer, 2008, 181.

20 Cf. *ibid.*, 180.

21 Because of the altitude, some patients had to negotiate steep mountain and foot paths to get to the valley. From there, in the case of Prettau, it was a further twenty-five kilometers to Sand and even forty kilometers to Bruneck.

22 For developments in the presence of physicians cf. Elisabeth Dietrich-Daum, 'Die Klage der Ärzte. Marktprobleme und Professionalisierung in der zweiten Hälfte des 19. Jahrhunderts in Österreich', in Alois Unterkircher, with Reinhard Bodner and Kathrin Sohm (eds), *Medikale Kulturen* [special issue: *Bricolage* 5 (2008)], 202–21.

23 Cf. K.K. Statistische Central-Commission 1882, XXIIIf., quoted in: Unterkircher, Konkurrenz, 2012, 165.

24 Cf. Elena Taddei, 'Bestellungsverfahren von Ärzten in Tirol in der zweiten Hälfte des 19. Jahrhunderts. Qualifikationen, Auswahlkriterien und Erwartungen', in Elisabeth *et al.*, *Arztpraxen* (2008), 221–37.

direct effect on Ottenthal's professional life: in the 1880s he was at times the only physician in the valley, his range of activities had greatly expanded and patient numbers had tripled. The official medical positions in the district – that of community and forensic physician in Sand and that of surgeon in Steinhaus – were often unoccupied for long periods of time due to illness of the post holders and difficulties with finding successors. As a consequence, it was no competition but came as a relief to Ottenthal when, in 1888, young Dr Kortleitner took on the position of community physician and Dr Pescota[25] that of surgeon. Ottenthal had had to stand in for his friend, community physician Dr Daimer Senior, who had suffered from poor health for years; and the position of forensic physician had been vacant for long stretches of time, which is why Ottenthal was employed – since 1861 occasionally and between 1884 and 1888 permanently – as locum community and forensic physician.[26] The position required the holder to treat the poor free of charge, examine the 'impotence of marital partners' and the mental state of persons in guardianship disputes, refer psychiatric patients to an institution, establish the cause and nature of physical damage and the cause of death in court trials, confirm fitness for military service, carry out post-mortem dissections, supervise 'lower healthcare staff' and pharmacies, carry out inoculations and report public health problems including animal diseases.[27] This wide range of duties provided Ottenthal with countless new patients, but also with an increase in office duties and contact with the authorities. The physician experienced this additional workload as particularly stressful. Often, and sometimes very openly,[28] he criticized the fact that physicians, in the words of Dr Josef Daimer Jr, 'had become like clerks, particularly in the country'.[29] In the end, the 66-year-old Ottenthal found the workload all too much and, in 1884, he applied to the authorities to be released from the 'vaccination business': '…the undersigned is 66 years of age, his performance therefore restricted, the possibility of illness due to disproportionate demands increased; his eyesight which is impaired due to age renders him also less suited for the vaccination business'.[30]

It was typical of his rural situation that Franz von Ottenthal, throughout his life, had to struggle with the excessive demands that resulted from high patient

25 Cf. Unterkircher, Jungen, 2014, 83–84.

26 Cf. Roilo, 1999, 59; Taddei, 2010, 75.

27 Cf. Dietrich-Daum *et al.*, 2006, 647; Oberhofer, 2008, 169, Taddei, 2010, 156.

28 Cf. *ibid.*, 157–66.

29 Cf. *ibid.*, 158.

30 Südtiroler Landesarchiv, Nachlass Ottenthal, Verordnungen und Dekrete, 19, quoted in: Taddei, 2010, 167.

numbers rather than with the problem of having too few patients.[31] This explains why he did not see his colleagues as rivals and hardly ever tried to promote himself or express criticism. Comments of that nature in the *Historiae* never concerned aspects outside the boundaries of professional activities, even though Ottenthal occasionally benefited from other physicians' lack of success and noted this down in his records.[32] His entries always appear neutral and never sound malicious, which was probably not just due to his noble upbringing, but part of the 'collegiality' typical among physicians.[33] 'Physicians shall not judge physicians', had been the concluding thesis in Ottenthal's dissertation.[34] It can generally be said that Ottenthal was hardly ever critical of colleagues in the *Historiae*. As one example of many, an entry shall be mentioned that concerned the unsuccessful therapy of an eye condition in a boy from Ahornach. Ottenthal's laconic comment was: [*M*]*edic. Collegae Kortleitn nil juvit.*[35] A failed attempt at curing a patient of diarrhoea was commented on by Ottenthal in 1853 with the words: *sumsit* [sic] *medicamen Dr Daimer et post hoc multus* [sic] *dolor lateris magnaque debilitas.*[36] Ottenthal and the Bruneck physician Dr Vorhauser apparently even warned each other of particularly difficult patients.[37] This and numerous similar entries show that Ottenthal liaised, partly very closely, with his colleagues, referring to these contacts in the *Historiae* as *colloquium*, *consultio* or *consilium*. Just under 0.2 per cent of the case histories contain references to previously or simultaneously treating *collegae*.[38] There are also entries regarding mutual referrals[39] and joint house calls. With Josef Daimer Sr he not only worked for many years on coroner's inquests and forensic or psychiatric statements,[40] the two men also enjoyed a

31 Cf. Unterkircher, Jungen, 2014, 79–85.

32 Cf. Unterkircher, 'Konkurrenz', 2012, 172–74; Taddei, 2010, 129.

33 Cf. Claudia Huerkamp, *Der Aufstieg der Ärzte im 19. Jahrhundert. Vom gelehrten Stand zum professionellen Experten. Das Beispiel Preussens* (Göttingen 1985), 130.

34 Cf. Taddei, 2010, 129.

35 'The medicament of colleague Kortleitn did not help': *Historiae Morborum* 1780/1891 (22 Dec. 1891), quoted in: Unterkircher, 'Konkurrenz', 2012, 173.

36 'Took a medicament from Dr Daimer and experienced great pain in his side and great weakness afterwards', cf. Taddei, 2010, 128.

37 Cf. *ibid.*

38 Cf. Oberhofer, 2008, 170.

39 Cf. Unterkircher, 'Konkurrenz', 2012, 174.

40 Cf. Elisabeth Dietrich-Daum and Elena Taddei, 'Curare – segregare – amministrare. L'assistenza e la gestione dei "mentecatti" in un contado del Tirolo. L'esempio del medico generico Franz von Ottenthal (1818–1899) di Campo Tures', in Elisabeth Dietrich-Daum and Rodolfo Taiani (eds), *Psychiatrielandschaft / Oltre il manicomio* [special issue: *Geschichte und Region / Storio e Regione* 17, no. 2 (2008)], 83–102; Taddei, 2010, 128.

warm and lasting friendship that also found expression in Ottenthal asking Daimer to be godfather to his children.[41] In 'questions of insanity', Ottenthal occasionally also contacted the priests, assuming that they knew more about individual family situations and local conflicts than he did.[42] Even though only fragments of Ottenthal's correspondence have been preserved, the *Historiae* entries as such reveal that he was part of a close-knit collegial and, due to his official appointments by the regional authorities, also a local communication network.

Unfortunately, little is known of his relationship with the surgeon who was mostly responsible for the copper miners in Steinhaus.[43] Oberhofer mentions occasional therapeutic consultations between Ottenthal and the surgeon whom the academic physician also credited with diverse healing successes.[44] The range of licensed medical care providers in the valley was complemented by twelve, later eighteen, qualified midwives who received their salaries from the court cashier in Taufers. Ottenthal fostered – within the margins of the medical hierarchies – a constructive relationship with the local midwives, achieving a relatively smooth and workable separation of labour without major tensions or competition. The physician did not act as *accoucheur*, but left the obstetrics, if there were no complications, to the female experts. Only in emergencies and when asked directly by the midwife in question, would Ottenthal interfere, using medication or instruments, such as forceps, and providing perinatal medical care. Especially his relationship with the local midwife in Sand, Magdalena Auer, appears to have been based on mutual respect and trust, for Ottenthal entrusted her with all nine deliveries of his wife, Katharina von Preu.[45] The midwife, on the other hand, consulted the physician for over thirty years: the *Historiae* reveal evidence of 134 consultations for various ailments.[46]

41 Cf. *ibid.*, 49–50, 128.

42 Cf. Dietrich-Daum *et al.*, 'Curare', 2008.

43 On the social group of miners in Ottenthal's practice cf. Andreas Oberhofer, 'Häuer – Holzer – Schmelzer. Unfälle und Krankheiten im Bergbau Prettau. Eine Auswertung der Krankengeschichten des Südtiroler Landarztes Dr Franz von Ottenthal (1818–1899)', in Wolfgang Ingenhaeff and Johann Bair (eds), *Bergvolk und Medizin. 3. Internationales Bergbausymposium Schwaz 2004* (Innsbruck 2005), 251–68.

44 Cf. Oberhofer, 2008, 170–71.

45 Cf. Marina Hilber, 'Der Landarzt als Geburtshelfer – Dr Franz von Ottenthal und der medizinische Markt in Südtirol (1860–1869) [Countryside obstetrics – Dr Franz Ottenthal and the South Tyrolean Medical Market]', in Lina Gafner *et al.*, 2012, 149–50.

46 From 1849–1880 cf. e.g. *Historiae Morborum* 944/1849, 1245/1854, 1229/1868, 1419/1879 or 315/1880.

If midwives ever exceeded their remit and administered internal medicines, Ottenthal would note this down in a dispassionate way. In the case, at the very beginning of his career, of a 68-year-old man from St Peter in the Ahrn Valley, who had consulted him several times in the same year, Ottenthal noted that a purge administered by the midwife had evoked acute fever and haemorrhagic diarrhoea.[47] It is possible that Ottenthal did not want to compromise the success of his practice by embarking on an open conflict with the midwife. But there is evidence of a similar case at the end of his career, when a midwife gave a laxative to an infant who had grazes in the mouth and therefore no appetite. Only when the treatment showed no effect but produced new symptoms, was the physician finally consulted.[48] Even if his tone tended to remain neutral, some curative attempts of lay people or, occasionally, of a midwife seem to have astonished if not annoyed him. In 1890 he noted down the case of a 2-year-old girl whom 'quacks' and 'know-all' women (*agyrtae et mulieres superintelligentes*) had wrongly diagnosed with rickets (*rachitidem invenerunt*), a conclusion that struck Ottenthal as unfounded and laughable.[49] Since these offences were not reported to the health authorities – there is no evidence of such reports – Ottenthal was apparently not a stickler for standards in healthcare. This is the more remarkable because, apart from his position as forensic physician, he held a number of prominent offices in the local administration (court cashier, school fund administrator) and for almost twenty years served as a member of the local parliament.[50] One would certainly expect such an office holder and delegate to support public order. His noticeable reticence in this respect – despite the fact that he was officially obliged to report – was apparent also in his contact with health-care providers who were not part of the medical hierarchy. Although Franz von Ottenthal noted down failed attempts at curing patients by diverse local lay healers and itinerant dealers,[51] he never seems to have gone as far as reporting them. On 11 July 1858, for instance, Ottenthal wrote down the story of a man of around fifty-four years of age who suffered from *dolore ischiadico ex 3 septimanis* and had been given ineffective remedies by a lay healer before he consulted the physician.[52] One also finds indications in the *Historiae* that the plasters, laxatives or purges administered by such healers who usually remained

47 *Ibid.* 1363/1847 (15 Oct. 1847).

48 *Ibid.* 998/1896 (22 July 1896), quoted in: Unterkircher, 'Konkurrenz', 2012, 171.

49 *Ibid.* 1283/1890 (5 May 1890), quoted from: Unterkircher, 'Konkurrenz', 2012, 171.

50 Cf. Taddei, 2010, 155.

51 Cf. Unterkircher, 'Konkurrenz', 2012, 167; Taddei, 2010, 84–86.

52 *Historiae Morborum* 1012/1858 (11 July 1858).

anonymous, had occasionally also aggravated the condition,[53] as in the many cases mentioned in the *Historiae* of inflammations ensuing after the inexpert pulling of teeth.[54] In the case of the vicar of Luttach, who appears to have been famous for his heroic laxatives, Ottenthal's failure to register the irregular healthcare behaviour might be explained by the elevated social status of the 'charlatan'. There is also no evidence of Ottenthal reporting a chaplain who treated a 19-year-old man for diarrhoea, whose symptoms grew worse after he had taken the 'spiritual laxative'.[55] *Laxans a parocho propinatum nil juvit*, Ottenthal wrote in the case of a 13-year-old girl who had been treated by the vicar of Lappach, but he did not report this case either. Once, in 1891, Ottenthal even noted that he had examined the 'stock of a quack doctor' and found that the man's medical activities did not need reporting.[56] Ottenthal's commitment to the fight against charlatanism, which the health authorities so firmly demanded, was clearly rather limited. His apparent omissions in this respect might have been due to his wish to 'not cause a stir', which led him to largely tolerate traditional fellow-healers in his catchment area.

A case of special interest is that of a bandage seller in Salzburg, Austria, whom Ottenthal, in his position as locum forensic physician, was ordered to examine and inspect. The imperial district commissioner instructed Ottenthal to ascertain,

> ...whether the bandage seller in question, apart from offering bandages for sale, has carried out acts that fall within the remit of physicians or surgeons, such as applying bandages, taking measurements, repairing abdominal hernias, administering unguents, etc. and inform me of the result, if possible, within eight days'.[57]

Three days later Ottenthal wrote back: '...not found anything beyond his having been there'.[58] He seems to have neither strictly condemned such healing

53 The identity of these *agyrtae* remains secret in almost all cases. Only in the case of a man from St Johann, who was himself treated by Ottenthal, the latter noted down the term *agyrta* as job title. *Historiae Morborum* 892/1867 (5 June 1867), quoted in: Unterkircher, 'Konkurrenz', 2012, 167.

54 Cf. for instance *Historiae Morborum* 603/1855 (1 May 1855), *ibid.* 710/1873 (6 May 1873), or *ibid.* 456/1883 (22 Feb. 1883).

55 *Historiae Morborum* 992/1861 (14 Sept. 1861); see the similar case in *ibid.* 1003/1867 (11 July 1867).

56 *Ibid.* 1461/1891 (17 Oct. 1891), quoted in: Oberhofer, 2008, 171.

57 Südtiroler Landesarchiv, Nachlass Ottenthal, 252 files on individual cases, No. 5440, the royal-imperial district commissioner to Franz v. Ottenthal on 23/7/1886, quoted from: Taddei, 2010, 169.

58 Taddei, 2010, 169.

attempts nor feared their competition, having, as he did, more than enough patients. Even if, as an academically trained physician, he privately mistrusted the cures of lay-healers or was wary of them, he accepted the presence of non-licensed therapists in his medical territory and, above all, respected the choice of his patients.[59]

The Country Doctor's Patients

Ottenthal's practice was a typical country doctor's practice which presented him with a wide variety of tasks. In keeping with his training, he ministered to inner, organic diseases, carried out minor surgery such as pulling teeth or setting fractured bones and, in cases of emergency, acted as obstetrician if a midwife reached the end of her official competences or thought it wise to consult a physician.[60] There is no evidence that he specialized in particular syndromes or favoured a particular approach to treatment. Concerning the structure of his clientele, Ottenthal's country practice was in no way out of the ordinary either: the *Historiae* reveal the classic picture, characterized by a slight predominance of female patients, not as a result of his obstetric interventions, but due to his attending women suffering from menstrual or gestational problems.[61] There is no noticeable prevalence of a particular age group among his patients either, since all ages from babies to the elderly are included.[62] Patient numbers grew continuously, reaching a peak in the 1880s during the period when Ottenthal served as community and forensic physician. After this busy period the number of cases treated per year decreased steadily, settling in, after 1890, at the level of the first ten practice years. The *Historiae* mention an average of almost 4,000 physician–patient contacts per year, corresponding to an average of ten consultations per day.[63] Consultations include house calls (*visita*), consultations by letter, consultations by messenger (*nuntia* or *nuntius*), as well as face-to-face contact in the physician's house. Even though Ottenthal did not hesitate, in cases of emergency, to visit patients at night time (*noctu*) anywhere in the Taufer Ahrn Valley, the majority of consultations took place in his own rooms. According to Oberhofer,[64] Ottenthal

59 Cf. Unterkircher, Konkurrenz, 2012, 168–70.

60 Cf. the definition of country physician in Huerkamp, 1985, 185.

61 Cf. Hilber, 2012, 151.

62 Particularly for his deployment in cases of childhood illnesses cf. Unterkircher, 'Konkurrenz', 2012, 162–64. More detailed information on the gender ratio and on age-specific deployments of the physician can be found in Unterkircher, Jungen, 2014.

63 Cf. Oberhofer, 2008, 180.

64 Cf. *ibid.*, 176–77, 184–85.

tried to avoid long absences for house calls or vaccination rounds, probably due to the size of his catchment area and his duties as a forensic physician.

Although house calls were more profitable – Ottenthal charged thirteen kreutzer per kilometre for *tour et retour*[65] – Ottenthal would apply to be released from this duty, especially towards the end of his medical career:

> The Ebnerhof in Mühlwald, which lies at a high altitude, is a two-and-a-half to three-hour-walk from Sand. Walking there – driving is not an option – will have obvious adverse effects on a 77-year-old man considering the severity of the winter and the poor road conditions… The esteemed administrators, when issuing the decree regarding the physicians' duty to report, might not have sufficiently considered the greatly differing conditions pertaining to town and country practices, or the old wisdom that "the work is as good as the wages".[66]

Ottenthal's consultation habits might also reflect his awareness of the status due to him as an aristocrat and physician. Patients' going to see their doctor tend to underline the latter's superior social position. Ottenthal's patients were, for reasons of cost alone, compelled to see him in Sand, in the manor house he had inherited after his father's death. For the people of the valley, most of them peasants, who had to undertake long and arduous walks to see the doctor, the grand house must have seemed rather imposing. The topographer and statistician J.J. Staffler described Castle Neumelans in 1844 as 'nobler than any other edifice, with a family chapel and rounded walls, and with a lobby attached'.[67] While the set-up of waiting and ordination rooms or the storage rooms for medicines can unfortunately no longer be reconstructed, there is plenty of evidence of the doctor-patient contacts taking place there.

The Doctor–Patient Relationship

Despite his position as academic physician, medical expert and local figure of authority, Ottenthal's demeanour towards his patients was never authoritarian or based on hierarchical thinking. This is most obvious in cases where he took account of the patients' own medical ideas and therapeutic interests. 'As a private practitioner who did not draw a state salary, Ottenthal catered to the

65 Cf. *ibid.*, 176.

66 *Ibid.*

67 Johann Jakob Staffler, *Das deutsche Tirol und Vorarlberg, topographisch, mit geschichtlichen Bemerkungen II* (Innsbruck 1847), 246.

wishes of his patients in a particular way and prescribed what they came for or asked for in their letters or through their messengers'.[68] We find proof of this in the rising number of requests for 'laxative cures' from patients in spring and summer.[69] *Laxans desiderat quia in Mühlbach Balneum ire vult,* Ottenthal wrote in 1855 about the therapy requested by a man from Uttenheim, who intended to use the laxatives to prepare himself for a spa treatment.[70] The fact that patients came to the practice with clear ideas of their own regarding the treatment of their complaints is also apparent from the case of Wallburga P., probably a servant, who consulted Ottenthal for the first time on 2 February 1847 and who, on the very next day, demanded a more powerful medicine (*Plane non operavit laxans ergo magis energicam desiderat medicinam*).[71] Over the next eight years, the young woman consulted Ottenthal seventeen times, insisting in nine of those consultations on being prescribed laxatives for the most varied conditions (*ob dolores rheumaticos, ob dolorem sacralem*).[72] Franz von Ottenthal only rarely rejected his patients' wishes regarding treatment, but there is evidence that he had practical reasons for his attitude. His medical views were strongly informed by the notions of humoral pathology. He was not a modern physician. Like his patients he assumed that many illnesses were caused by the wrong diet. Ottenthal accepted the forms of self-medication that were wide-spread in rural areas. Facilitating them meant that he had at least some kind of control over them. For a 68-year-old man, for instance, who had tried to cure his *sciatica* by applying cow dung, he prescribed opium without openly condemning the man's prior attempt at self-medication.[73] In the case of a 49-year-old woman from Rein, who was plagued by an acute gastro-intestinal disorder accompanied by vomiting, diarrhoea and pain, Ottenthal was astonished by the chosen form of self-medication: 'she ate horse dung soup'.[74] Ottenthal was not in the habit of moralizing. He knew what people were like and accepted them, well aware that his influence was limited. His dedication to the patients is also apparent from his activities as assessor. Especially in cases of mental illness, the physician stood up for the personal rights of his patients. The numerous occasions, on which attempts to have a patient put

68 Taddei, 2010, 100.

69 Cf. Alois Unterkircher, 'Die Praxis des Südtiroler Landarztes Franz von Ottenthal. Krankheitsspektrum und Gesundheitsverhalten seiner Patientenschaft um 1860', in Rainer Alsheimer *et al.* (eds), 2005, 226; Oberhofer, 2008, 182.

70 *Historiae Morborum* 375/1855 (8 July 1855).

71 *Ibid.* 583/1847 (3 Feb. 1847).

72 Cf. e.g. *Historiae Morborum* 591/1852 (1 May 1852), *ibid.* 862/1854 (5 June 1854).

73 *Ibid.* 1198/1860 (15 Aug. 1860), quoted in: Taddei, 2010, 101.

74 *Ibid.* 632/1866 (2 June 1866). In the original German: '*Rossdrecksuppe hat sie gefressen*'.

under the care of a guardian, were frustrated by Franz von Ottenthal's clinical assessments, cast light on his protective and patriarchal attitude towards many of his patients that was not based on authority or guided by the wish to exert power.[75] Having said this, his assessments regarding fitness for military service stand out due to their exceptional rigour.[76] Ottenthal's consideration of patient interests also applied to the times patients could consult him. He received patients every day, including Sundays and holidays and, in cases of emergency, even at any time of day or night. On Sundays and market days the physician saw a particularly large number of patients,[77] a clear sign that he adapted to the rhythm and needs of a rural population. The stream of patients accordingly receded during harvest time.[78] Ottenthal's efforts to take into account the needs of his clients are also apparent from his habit of sending patients advice or medicine by messenger or of posting requested remedies without asking for the money in advance,[79] as illustrated by several letters enclosed in the *Historiae*. Moreover, urine samples delivered by messenger from remote patients were accepted by the physician even if he was occasionally dissatisfied with the inferior quality of the sample due to transport and weather conditions, as is apparent from this note: *urina in transportu turbida facta per frigus*.[80] Lastly, Ottenthal was relatively lenient as regards fee payments: he would grant patients respite, accept payment in kind or even waive his fee altogether: 'Partial payment 1fl [florin] 28 x [kreutzer], still owing 42 x [kreutzer] balance paid in fish' or accepted 'as pledge 1 sm[all] opera glass'.[81] He also lent his patients items for the proper administration of treatments: 'loan of clay bedpan', 'lent pair of scissors', 'lent glass syringe' or 'lent cataract glasses and lenses to Bozen'.[82] Ottenthal's support of the poor exceeded his purely medical duties, when, in 1867, he gave a woman patient '10 xr [kreutzer] to buy bread'.[83] Many patients were indebted to him, not only because of unpaid fees or medicines, but because he lent them money at an annual interest rate of 4 per cent.

---

75 Cf. Elisabeth Dietrich-Daum and Elena Taddei, 'Psychiatrische Versorgung am Land. Die Rolle der praktischen Ärzte am Beispiel der Südtiroler Landarztpraxis des Franz von Ottenthal (1818–1899)', *Virus. Beiträge zur Sozialgeschichte der Medizin* 8 (2009), 27–22, 34–36. See also: Dietrich-Daum *et al.*, 'Curare', 2008.

76 Cf. Unterkircher, Jungen, 2014, 265–268.

77 Cf. Oberhofer, 2008, 184.

78 Cf. *ibid.*, 183–84; Taddei, 2010, 92.

79 Cf. Oberhofer, 2008, 178.

80 *Historiae Morborum* 205/1897 (3 April 1897), quoted in: Oberhofer, 2008, 176.

81 Cf. Taddei, 2010, 124.

82 *Ibid.*

83 *Ibid.*

His main debtor was the Taufer communal road administration, the *Straßen-concurrenz*, that owed him the sum of 9,500 florins for road construction works.[84] The reason why the physician was able to grant such high loans – in this case the sum was eight-times his practice income of 1874[85] – was his considerable family property, his income as land-owner, interest from capital investments and lending as well as his allowances as local MP, as locum community and forensic physician and as court cashier. It was therefore no problem for him to write off one or the other patient fee despite the fact that his large household with three to six servants and the education and support of his sons must have incurred considerable expenses. Ottenthal could clearly afford to be generous.

An Integrated Country Practice?

In November 1890 the then 71-year-old physician wrote to the district court in Taufers asking to be released from the office as health assessor because of his age and failing health. Nine years later he began to wind down his private practice. In a letter dated 19 January 1899 he announced his retirement in February of that year. But Ottenthal died sooner than expected, on 29 January 1899, the last day of his professional life as a medical doctor, having treated patients right up to the day before his death.[86]

Apart from his legal heirs, his six servants inherited money from him, as was customary for an aristocratic employer. The local school fund received fifty, the poor fund one hundred and the hospital in Taufers also one hundred florins as well as Ottenthal's instruments and pharmacy.[87] These legacies show how firmly he was part of the local community and are proof of the protective and patriarchal attitude he maintained towards the community. As a physician, he occupied a distinct place within the healthcare network of the region; his services were indispensable. Even though his practice was rather expensive for his rural patients, his position within this network was neither marginal nor exceptional, but very central. Ottenthal was recognized and valued as a part of the medical market even by groups who were traditionally seen as 'avoiding the medical profession'. He largely succeeded, moreover, in arriving at consensual therapeutic decisions with his patients. In reconstructing his

84 Cf. *ibid.*, 65.

85 Cf. *ibid.*, 125–26.

86 Cf. *ibid.*, 213–14.

87 Cf. *ibid.*, 219.

everyday practice, one finds a striking absence of public conflicts. This low conflict level had, on the one hand, entirely individual reasons: thanks to this personal independence, Ottenthal was not in the permanent economic and professional competition with other health providers that other representatives of the rapidly professionalizing healthcare system were in. On the other hand, there may have been other, more general or structural reasons, too: it was essential for the functioning of the healthcare system in this rural area that physician and other medical providers worked together. The situation was presumably similar for the patients. The relative smoothness of the interactions does not mean that there were no tensions. Conflicts are ubiquitous in the *Historiae*, but they did not tend to escalate. While this conclusion might meet with scepticism, there is no evidence whatsoever of any such escalation. Instead of arguing with his patients or colleagues, the liberal physician and politician preferred to argue with the health authorities. Right up until his final years in practice, Ottenthal objected to the severe and sharp control measures implemented by the health authorities. His critical attitude in this respect may, however, have had more to do with his own socialization than with his profession or his position as a physician.

CHAPTER 13

A Special Kind of Practice? The Homeopath Friedrich von Bönninghausen (1828–1910)

Marion Baschin

More than 11,500 people consulted the physician Friedrich von Bönninghausen between 1864 and 1889. He noted down details of these consultations in his case journals, devoting a page to each patient. Today, these journals provide documentation of a medical practice in the second half of the nineteenth century in the Westphalian city of Münster.[1] On the basis of Bönninghausen's journals it is possible to reconstruct his activities against the general historical context.

Bönninghausen was, in fact, no 'orthodox' physician: although he had graduated from medical school and been sworn in by the authorities of the city of Münster as 'physician, surgeon and obstetrician' in 1862, he differed in one essential aspect from other physicians offering their services in the city at the time. He was a homeopath and treated his patients according to principles put forward by Samuel Hahnemann (1755–1843), who had founded this medical approach in the early nineteenth century.

Did this make his practice special and, if so, in what way was it different from that of other physicians? These are the questions that this paper seeks to investigate. We begin by introducing Bönninghausen. His case journals form part of this process because they constitute the source on which this research is based. We will then present one case history to illustrate Bönninghausen's mode of treatment. Next, Bönninghausen's homeopathic practice – as established from the journals – will be briefly outlined: its size, its clientele and the range of complaints treated. In conclusion, we will offer an answer to the question as to whether and to what extent Bönninghausen's practice was 'special'.[2]

1 The total number of patients treated was calculated by adding up the journal entries. The figure constitutes a bottom line, cf. notes 20 and 52. The case journals are kept at the Institute for the History of Medicine of the Robert Bosch Foundation (hereinafter: IGM), P 116–49.

2 For an overview of medical practices previously examined see Martin Dinges, 'Immer schon 60% Frauen in den Arztpraxen? Zur geschlechtsspezifischen Inanspruchnahme des medizinischen Angebotes (1600–2000)', in *idem* (ed.), *Männlichkeit und Gesundheit im historischen Wandel 1850–2000* (Stuttgart 2007), 295–322; Martin Dinges, 'Arztpraxen 1500–1900. Zum Stand der Forschung', in Elisabeth Dietrich-Daum, Martin Dinges, Robert Jütte and Christine Roilo (eds), *Arztpraxen im Vergleich. 18.–20. Jahrhundert* (Innsbruck 2008), 23–61; Marion Baschin, *Wer lässt sich von einem Homöopathen behandeln? Die Patienten des Clemens*

 | DOI 10.1163/9789004303324_015

Friedrich von Bönninghausen and the Case Journals

Friedrich Paul Joseph Baron von Bönninghausen was born on 14 April 1828 in Münster, Westphalia.[3] His early professional career did not seem to indicate that he would follow in the footsteps of his father, the lay homeopath Clemens von Bönninghausen (1785–1864). Like his father, he first read law before switching to medicine in 1855, although still under the age of 30. He studied at the universities of Bonn and Berlin, graduating cum laude in 1859 with a dissertation on diabetes (De diabete mellito). Once he had gained his medical license and had been sworn in, Bönninghausen could start practising in Münster.[4] As one of thirty-four physicians, he devoted himself to the needs of the sick there,[5] working at first as his father's assistant rather than setting up his own practice.[6]

Bönninghausen had come into contact with homeopathy at an early age: his father had treated him from childhood with globules and tinctures. Nothing is known about his motivation for taking up medicine. One of his brothers also chose to become a homeopathic physician, but did not practise in Münster.[7] It was probably Clemens von Bönninghausen himself who introduced his two sons to Hahnemann's doctrine. Both adopted their father's approach to drug administration. This is remarkable in that Clemens von Bönninghausen favoured the high

---

Maria Franz von Bönninghausen, 1785–1864 (Stuttgart 2010); for an overview of the research into the history of homeopathy cf. Robert Jütte, 'The Historiography of Homoeopathy in Germany', *Orvostörténeti Közlemények. Communicationes de Historia Artis Medicinae* 186/187 (2004), 123–30; Martin Dinges, 'The Current State of Research on the History of Homeopathy', in Claudia Witt and Henning Albrecht (eds), *New Directions in Homeopathy Research. Advice from an Interdisciplinary Conference* (Essen 2009), 13–41.

3 Cf. Hermann Schnütgen, 'Sanitätsrath Dr Friedrich von Bönninghausen†', *Allgemeine Homöopathische Zeitung* 58 (1910), 351–52; Fritz Schroers, *Lexikon deutschsprachiger Homöopathen* (Stuttgart 2006), 16; for detailed information on the sources cf. Marion Baschin, *Ärztliche Praxis im letzten Drittel des 19. Jahrhunderts. Der Homöopath Dr Friedrich Paul von Bönninghausen (1828–1910)* [Stuttgart 2014].

4 For general information on the training of young physicians at the time cf. Claudia Huerkamp, *Der Aufstieg der Ärzte im 19. Jahrhundert. Vom gelehrten Stand zum professionellen Experten. Das Beispiel Preussens* (Göttingen 1985).

5 Cf. Marion Baschin, 'Warum besucht man einen Homöopathen? Eine Praxis in Münster im 19. Jahrhundert', in Lina Gafner, Iris Ritzmann and Katharina Weikl (eds), *Penning Patient's Histories – Doctors' Records and the Medical Market in the 18th and 19th Century* [special issue: *Gesnerus* 69, no. 1 (2012)], 126–40.

6 On Clemens von Bönninghausen cf. the biography by Friedrich Kottwitz, *Bönninghausens Leben. Hahnemanns Lieblingsschüler* (Berg 1985); on his patients cf. Baschin, 2010.

7 On the treatment of the Bönninghausen family cf. Baschin, 2010, 182; Marion Baschin, 'Carl von Bönninghausen. Ein vergessener Homöopath und seine Lernzeit', *Medizin, Gesellschaft und Geschichte* 28 (2009), 237–80: 241; on Carl the information in Schroers, 2006, 15, cf. also Baschin, 2009.

potencies which, with very few exceptions, he prescribed in his practice.[8] It is a well-known fact that homeopathic medicines are never given in their pure form, but in dilution or 'potencies'. Hahnemann had produced empirical evidence for this principle in experiments carried out in his own practice. He was, however, unable to provide an entirely plausible explanation for the fact that the effect of a substance was not reduced, but enhanced, the more one diluted it.[9] One differentiates stages of dilution that contain the original active agent in ratios of 1 in 10 up to 1 in 50,000.[10] Even in Samuel Hahnemann's lifetime the use of high potencies, in other words dilutions of D30 or C15, was controversial.[11] While some of Hahnemann's pupils turned to decimal potencies and, in doing so, laid the foundation for a 'scientific-critical' approach to homeopathy,[12] others followed Hahnemann and experimented with high potencies. The latter referred to themselves as the 'true' homeopaths.

Clemens von Bönninghausen began to use the '200th centesimal dynamization' in 1844.[13] His success convinced not just him, but obviously also his sons,

---

8 Cf. Marion Baschin, 'How Patients Built the Practice of the Lay Homoeopath Clemens von Bönninghausen. Quantitative and Qualitative Aspects of Patient History, *Dynamis* 31, no. 2 (2011), 475–95; Baschin, 2009, 265–66; Luise Kunkle, 'Von Bönninghausens Verschreibungspraxis', *Zeitschrift für Klassische Homöopathie* 52 (2008), 172–78.

9 Cf. J Robert Jütte, *Samuel Hahnemann. The founder of homeopathy 2012*) [http://www.igm-bosch.de/content/language1/downloads/SamuelHahnemannTheFounderofHomeopathy.pdf], Chapter 3, 19–20, Chapter 6, 20–21.

10 Cf. Robert Jütte and Martin Dinges, *Homoeopathy. A Medical Approach and its History* (Stuttgart 2009), 8; the potencies are referred to as: 1 in 10 = D-potency, 1 in 100 = C-potency, 1 in 50,000 Q- or LM-potency. Dilutions from D1/C1 to D12/C6 are low potencies. On the history of the high potencies cf. Ursula Jacobi, *Der Hochpotenzstreit. Von Hahnemann bis heute* (Stuttgart 1994); Cf. J Robert Jütte, *The LM Potencies in Homoeopathy. From their Beginnings to the Present Day* (Stuttgart 2008) [http://www.igm-bosch.de/content/language2/downloads/lm_potencies(1).pdf].

11 The term 'high potencies' was coined by Gustav Wilhelm Groß (1794–1847). In Hahnemann's time there was a degree of uncertainty in regard to this, which is why C30 was not consistently included with the 'high potencies', cf. Jacobi, 1994, 14, 33–34.

12 The division had other reasons too, the most important being the question as to what position one should adopt regarding allopathic medicine and Hahnemann's views on the origin of chronic diseases, cf. Robert Jütte, *Geschichte der Alternativen Medizin. Von der Volksmedizin zu den unkonventionellen Therapien von heute* (München 1996), 182–84; Kathrin Schreiber, *Samuel Hahnemann in Leipzig. Die Entwicklung der Homöopathie zwischen 1811 und 1821. Förderer, Gegner und Patienten* (Stuttgart 2002), 92–96; Samuel Hahnemann, *Die chronischen Krankheiten. Theoretische Grundlagen. Mit allen Änderungen von der 1. Auflage (1828) zur 2. Auflage (1835) auf einen Blick*, revised by Matthias Wischner (Stuttgart ³2006); Ulrich Fischer, *Die Chronischen Miasmen Hahnemanns. Grundgedanken zum Verständnis und zur Therapie chronischer Krankheiten aus homöopathischer Sicht* (Karlsruhe 1993).

13 Marion Baschin, 'Die Medikation in der Praxis Friedrichs von Bönninghausen', *Zeitschrift für Klassische Homöopathie* 55 (2011), 116–21: 117.

for they went on to use this potency regularly and almost without exception.[14] Even though this says little about Friedrich von Bönninghausen's 'theoretical' background, we know for certain that, despite his academic training, he treated his patients exclusively according to homeopathic principles. Even among the homeopaths he occupied a special position because he did not adhere to the scientific-critical approach to homeopathy, which was predominant in the second half of the nineteenth century. Instead he used the high potencies, which meant that he followed Samuel Hahnemann's instructions to the letter. It is therefore not surprising that Bönninghausen showed little interest in 'collegial' exchange. His method of formulating his prescriptions positioned him within a homeopathic minority, standing in clear opposition to the 'orthodox medicine' of his time. We have certainly no proof of any kind of involvement on his part with medical associations, homeopathic or allopathic. He never held a public office, although that would have been a likely step for him to take. Nor did he ever work in public healthcare, but preferred to concentrate on his private practice, which was subject to supervision by the medical authorities.[15]

After his father's death in January 1864, Bönninghausen took on the paternal practice in Münster. He also maintained his father's method of record-keeping, using the pre-printed case journals. Clemens von Bönninghausen, a former civil servant of the Prussian government, had been well versed in the administration and organization of written records. In 1863, he wrote in the homeopathic journal *Allgemeine Homöopathische Zeitung* about the record-keeping system he had invented.[16] He used pre-printed pages that were bound into a journal. Each patient was allocated one page in the journal.[17] His system differed from those of other homeopaths or allopathic physicians, in particular because of its focus on the patient's 'symptoms'.[18]

14 Cf. Baschin, 'Medikation', 2011; *eadem*, 2009.

15 Kay Peter Jankrift and Ruth Schilling, 'Medical Practice in Context: Religion, Family, Politics and Scientific Networks' (in this volume), Baschin, 'Homöopath', 2012.

16 Clemens von Bönninghausen, 'Das Krankenjournal', *Allgemeine Homöopathische Zeitung* 67 (1863), 114–16, 121–23, 129–31, 140–41, 147–49, 163–65.

17 If necessary, the patient's history was continued on the back of the page or on subsequent reverse sides, cf. Volker Hess and Sabine Schlegelmilch, '*Cornucopia officinae medicae*: Medical Practice Records and their Origins' (in this volume) and the description of sources (in the appendix of this volume).

18 Cf. Hess and Schlegelmilch (in this volume). For homeopathic case records in particular cf. Robert Jütte, 'Case Taking in Homoeopathy in the 19th and 20th Centuries', *British Homoeopathic Journal* 87 (1998), 39–47; Martin Dinges, 'Hahnemanns Falldokumentation in historischer Perspektive', *Naturheilpraxis* 63 (2010), 1356–62; in general cf. Jonathan

In the top half of the page, the patient's name, age, place of residence and, occasionally, occupation were entered, followed by extensive notes regarding the symptoms described by this patient. Patients were also asked about any health regimes they had used previously. The bottom half of the page was used to write down the date of the consultation, the medicines prescribed, and the further course of treatment envisaged. Clemens von Bönninghausen had filled 115 volumes of such pre-printed case journals. His son started his own entries on 23 January 1864 in the 116th journal and continued to enter regular and more or less detailed accounts up until 1889. Further entries, as well as his obituary, seem to indicate, however, that he continued to see patients right up to his death in 1910.[19] For the purpose of 'reconstructing' his medical practice, we selected a sample from the well-documented years of Bönninghausen's activity: journals from a sixteen-year period were used as a basis for the research. During that time Bönninghausen saw at least 6,832 patients.[20]

While the evaluation of the journals and the subsequent 'reconstruction' of Bönninghausen's practice required us to make a few assumptions, since the notes do not reveal every detail of each medical treatment, these assumptions are highly plausible considering the known conditions surrounding the practice. There are 'fundamentals' and instructions regarding the homeopathic treatment process – published by Samuel Hahnemann in his *Organon of the Healing Art* – that can help categorize the notes available to us without there being any guarantee that Bönninghausen always followed them to the letter. There is no doubt that he was familiar with these principles and that he also applied them, as we will see shortly.

Gillis, 'The History of the Patient History since 1850', *Bulletin of the History of Medicine* 80 (2006), 490–511.

19 He died in Münster on 6 August 1910. Individual case histories dating from after 1889 can be found in IGM p 149, the corresponding information in Schnütgen, 1910. Across all the journals we find barely legible notes on various patients, most of them dated later than 1880, which continue from older entries without obvious coherence. These unsystematic notes were not suitable for closer evaluation.

20 The years under examination are 1864–1867 (3,720 patients), 1872–1875 (1,901 patients), 1879–1882 (621 patients) and 1886–1889 (590 patients). 'At least' here refers to the fact that Bönninghausen treated other patients, also in the years under examination, whose histories are written down in the journals started by Bönninghausen senior and are therefore not included in the data base. For details regarding the database see Baschin, 2010, 44–49; *eadem*, 'Patients', 2011; *eadem*, *Ärztliche Praxis*, 2014.

Homeopathic Therapy – a Case Study

Catharina K., a 61-year-old woman from Münster, attended Bönninghausen's practice for the first time on 14 July 1865. On the basis of her patient file some 'special features' of homeopathic treatment can be demonstrated.

Catharina K. described her complaints[21] and the homeopath wrote down the following symptoms, probably still during the conversation: 'For some years, now aggrav[ated] again, gouty headaches, piercing pain; cough without sputum, with easterly or north-easterly wind; in thundery weather, after sleeping, in the evening, when resting; with toothache; cravings for sour foods, dark bread, spices'. This jumbled collection of symptoms may seem bewildering at first glance, but is in keeping with the requirements of homeopathic case-taking.

Samuel Hahnemann first offered an in-depth account of the principles of his medical approach in 1810 in his work *Organon of the Healing Art*.[22] Among the main characteristics of this approach, some of which contradicted the medical knowledge taught at universities at the time, are the simile principle, the proving of drugs on healthy subjects, the extensive individual case-taking and the way medicines are prepared.[23] The simile principle ('may like be cured by like') states that a sick person can be healed with the substance that produces, in a healthy subject, symptoms that are most similar to those experienced by the sick person in question. According to the homeopathic view, illness is a disorder that affects the whole person. It can be seen as the sum total of all the symptoms that can be observed and described.[24] The patients and their subjective experience of illness therefore form the centre of the homeopathic treatment. The patient's description, in conjunction with the physician's observations, provide the key for choosing the 'right' medicine that, according to the principles of homeopathy, should effect a 'fast, safe and gentle' cure.[25] Homeopathic physicians and healers are still expected to listen

21 IGM p 118 Fol. 142. In the original: '*Seit mehreren Jahren, jetzt wieder verschl.*[*immert*] *Kopfgicht, wie Stiche; Husten ohne Auswurf, bei Ost-Nordostwind; – bei Gewitterluft, nach dem Schlafe, Abends, in der Ruhe; mit Zahnweh; Verlang*[*en*] *nach Sauer, Schwarzbrod, Gewürze*'. Cf. Hess *et al.* and the description of sources (in this volume).

22 Cf. Jütte, *Samuel Hahnemann*, Chapter 4.

23 Cf. Jütte *et al.*, 2009, 4–9; Jütte, *Samuel Hahnemann*, Chapter 1, Chapter 4.

24 On homeopathic symptomatology cf. Andreas Wegener, 'Homöopathische Symptomenlehre', in Thomas Genneper and Andreas Wegener (eds), *Lehrbuch Homöopathie. Grundlagen und Praxis der klassischen Homöopathie* (Stuttgart [3]2011), 50–78.

25 Cf. Samuel Hahnemann, *Organon of Medicine. Edition 6B*, ed. by Kurt Hochstetter (Santiago–Chile 1977), § 2.

carefully to a patient's history and take extensive notes during the process. The homeopath should ask questions to obtain further relevant information only once the patient has finished his or her report. Hahnemann published clear instructions about individual case-taking in the *Organon*. These instructions have hardly changed across the various editions of this work.[26]

Hahnemann proposed that the choice of the right medicine should always be determined by the 'conspicuous, peculiar, unusual and idiosyncratic (characteristic)' symptoms.[27] The way in which an ailment developed also played an important part, as did the 'modalities', i.e. the circumstances that cause the symptoms to improve or deteriorate. If such modalities are considered, frequent and therefore common symptoms such as fatigue, which appears with almost any illness, can be indicators for the choice of a particular medicine.[28] Clemens von Bönninghausen had attached particular importance to these 'modalities' in choosing the right medicine.[29] Each symptom needed to be clearly defined. The kind of pain the patient experienced had to be specified: was it a dragging or rather a piercing pain? In the case of Catharina K. therefore, Bönninghausen wrote down what was probably the most painful symptom ('gouty headache') first. The patient specified that her pain was 'piercing'. She also suffered from cough without sputum, a dry cough in other words. Had there been sputum she would have been asked to describe it in great detail. The case history reveals that the complaints were aggravated in 'easterly or north-easterly wind', in 'thundery weather, after sleeping, in the evenings' and 'when resting'. These are the modalities mentioned above. Catharina K. also suffered from toothache and mentioned particular food cravings.

26 *Ibid.*, §§ 83–104, on earlier editions see Robert Jütte, 'Homöopathie. Das besondere Arzt-Patienten-Verhältnis', in Martin Momburg and Dietmar Schulte (eds), *Das Verhältnis von Arzt und Patient. Wie menschlich ist die Medizin?* (München 2010), 71–87: 80. For modern-day instructions and explanations for carrying out homeopathic case-taking, cf. Thomas Genneper, 'Die homöopathische Anamnese', in Thomas Genneper *et al.*, 2011, 80–106.

27 Hahnemann, 1977, § 153.

28 Cf. Wegener, 2011, 56–57.

29 Bönninghausen noted down, inside the cover of journal IGM P 117, a verse by his father to help him remember the full recording of a symptom. Andreas Wegener, 'Mittelfindung nach dem Therapeutischen Taschenbuch von Bönninghausen', in Thomas Genneper and Andreas Wegener (eds), *Lehrbuch Homöopathie. Grundlagen und Praxis der klassischen Homöopathie* (Heidelberg 2001), 139–51 (= Stuttgart ^{3}2011), 118–30; Bernhard Möller, 'Die Methodik Clemens von Bönninghausens. Dargestellt anhand seines Therapeutischen Taschenbuchs', *Die Bönninghausen- und Bognermethodik. Homöopathie-Zeitschrift*, Sonderheft 2002, 6–25; Genneper, 2011, 86–90.

Bönninghausen would, either during the conversation or afterwards, mentally compare these symptoms with the drug pictures of substances known to him from the *materia medica*. But the homeopathic *materia medica* includes a great number of substances. There are repertories that list substances according to categories to help homeopaths choose the right simile.[30] Apart from relying on his own expertise, Bönninghausen would have used the repertories his father had published: the two-volume *Systematic, Alphabetic Repertory of Homeopathic Medicines* of 1832/35 and the 1846 *Therapeutic Pocket-Book for Homeopaths*. The latter was restricted to 125 remedies that were considered well proven in Clemens von Bönninghausen's lifetime.[31] Finding the 'right' remedy can nevertheless be a challenge, even for experienced homeopaths. Symptoms might be ambiguous and not allow for immediate identification of one particular substance as the suitable simile. Catharina K.'s case history shows that Bönninghausen decided to give sulphur first, but changed his mind even before the consultation was over and gave the woman four little sachets of crushed homeopathic globules. Three of the sachets contained lactose and the fourth Spigelia as the only active agent, in C200, Bönninghausen's preferred potency. He also advised his patient on how to take the remedies. At particular intervals, which are not specified for Catharina K., she had to dissolve one of the powders in water and drink it.[32]

Bönninghausen did not often use Spigelia.[33] It is not clear from the case history what kind of considerations led him to decide in favour of its use on that occasion.[34] He may have asked his patient whether her pain was more to the left or to the right side of the head or whether she could describe other symptoms in more detail.[35] Unfortunately he omitted to write down this information. This is one of the weak points of the patient journals. It is also known that the homeopath did not only rely on his patient's oral account. There are

30 For an overview of the various repertories and their usage cf. Klaus Holzapfel, 'Repertorien und Repertorisation', in Thomas Genneper *et al.*, 2011, 143–69; for the repertory according to Bönninghausen mentioned below cf. *ibid.*, 147–49 as well as Clemens von Bönninghausen, *The Bönninghausen Repertory. Therapeutic Pocketbook Method*, ed. by George Dimitriadis (Sydney [2]2010).

31 Cf. Baschin, 'Medikation', 2011, 117. Bönninghausen generally referred to his father's publications as we know from copies inside the journal covers.

32 For these instructions regarding drug usage of which only a single detailed one is known, cf. Baschin, 'Medikation', 2011.

33 On medication see *ibid.*

34 For this problem see *ibid.*

35 Cf. Hahnemann's 'drug picture' of Spigelia from 1881, 479–508, where he spoke of stabbing sensations, dry cough and accompanying toothaches. According to Bönninghausen, 2010, 183–84, Spigelia had proved to be particular effective in cases of 'stabbing' pain.

other case histories that mention examinations. Certain remarks about regular or irregular heart tones indicate that Bönninghausen carried out chest auscultations, a technique he had acquired as part of his 'orthodox medical training'.[36]

Homeopathic therapy also included dietary instructions.[37] The diet was not only meant to guide the patient 'back to a natural lifestyle', it would also ensure that the effect of the homeopathic medicine was not 'disturbed by other medicinal stimuli'.[38] This kind of interference was said to be expected from a number of foods and stimulants. Bönninghausen would probably have advised Catharina K. to stay clear of alcoholic drinks as well as tea or coffee while she was taking the powders. The same applied to strong smells and spices such as camphor or camomile.[39] Another patient from the Münster area had consulted Bönninghausen specifically because of a 'stomach upset' she had suffered from after eating 'cucumber salad'.[40] Even though the patient would have been told to adhere to the particular diet indicated in her case and that it would be wise to consume certain foodstuffs only with care, she kept coming back to the homeopath, complaining that she had 'corrupted' her stomach with coffee, beer or Madeira wine, cream cakes and cheese. Catharina K. did not always observe the dietary instructions she was given either. When she consulted Bönninghausen on 9 August 1865, probably by messenger, because she had a toothache, he wrote down 'n[on] v[idi]', which meant that he had not seen the patient on this occasion. A note indicates that the patient had 'drunk cam[omile]'.[41] This made it more difficult

36 Among Bönninghausen's teachers was the surgeon Ludwig Traube (1818–1876), a representative of the scientific-critical school of medicine, who was known for his auscultation and percussion courses, cf. August Hirsch (ed.), *Biographisches Lexikon der hervorragenden Ärzte aller Zeiten und Völker*, vol. 5 (München ³1962), 625–26; cf. Annemarie Kinzelbach, Stephanie Neuner and Karen Nolte, 'Medicine in Practice: Knowledge, Diagnosis and Therapy' (in this volume).

37 Cf. Hahnemann, 2006, 68–72. In 1833 Clemens von Bönninghausen had written an instructive paper on essential dietary rules (Clemens von Bönninghausen: *Die homöopathische Diät und die Entwerfung eines vollständigen Krankheitsbildes behufs homöopathischer Heilung für das nichtärztliche Publikum* (Münster 1833)) which was also used by Samuel Hahnemann, cf. Jütte, *Samuel Hahnemann*, Chapters 6 and 12; on the patients' attitudes towards such instructions cf. Bettina Brockmeyer, *Selbstverständnisse. Dialoge über Körper und Gemüt im frühen 19. Jahrhundert* (Göttingen 2009), 117–32; Baschin, 2010, 358–61.

38 Bönninghausen, 1833, 6.

39 Bönninghausen, 1833, 13–18.

40 IGM P 118 Fol. 398.

41 *Ibid.* Fol. 142. In the original German: '*Nun seit heute Nacht Zahnweh, nach Warm schl*[*immer*], *Warmwerden in Bette; Waß*[*er*] *beß*[*essert*] *n*[*on*] *v*[*idi*] *Cham*[*ille*] *getrunk*[*en*]'. We find such notes about patients ignoring dietary instructions frequently with both Friedrich and Clemens von Bönninghausen, cf. Baschin, 2010, 358–61.

for the homeopath to choose a remedy for the toothache. Catharina K. had obviously not complied with his instructions to avoid this herb. Bönninghausen gave Pulsatilla, a choice he often made when a patient complained about toothache.[42] He either handed the preparation over to the messenger or sent it by post. Since the previous complaints were no longer mentioned on this occasion, we are left wondering whether the administration of Spigelia had alleviated the 'gouty headache'.

What else can we learn about Bönninghausen's practice from the case history of Catharina K.? His note regarding previously tried cures is of interest. Under the heading 'allopathic treatments' he wrote, in the case of this patient, that she had used 'mustard-ant-spirit' and that she had consulted 'Dr Riefenst[ahl]'. Dr Riefenstahl was one of many rival allopathic physicians practising in Münster.[43] Like Catharina K., the majority of patients had tried other therapies before deciding to see the homeopath.[44] The patient who suffered from constant stomach problems had used unspecified 'home remedies' before she consulted Bönninghausen for the first time. Before she came to see him for the second time, she had also tried 'home remedies' against toothache, but it is not apparent what they were. It seems that the toothache was not her reason for seeing the homeopath this time. She was suffering from 'obstruct[iones]' for which he prescribed Nux vomica and Opium in the C200 potency. This shows that even patients who saw Bönninghausen repeatedly did not feel obliged to put all their trust into homeopathy.[45] It needs pointing out, however, that allopathic physicians also had the problem that their patients saw other medical practitioners as well. They clearly did not feel obliged to remain with one practitioner or follow his advice over a longer period of time.[46]

42 Cf. Baschin, 'Medikation', 2011, 118; 40 per cent of patients with toothache were given Pulsatilla.

43 Baschin, 2010, 128 on that physician.

44 Baschin, 'Homöopath', 2012; with 3,623 of the 6,832 patients (53 per cent) we find comments on previous allopathic treatments.

45 We find evidence for this in Baschin, 'Homöopath', 2012; *eadem*, 2010, 361–62, with further reading suggestions as well as Schreiber, 'Samuel Hahnemann', 2002, 135; cf. also Samuel Hahnemann, *Krankenjournal D 38 (1833–1835)*, ed. with a commentary of the transcription by Monika Papsch (Stuttgart 2007), 63–64; Brockmeyer, 2009.

46 Cf. Marion Baschin, 'The Patients' Choice. How and Why Sick People Used Homeopathy in 19th Century Münster', in Martin Dinges (ed.), *Medical Pluralism and Homoeopathy in India and Germany (1810–2010). A Comparison of Practices* (Stuttgart 2014), 149–65; cf. with further reading, e.g. Mary Lindemann, *Health and Healing in Eighteenth-Century Germany*

Much can be learned about a physician's approach from a single – or just a few – patient histories. The case presented gives insight into the 'special features' of homeopathic therapy. These are primarily the prescriptions based on the simile principle and the use of potentized medicines. The in-depth case-taking at the first consultation without establishing a diagnosis in the usual sense, is another aspect that is distinctive of homeopathy. The fact that allopathic physicians in the second half of the nineteenth century relied equally on the history of their patients as a basis for deriving a diagnosis and treatment regimen[47] might give the impression that the patient-focused approach typical of homeopathy was not all that different from the practice of other physicians at the time. But allopathic physicians tended to attach less importance to the subjective descriptions of their patients when establishing a diagnosis and treatment plan as case-taking tended to consist of patients answering the physician's questions.[48] The way homeopaths 'listened' to their patients and the value they attached to the patients' subjective sensations were therefore, even in those days, 'distinctive features' of homeopathy.[49] Further aspects of the homeopathic practice emerge when a number of case histories are evaluated, rather than only one.

(Baltimore 1996), 365; Lucinda McCray Beier, *Sufferers and Healers. The Experience of Illness in Seventeenth-Century England* (London 1987), 57.

47 Cf. Martin Dinges and Klaus Holzapfel, 'Von Fall zu Fall. Falldokumentation und Fallredaktion Clemens von Bönninghausen und Annette von Droste-Hülshoff', *Zeitschrift für klassische Homöopathie* 48 (2004), 149–67: 149–50; Volker Hess, 'Samuel Hahnemann und die Semiotik', *Medizin, Gesellschaft und Geschichte* 12 (1993), 177–204; Volker Hess, *Von der semiotischen zur diagnostischen Medizin. Die Entstehung der klinischen Methode zwischen 1750 und 1850* (Husum 1993). Ottenthal also often refrained from establishing a diagnosis, cf. Marion Baschin, Elisabeth Dietrich-Daum and Iris Ritzmann, 'Who Consulted the Doctor? The Patients and their Motives' and Elisabeth Dietrich-Daum, Marina Hilber and Eberhard Wolff, 'An Alpine Doctor's Private Practice: Franz von Ottenthal (1847–1899)' (both contributions in this volume).

48 For the relevant examples cf. Jütte, *Samuel Hahnemann*, Chapters 1, 4, 6, and 12; Lindemann, 1996, 300–03; Gillis, 2006; Jens Lachmund and Gunnar Stollberg, *Patientenwelten. Krankheit und Medizin vom späten 18. Jahrhundert bis zum frühen 20. Jahrhundert im Spiegel von Autobiographien* (Opladen 1995), 203–09.

49 Michael Stolberg, 'Krankheitserfahrung und Arzt-Patienten-Beziehung in Samuel Hahnemanns Patientenkorrespondenz', *Medizin, Gesellschaft und Geschichte* 18 (1999), 169–88; Martin Dinges, 'Patients in the History of Homoeopathy', introduction to *idem* (ed.), *Patients in the History of Homoeopathy* (Sheffield 2002), 1–32, 4–6; Brockmeyer, 2009, 72–90; Jütte, 2010.

The Practice – The Journals Tell Their Stories

Leafing through case journal P 118 reveals that, on the day when Catharina K. first attended the Baron's practice, the histories of three other first-time patients were also taken. Catharina was therefore not the only person to consult the homeopath on Friday, 14 July 1865. Apart from these 'newcomers', four other patients came to see Bönninghausen for follow-up appointments, bringing the total number of patients up to eight for that day. There was certainly enough time for taking the history of each new patient or to give careful attention to new symptoms.[50] Over the years, the number of first-time consultations varied considerably. While in the early years an average of 930 patients saw Bönninghausen for the first time, there were only 155 in the final years of his practice.[51] In the sixteen years investigated more closely, an average of just 430 new patients consulted the Baron per year. The overall number of consultations was higher because many patients came more than just once. During the years under investigation, 22,537 encounters with patients were documented, corresponding to an average of 1,409 treatments per year.[52] Between 1864 and 1889, the number of actual doctor–patient encounters also decreased. Bönninghausen's overall workload was relatively modest at that time: he treated five patients per working day, two of which were first-time patients.[53]

This 'low' workload did not necessarily distinguish Bönninghausen's practice from that of other homeopathic or allopathic physicians.[54] It is remarkable, however, that Bönninghausen gained a large number of patients especially in the early years of his practice, and that he was able to increase the number of new patients. Beginners in the profession usually had to work hard to set up a 'customer base' and acquire a good reputation by curing patients

---

50 Nothing is known about consultation hours. If Bönninghausen devoted five hours a day to his practice, as Hahnemann and his father used to do, he would have had forty-five minutes per patient, cf. Jütte, 2010, 83; Baschin, 2010, 332–33.

51 Cf. Philipp Klaas, Hubert Steinke, Alois Unterkircher: 'Daily Business: the Organization and Finances of Doctors' Practices' (in this volume).

52 Since Bönninghausen continued to see former patients of his father's and their first case-taking is not included in the data-base, this figure represents a minimum. More patients came in later years whose first consultations took place outside the period of time under investigation and who are therefore not included in the evaluation.

53 Regarding the workload during the individual practice periods under investigation, cf. Klaas *et al.* (in this volume).

54 Cf. Klaas *et al.* (in this volume).

successfully.[55] Bönninghausen probably benefited from being able to take over a well-established practice. If one directly compares the number of actual physician–patient encounters and the number of first-time callers per day with those of the father it, appears that Bönninghausen Jr was considerably less busy as a physician.[56] For Samuel Hahnemann the number of patients varied between four and ten per day.[57] The Belgian homeopath Gustave van den Berghe (1837–1902) also saw two to three new patients per day between 1869 and 1902 and similar numbers apply for the French homeopath Leon Vannier (1880–1963) who practised in Paris from 1930 to 1954.[58] The physician Johann Friedrich Glaser (1707–1789), who practised in the Harz region of Germany, carried out an average of three consultations per day in 1750.[59] Overall, Bönninghausen's workload compares to that of the Canadian country doctor James Miles Langstaff (1825–1889), whose average number of patients grew from only one per day at the beginning to nine, but the development was

55 Cf. Michael Stolberg, 'Patientenschaft und Krankheitsspektrum in ländlichen Arztpraxen des 19. Jahrhunderts', *Medizinhistorisches Journal* 28 (1993), 3–27: 15. The physician, who, with 328 new patients per year, had the smallest practice there, was new to the job, cf. Andrea Thümmler, 'Rekonstruktion des Alltags eines thüringischen Arztes im 18. Jahrhundert anhand seines Praxistagebuchs 1750–1763' (Med. diss., University of Berlin, 2004), 39; Andreas Oberhofer, 'Eine Landarztpraxis im 19. Jahrhundert am Beispiel der Ordination des Dr Franz von Ottenthal (1818–1899)', in Dietrich-Daum *et al.*, *Arztpraxen*, 2008, 167–91: 180. The same applied to Hahnemann after one of his relocations, cf. Schreiber, 'Samuel Hahnemann', 2002, 141–42; Kathrin Schreiber, 'Was Hahnemann Driven out of Leipzig? The Leipzig Practice and Why Hahnemann Moved to Köthen in 1821. Patient Numbers and Polemics', in Dinges, 2002, 53–63: 56–60.

56 Cf. Baschin, 2010, 321. Clemens von Bönninghausen saw five to eleven patients per day.

57 Cf. Robert Jütte, 'Samuel Hahnemanns Patientenschaft', in Martin Dinges (ed.), *Homöopathie. Patienten, Heilkundige, Institutionen. Von den Anfängen bis heute* (Heidelberg 1996), 23–45: 27–29, 37. (Between the years of 1801 and 1831, Hahnemann had between four and nine consultations per day), cf. Dinges, 2008, 45; for 1834/35 cf. Papsch, 2007, 27, only speaks of three treatments per day.

58 Cf. Anne Hilde van Baal, 'In Search of a Cure. The Patients of the Ghent Homoeopathic Physician Gustave A. van den Berghe (1837–1902)' (PhD diss., University of Amsterdam, 2004; published under the same title: Rotterdam 2008), 51; Olivier Faure, 'La clientèle d'un homéopathe parisien au XXe siècle. Recherche sur les patients de L. Vannier 1928 1948', in *idem* (ed.), *Praticiens, patients et militants de l'homéopathie aux XIXe et XXe siècles (1800–1940)* (Lyon 1992), 175–196: 181.

59 Cf. Thümmler, 2004, 39. The physician Soemmerring (1755–1803) also only carried out two to three consultations per day in 1796, even though there could be more than ten on occasion, cf. Franz Dumont, 'Nicht nur Hölderlin. Das ärztliche Besuchsbuch Soemmerings als Quelle für sein soziales Umfeld in Frankfurt am Main', *Medizinhistorisches Journal* 28 (1993), 123–54: 127.

reversed in the case of the homeopath.[60] Other physicians had much more work in comparison. The Bochum physician Carl Arnold Kortum (1745–1824), for instance, saw around ten patients a day.[61]

It cannot be said with any certainty why the number of patients who came to Bönninghausen for treatment declined. On the one hand, this kind of decline has been observed with other practices too, when physicians with advancing years deliberately reduced their working hours,[62] but Bönninghausen was only fifty-one years old in 1879. It is conceivable, on the other hand, that a 'saturation point' had been reached and the number of potential clients was exhausted after some years.[63] Moreover, competitors in Bönninghausen's special field of homeopathy had appeared on the horizon in 1878.[64] Patients keen on trying out Hahnemann's method now had a choice. The discovery of various pathogens and the resulting euphoric celebration of 'orthodox medicine' might have contributed to a decline in the demand for homeopathic cures from the 1880s onwards.[65] The fact that the Baron was not under contract with the sickness funds might also have kept patients away. It is known, however, that Bönninghausen, despite his relatively low number of physician–patient contacts, had no financial worries.[66] Thus, the decline in numbers could be the result of a deliberate decision on his part to accept fewer patients.

---

60 Cf. Jacalyn Duffin, 'A Rurale Pratice in Nineteenth-Century Ontario. The Continuing Medical Education of James Mile Langstaff', *Canadian Bulletin of Medical History* 5 (1988), 3–28: 6–7.

61 Cf. Wolfgang Balster, 'Medizinische Wissenschaft und ärztliche Praxis im Leben des Bochumer Arztes Karl Arnold Kortum (1745–1824). Medizinhistorische Analyse seines Patiententagebuches' (Med. diss., University of Bochum, 1990), 131. The number was slightly higher for Grotjahn (1794–1872), a physician in the Harz region of Germany. Regina Engel, 'Das Patientengut eines praktischen Arztes um die Mitte des 19. Jahrhunderts, gezeigt am Beispiel des Patientenjournals von Dr Heinrich Grotjahn in Schladen am Harz' (Med. diss., University of East-Berlin, 1978), 53, 94.

62 The phenomenon was observed in his father's practice as well as with the Canadian physician Langstaff, the Tyrolean physician Ottenthal (1809–1899) and the French homeopath Vannier, cf. Baschin, 2010, 318; Jacalyn Duffin, *Langstaff. A Nineteenth-Century Medical Life* (Toronto 1993, repr. 1999), 36, Tabelle 2.2; Oberhofer, 2008, 181; Faure, 1992, 181.

63 Schreiber, 'Samuel Hahnemann', 2002, 144–45, observed a similar phenomenon for Hahnemann's practice in the years 1815 and 1816. She concluded that the drop in new patients was due to 'a decrease of public interest in Hahnemann'.

64 Cf. Baschin, 'Homöopath', 2012; they also worked on a contractual basis with the sickness fund and followed the scientific-critical approach to homeopathy.

65 Patient numbers also declined for the Belgian homeopath van den Berghe after 1881, cf. Baal, 2004, 49, Fig. 2, 51.

66 See the obituary, which mentions that Bönninghausen and his wife were very wealthy, cf. Schnütgen, 1910. One needs to consider that the regular entries stopped in 1889 and that,

The homeopath's clientele hardly differed from that of other physicians.[67] The majority of patients were – like Catharina K. – female adults. Bönninghausen also treated many children. Indeed, children from birth to the age of five constituted his largest patient group. As time passed, the average age of the patients increased. Most of his adult patients were between twenty-six and thirty years old when they first consulted him. The majority of his patients came from the immediate surroundings of the city of Münster and few had to travel more than fifty kilometres. While members of all social strata were represented, the occupations of most patients indicate that they belonged to the lower social classes. Homeopathic treatment was nonetheless, up until 1889, increasingly sought by members of the middle classes. These findings are comparable to those on other physicians' practices and are therefore not indicative of a fundamental difference between homeopaths and other doctors. We certainly do not find any evidence, in the case of this Münster doctor's practice in the second half of the nineteenth century, of such differences between the composition of homeopathic and allopathic patient clienteles as we see today.[68]

The complaints that led patients to consult Bönninghausen did not differ considerably from those of patients in other doctor's practices either. The homeopath's patients neither suffered particularly frequently from chronic diseases nor were their afflictions on the whole more difficult to treat.[69] Overall, the range of illnesses treated by Bönninghausen tended to be of a 'general medical' nature. Conditions such as 'fevers'[70] or gastro-intestinal complaints were frequent, with patients' descriptions typically focusing on excretions, headaches and coughs.[71]

---

from 1880 at the latest, patient histories were added randomly to all the journals. These were not included in the evaluation (see note 19). The apparent decline in patient and consultation numbers has therefore also source-related causes.

67 For more detailed information and the corresponding categories cf. Baschin *et al.* (in this volume).

68 For more information and further reading suggestions cf. Martina Günther and Hans Römermann, 'The Homoeopathic Patient in General Practice. Findings of a Comparative Poll of Patients in Conventional Medical Practices and Homoeopathic Private and Health Insurance Scheme Practices', in Dinges, 2002, 281–99; Ursula Sharma, *Complementary Medicine Today. Practitioners and Patients* (London ^{2}1995), 18–24.

69 Cf. Baschin, 'Homöopath', 2012; nowadays, patients with such complaints constitute the majority of the homeopathic clientele, cf. Günther *et al.*, 2002, 288–89; Sharma, 1995, 24–26.

70 This included any unusual incidence of chilliness, warmth, shivering or sweating, cf. Bönninghausen, 1833, 38.

71 Cf. Baschin *et al.* (in this volume). The same result was arrived at for the practice of the father and of the Belgian homeopath, cf. Baschin, 2010, 310–13; Baal, 2004, 198.

It is apparent that a greater number of patient histories need to be evaluated before essential aspects of the practice can be derived. Even if quantification can prove difficult with some of the characteristic properties, it is a necessary step, without which, important information about the practice would be excluded.[72] One must also not forget that the basis of a doctor's practice is the sum of the individual patients' decisions to consult a physician – in this case the homeopath Bönninghausen – at least once. An investigation into a physician's everyday practice therefore needs a combination of qualitative and quantitative approaches.

A Special Kind of Practice?

The patients who consulted Friedrich von Bönninghausen were certainly aware of his homeopathic approach.[73] Providing medical care according to homoeopathic principles was, however, the only aspect that distinguished his practice from others. His approach included the prescription of potentized medicines based on the simile principle and focusing on the patients and the symptoms they described. While it is true that allopathic physicians also based their diagnosis and treatment on patients' statements, the kind of attention devoted by a homeopath to his patients and their descriptions was 'special' even then.

Yet the 'special' circumstance of Bönninghausen offering homeopathic treatment had little effect on the organization and everyday routine of the practice. His services were as much in demand as those of other physicians already investigated. The homeopath's clientele was not different from that of allopathic doctors. The patients behaved in a confident manner: they did not always feel obliged to follow doctor's orders and used a variety of medical approaches if necessary. The range of illnesses treated was the same as that of a 'general practitioner'. The number of consultations Bönninghausen carried out and the number of patients he saw are comparable with those of other physicians. The unusual decrease in medical activity is best explained by personal and general circumstances, such as increasing 'homeopathic competition'. But these are factors that could well apply to allopathic doctors, too. Therefore, there is good reason to conclude that, in the second half of the nineteenth century, a homeopathic doctor's practice hardly differed from that of other physicians.

---

72 For more detail on these problems cf. Baschin, 2010; *eadem*, 'Patients', 2011; Dinges, 2008.

73 This particular qualification was mentioned in the telephone directory. It was also known in Münster that his father was a famous homeopath, cf. Baschin, 'Homöopath', 2012.

The Sources

Johannes Magirus: *Diarium Medicum* (1647–1656)

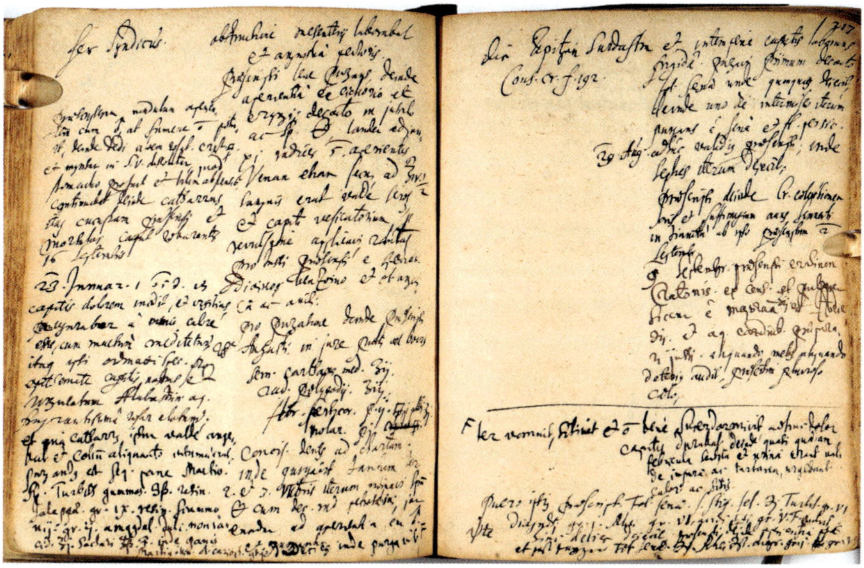

FIGURE 14.1 *Johannes Magirus,* Diarium Medicum.
UNIVERSITY LIBRARY MARBURG.

Description

One volume in quarto-format of 620 pages, 540 of which are used for the *Diarium.* It contains, in chronological order, case histories from Berlin (from September 1647 until 1650) and Zerbst (up to April 1656), and, in part, the costs of prescriptions and settlement thereof (Berlin). The *Diarium* was pre-bound and paginated by Magirus (with errors). All pages are folded in the middle to create two columns, the entries are written in commonplace style (name or occupation | observation). Individual treatments in the consecutive notes (the ink is sometimes changing) are usually not dated. At the end of each page the notes change from the right to the left column. If space runs out, the case history is continued on the next empty page. The *Diarium* contains an (incomplete) name index and references to earlier diaries of which there must have been at least three. Special entries: two cost reports (trade with herbal wine and beer), draft letters, consilia, a dissection report, a list of books he lent out. Language: German/Latin.

 | DOI 10.1163/9789004303324_016

Further records from this practice: one volume of *Loci communes*, three notebooks.

Johann Christoph Götz: *Observationes et curationes Nurimbergenses* (1716–1726)

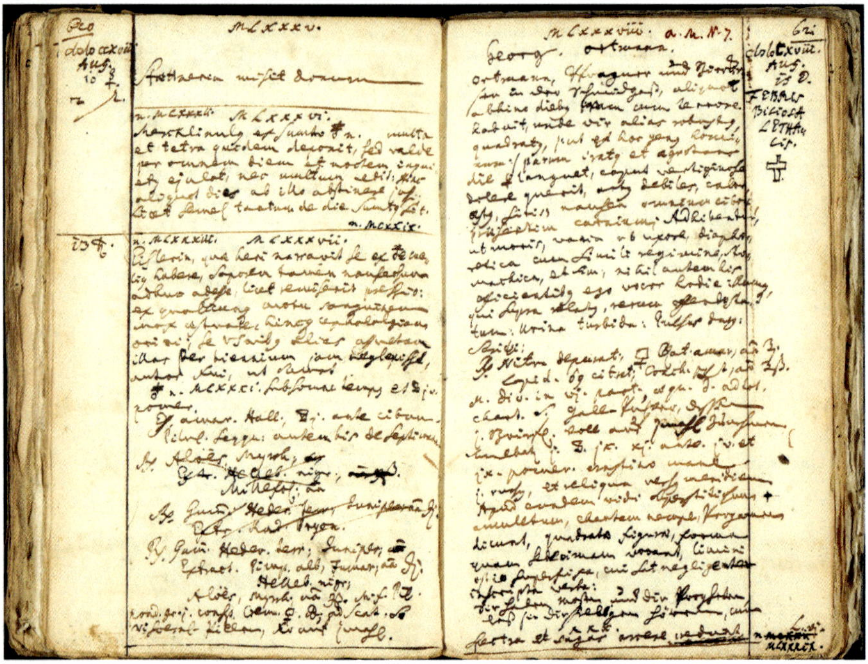

FIGURE 14.2 *Johann Christoph Götz,* Observationes et curationes Nurimbergenses. UNIVERSITY LIBRARY ERLANGEN, MS. 1200/3, 620-I.

Description

Seven octavo volumes (10×16 or 18×22 cm), up to 600 pages from the period between 1716 and 1726. The chronological records were divided into days, with serial numbers for the patients treated that were extended in alphabetical order for each follow-up consultation. Entries were made several times a day (changing ink). The *Observationes* include information on payments, third-person entries regarding prescriptions, references and indices. Some letters or draft letters are included. Some case histories were used for a publication. Language: Latin.

Further records from this practice: one account book (1720).

Johann Friedrich Glaser: *Medizinisches Register* (1750–1763)

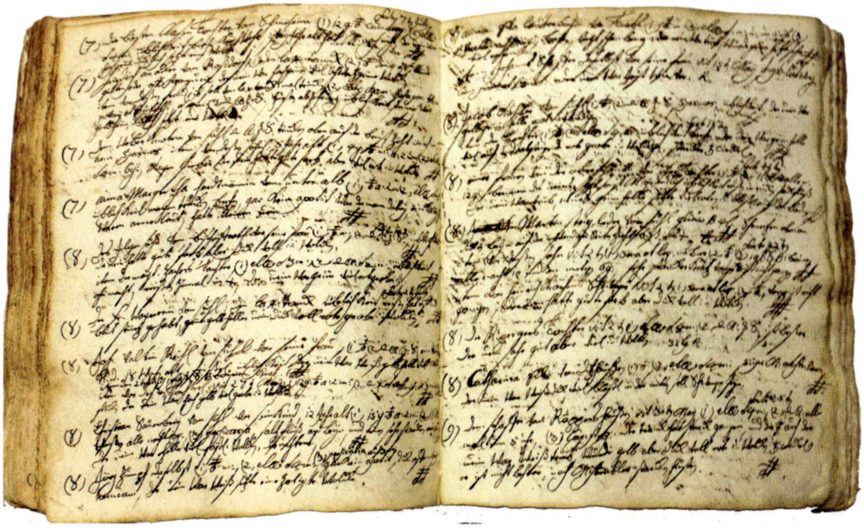

FIGURE 14.3 *Johann Friedrich Glaser,* Medizinisches Register.
INSTITUT FÜR GESCHICHTE DER MEDIZIN DER CHARITÉ, BERLIN.

Description

One volume (probably of several) in octavo (17×21 cm), with 1,200 pages from the period between 1750 and 1764. The book consists of smaller journals that were (presumably subsequently) bound together. It contains records of consultations in chronological order noting name, profession and place of origin of the visitors, the complaints, prescriptions, diagnoses and comments regarding payment. All entries are written in German script (with the exception of technical terms in Latin). Only a few enclosures (prescriptions). No index.

Further records from this practice: *Observationes* prepared for publication.

Gottfried Wachter: *Chirurgisches Operation Büchli* or *Tagebuch für innerliche und eußerliche Kranckheiten* (1803–1843)

Description

Twelve volumes in small octavo format or B6 or 1/32-sheet (11 cm×18 cm) with a total of 643 pages (23 to 79 pages per journal). The journals are from the period between 1803 and 1843, their original number is not known. The case histories were mostly kept in chronological order; further history was added – as

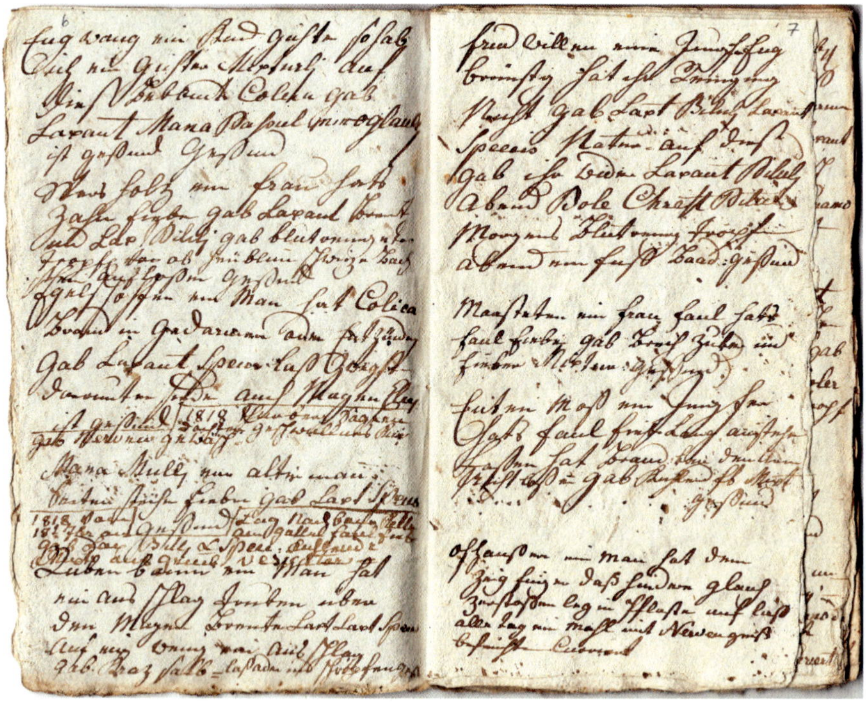

FIGURE 14.4 *Gottfried Wachter,* Chirurgisches Operation Büchli or Tagebuch für innerliche und eußerliche Kranckheiten.
ARCHIV FÜR MEDIZINGESCHICHTE, UNIVERSITÄT ZÜRICH, MS H 17.3, PP. 6–7.

long as there was space. A second patient entry was only added when the empty space was used up before the next entry. Two of the volumes were penned twice by filling the gaps between the elder entries. Here the respective year is noted; apart from this, dates are rarely given, accounting is separate, no visible signs of revision. Language: German.

Further records from this practice: one account book (1837–54).

Conrad Heinrich Fuchs' Policlinic: *Krankheitsgeschichten* (1833–1842)

Description

Eleven (of originally twelve) octavo volumes, paperback, with 160–80 pages each, selected patient reports from the period between 1833 and 1842. The

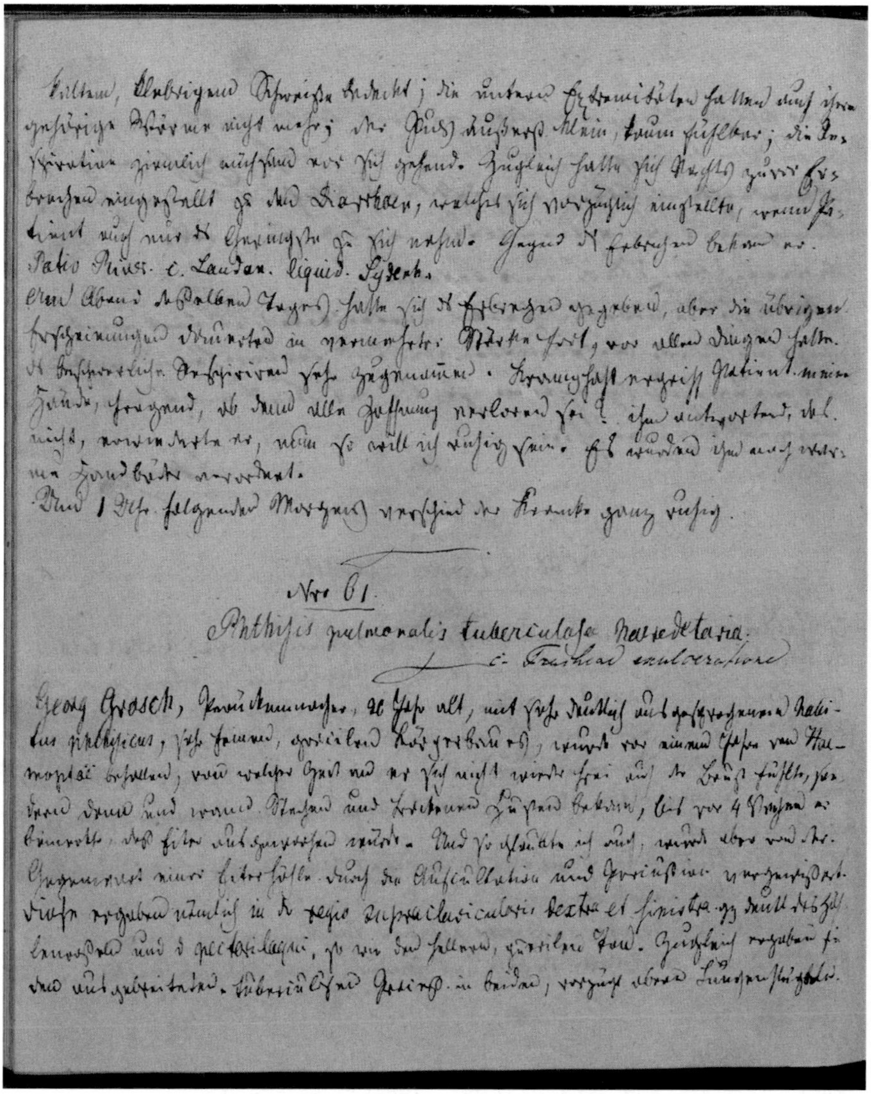

FIGURE 14.5 *Conrad Heinrich Fuchs's policlinic,* Krankheitsgeschichten. NIEDERSÄCHSISCHE STAATS- UND UNIVERSITÄTS BIBLIOTHEK GÖTTINGEN, HSD 8 COD MS H NAT 61:VII, NO. 61.

entries were copied out in literary style in various handwritings and served for clinical training and research. With indices (of patients and diagnoses). Numerous cross-references within the case histories and to the separate five dissection books. The entries are preceded by a tabulated journal sheet of the time allocated per patient.

Caesar Adolph Bloesch: *Tagebuch ärztlicher Beobachtungen* (1828–1863)

Description

Fifty-five (of originally fifty-seven) volumes in folio form (c.23×35 cm) of between 200 and 800 pages each, from 1828 through to the author's death in 1863. Commercially available notebooks. Entries in strict chronological order (follow-up

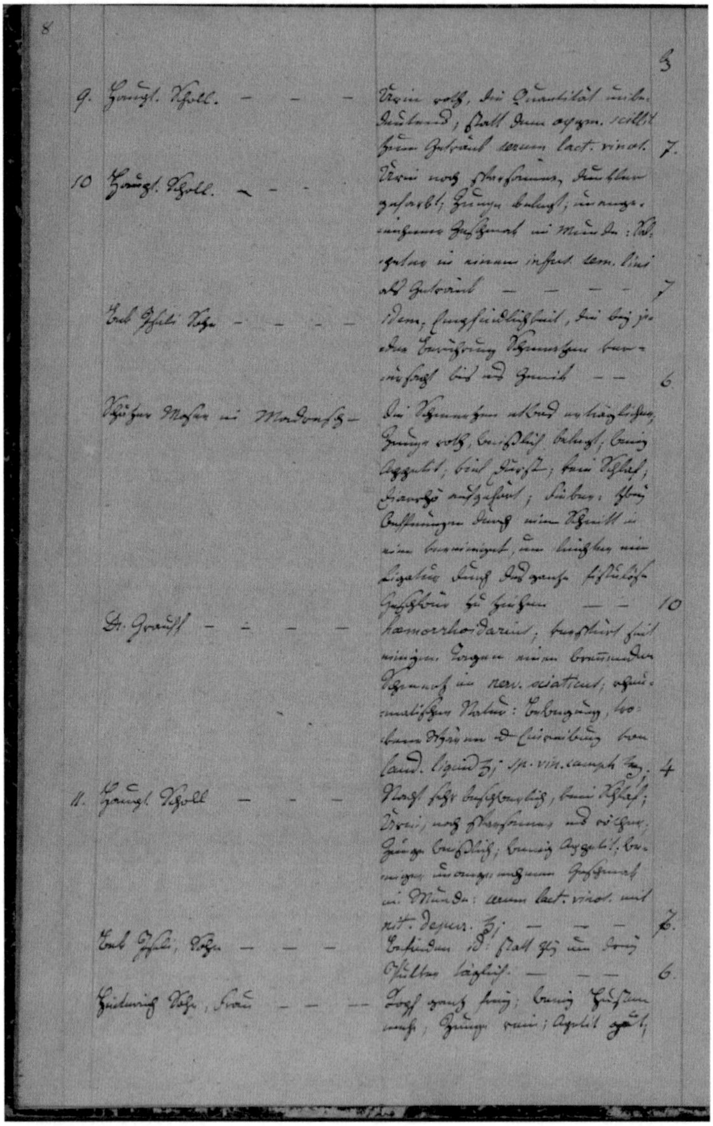

FIGURE 14.6 *Caesar Adolph Bloesch,* Tagebuch ärztlicher Beobachtungen.
STADTARCHIV BIEL, 4 A 1, NACHLASS C. A. BLOESCH, BD. 3, 8.

consultations are entered on the day when service was provided); in four columns drawn with pencil (date, name, observation, fee). The number of treatments, accrued fees and their settlements are added up monthly. Among the daily entries are insertions of draft letters, expert reports and references, occasional tabular vaccination lists for passing on to the health authorities. Alphabetical index at the end of each volume. No signs of revision, neatly written at the 'end of the day', second handwriting (wife?). German, French expressions in Latin script.

Friedrich von Bönninghausen: Patient Journals (1864–1889)

Description

Thirty-two octavo volumes (17×20 cm) of 125 to 479 pages, from January 1864 to October 1889 from whole-page pre-printed journal sheets that were prepared for each patient (and continued on the back or further sheets when space ran out). The format corresponds to the contemporary hospital journals, and consisted firstly of a headline with a large empty space for the patient's history and present condition (the *Krankheitsbild*, i.e. the described and perceived symptoms), and secondly, of a tabular course of treatment in four columns (date | prescription | dosage | success and new symptoms) that was probably noted down during consultation. Occasional recordings of costs; index and account book were kept separately.

Further records from this practice: One account book (*Conto-Buch*) that lists open accounts and their settlement.

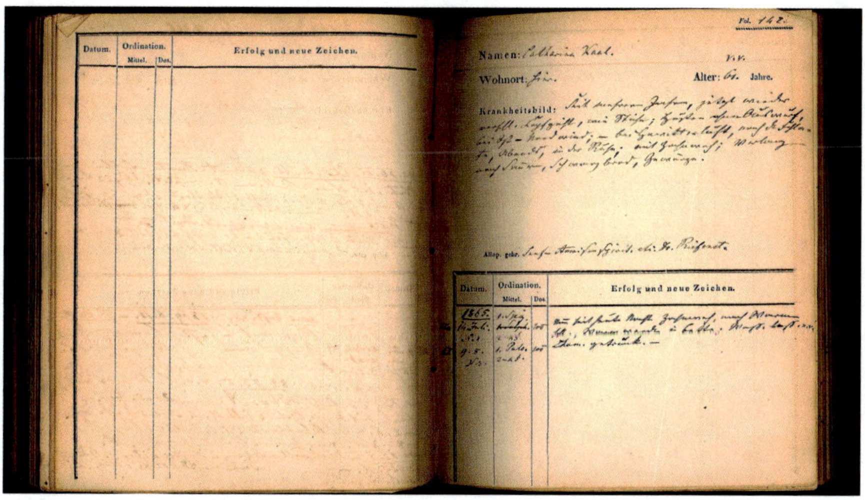

FIGURE 14.7 *Friedrich von Bönninghausen, Patient journal*
INSTITUT FÜR GESCHICHTE DER MEDIZIN ROBERT BOSCH STIFTUNG STUTTGART, P 118, FOL. 142

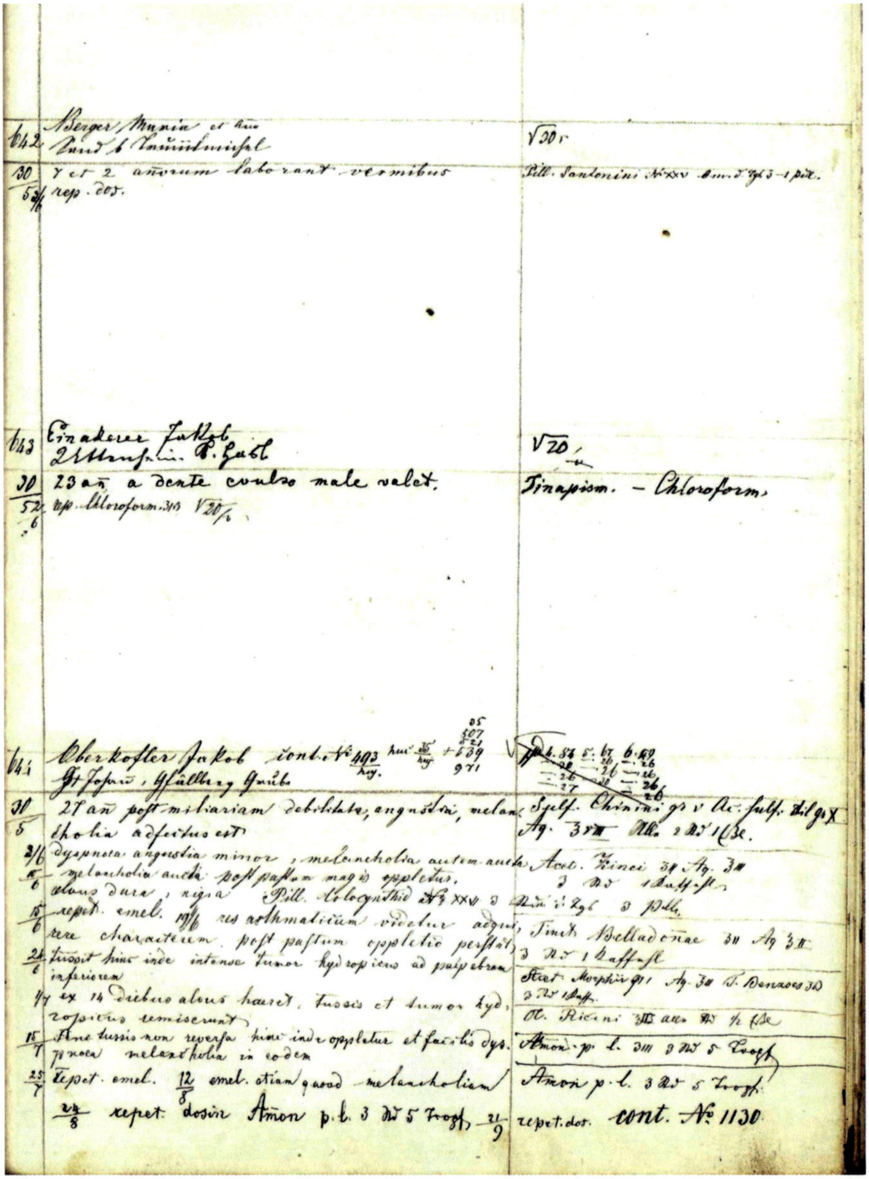

FIGURE 14.8 *Franz von Ottenthal,* Historiae morborum.
SÜDTIROLER LANDESARCHIV / ARCHIVIO PROVINCIALE DI BOLZANO, NACHLASS FRANZ V. OTTENTHAL, LFD. NR. 97, HISTORIAE MORBORUM 1866, HEFT 2.

Franz von Ottenthal: *Historiae morborum* (1844–1899)

Description

244 journals in quarto (22×28 cm) of 20 to 100 pages each, covering the period between 1844 and 1899, handwritten in strongly differing scripts, obviously also by a third person. Chronological, hospital-style records, in three hand-drawn columns. Two lines per patient, one small headline (running number | name and place of origin| fee) and one wide line that was kept empty for follow-up consultations (date | observations | prescriptions). Double-chronological structure with cross-references between the entries and one index at the end of each journal. Occasional explanatory drawings (findings, scribbles); patient letters inserted in the appropriate places. Frequent corrections, presumably transferred from slips of paper or notebooks.

Bibliography

I Primary Sources (Printed before 1900)

Adelung, Johann C., *Versuch eines vollständigen grammatisch-kritischen Wörterbuches der hochdeutschen Mundart* (Leipzig 1786), vol. 5, 1 (W–Z).

Adelung, Johann C., *Grammatisch-kritisches Wörterbuch der hochdeutschen Mundart. Zweyte vermehrte und verbesserte Ausgabe* (Wien 1811), vol. 4 (Seb–Z).

Anonymous, *Die Nürnbergischen Adress- und Schreibkalender: Das jetzt lebende Nürnberg: bey einem Calender auf das Jahr 1705/06. Verzeichnus der Republic Nürnberg Regenten, Beamten und Bedienten* (Nürnberg 1705) [microfiche edition, Erlangen 2000].

Anonymous, 'Anzeige neu entworfener Krankentabellen, zum Gebrauche für Ärzte und Wundärzte zu genauer und bequemer Führung ihrer medizinischen Tagebücher', *Medicinische Nationalzeitung für Deutschland und die mit selbigem zunächst verbundenen Staaten: Intelligenz-Blatt* 10 (1799), 39–40.

Anonymous, 'Rezension: Kranken-Diarium oder Medicinisches Taschenbuch zu fortgehender Einzeichnung der in der täglichen Praxis vorkommenden Krankheitsfälle und die diesfalls getroffenen medicinischen Anordnungen wie auch des besorglichen und des endlichen Ausgangs der Krankheit', *Heidelbergische Jahrbücher der Literatur* 3 (1810), 223.

Anonymous, 'Rezension: Kranken-Manual oder medicinisches Hauptbuch zu angemessener Aufzeichnung der in der medicinischen Praxis vorkommenden bemerkungswerthen Krankheitsfälle nach ihrem Verlaufe, dem Heilverfahren dabey und dem besorglichen wie dem wirklichen Ausgange der Krankheit', *Heidelbergische Jahrbücher der Literatur* 3 (1810), 223.

Anonymous, 'Rezension: Medicinisch-praktischer Geschäfts- und Address-Kalender auf das Jahr 1809. Für praktische Aerzte, Chirurgen und Apotheker. Herausgegeben von Dr. Carl Heinrich Ludwig Schulz. Nebst 12 Monatstafeln. Leipzig, bey Heinrich Gräff', *Heidelbergische Jahrbücher der Literatur* 3 (1810), 221–2.

Anonymous, 'Rezension: Medicinisch-praktisches Geschäfts- und Adress-Buch auf das Jahr 1811 für praktische Ärzte, Chirurgen und Apotheker. Herausgegeben von Dr. Carl Heinrich Ludwig Schulz. Nebst 12 Monatstafeln und einem Bildnisse des Hn. Geh. Hofr. Dr. Wendt. Erlangen, bey Palm', *Jenaische Allgemeine Literaturzeitung* 9 vol. 2, no. 69 (1812), col. 29–31.

Anonymous, *Die ärztliche Besuchs-Anstalt der erkrankten Armen und die ambulante Klinik in Würzburg* (Würzburg 1820).

Anonymous, 'Witterungs- und Gesundheits-Constitution von Berlin im September', *Journal der practischen Heilkunde* 55 (1822) 6, 129–33.

Anonymous, 'Schema zu zweckmässigen, in Folio anzulegenden Krankentabellen für die Praxis (das aber noch jeder Arzt beliebig modificiren kann)', *Medicinischer Almanach für das Jahr 1836* 1 (1836), 72.

Avoda Sara: Götzendienst, ed. by Michael Krupp (Ein Kerem 2002).

Berends, Carl August W., *Vorlesungen über praktische Arzneiwissenschaft* (Berlin 1827), vol. 1 (*Semiotik*).

Blanckaert, Steven, and Schulze, Heinrich, *Lexicon medicum. Ioann. Henrici Schulzii Opera Insigniter Auctum Et Emendatum, Nunc Denuo Recognitum Variisque Accessionibus Locupletatum Curante Mich. Gottl. Agnethlero. Praefationem Praemisit Andreas Elias Büchner* (Halle 1748) [online access: http://www.uni-mannheim.de/mateo/camenaref/blanckaert.html, accessed 11.08.2014].

Bloesch, Caesar A., *Beytrag zur Pathologie und Therapie der gastrischen Krankheiten* (Bern 1832).

Bloesch, Caesar A., 'Gutachten über einen gerichtlich-psychologischen Fall', *Schweizerische Zeitschrift für Medizin, Chirurgie und Geburtshülfe* 1 (1842), 369–80.

Bloesch, Cäsar A., 'Die Nervenfieber Epidemie von 1842 in Biel', *Schweizerische Zeitschrift für Medizin, Chirurgie und Geburtshülfe* 2 (1843), 241–52.

Bloesch, Cäsar A., 'Einige Rückerinnerungen aus dem Jahr 1846', *Schweizerische Zeitschrift für Medizin, Chirurgie und Geburtshülfe* 7 (1848), 345–56.

Bloesch, Cäsar A., 'Einige Bemerkungen über pantheistische Weltanschauung', *Schweizerische Zeitschrift für Medizin, Chirurgie und Geburtshülfe* 9 (1850), 29–48.

Bloesch, Cäsar A., *Bern und Biel. Eine Erinnerung an die grosse Vorzeit* (Biel 1853).

Bloesch, Cäsar A., 'Einige Bemerkungen über die Choleraepidemie in Aarau im Herbste 1854', *Schweizerische Zeitschrift für Medizin, Chirurgie und Geburtshülfe* 14 (1855), 1–16.

Bloesch, Cäsar A., *Geschichte der Stadt Biel und ihres Panner-Gebietes*, 3 vols. (Biel 1855–1856).

Bloesch, Caesar A., 'Allgemeine Grundsätze der theoretischen und der praktischen Medizin, ausgezogen aus meinem Tagebuch ärztlicher Besuche und Konsultationen', *Mitteilungen der medizinisch-chirurgischen Gesellschaft des Kantons Bern* 1 (1866), 1–150.

Bock, Carl Ernst, *Lehrbuch der pathologischen Anatomie und Diagnostik*, 2 vols. (Leipzig 1853), vol. 2 (*Lehrbuch der Diagnostik*).

Bönninghausen, Clemens von, 'Das Krankenjournal', *Allgemeine Homöopathische Zeitung* 67 (1863), 114–6, 121–3, 129–31, 140–1, 147–9, 163–5.

The Bönninghausen Repertory. Therapeutic Pocketbook Method, ed. by George Dimitriadis (Sydney [2]2010).

Busch, Dietrich Wilhelm Heinrich, Dieffenbach, Johann Friedrich *et al.* (eds), *Encyclopädisches Wörterbuch der medicinischen Wissenschaften* (Berlin 1842), vol. 24.

Carl, Johann Samuel, *Specimen Historiae Medicae Ex Solidae Experientiae Documentis, Maxime Vero Monimentis* [*!*] *Stahlianis, In Syllabum Aphoristicum Redactum. Ut Ita*

Ex Morborum Circumstantiis Perpetuis Essentialibus Et Extraessentialibus Universa Oeconomia Medicinae Internae, Indeque Per Se Fluens Vera Et Constans Pathologica Et Therapeutica Consideratio Patescat (Halle 1719).

Carl, Johann Samuel and de Cilano, Georg Christian, *Erfahrungs-Gründe von des Blutlassens wahren Gebrauch und Mißbrauch* (Flensburg 1739).

Choulant, Ludwig, *Der junge Arzt am Krankenbette* (Leipzig 1823).

Collegium Pharmaceuticum Nürnberg (ed.), *Versuch einer Geschichte des Apothekenwesens in der freyen Reichsstadt Nürnberg* (Nürnberg 1792).

Conradi, Johann Wilhelm, *Einrichtung der Medicinischen Klinik im Academischen Hospitale zu Heidelberg* (Heidelberg 1820).

Commercium litterarium ad rei medicae et scientiae naturalis incrementum institutum quo quicquid novissime observatum agitatum scriptum vel peractum est succinte dilucideque exponitur (Nürnberg 1731–1745).

Christian Ernst, Fischer, *Versuch einer Anleitung zur medizinischen Armenpraxis* (Göttingen 1799).

Erzstein, *Ertappter Briefwechsel von der Zauberey, Schröpfers Künsten, Nativitätstellen, Sympathie, Gespenstern u.d.g* (Leipzig 1777).

Frank, Johann Peter, *Ankündigung des klinischen Instituts zu Göttingen wie solches bey seiner Wiederherstellung zum Vorteil armer Kranken und zur Bildung praktischer Ärzte eingerichtet werden sollte* (Göttingen 1784).

Frank, Johann Peter, *System einer vollständigen medicinischen Polizey* (Mannheim 1788).

F.R., 'Bemerkungen über die Erkenntniß der Krankheiten nebst einer Anleitung zur Führung einer geordneten Kranken-Controle. Mit einer Tabelle. Von Dr. Heinrich Küchler, Arzt in Darmstadt. 1844', *Neue medicinisch-chirurgische Zeitung*, Neue Folge 2 (1844), 301–2.

Fürchtegott Gellert, Christian, *Werke* (Frankfurt am Main 1979).

Fuchs, Conrad Heinrich, 'Über die Sterblichkeit der Stadt Würzburg vom 1ten Juli 1819 bis zum 30sten Juli 1829', *Zeitschrift für Staatsarzneikunde* 25 (1833), 368–403.

Fuchs, Conrad Heinrich, 'Ueber den Einfluß der verschiedenen Gewerbe auf den Gesundheitszustand und die Mortalität der Künstler und Handwerker in den Blüthejahren nach den Tabellen des Instituts für kranke Gesellen zu Würzburg 1786–1834. Ein Beitrag zur medicinischen Statistik', *Wissenschaftliche Annalen der gesammten Heilkunde* 32 (1835), 385–419.

Fuchs, Conrad Heinrich, *Bericht über die Vorgänge an der Poliklinik zu Würzburg im Jahre 1835. Mit einigen Bemerkungen über die Krankheits-Constitution dieses Jahres* (Würzburg 1835).

Fuchs, Conrad Heinrich, *Bericht über die Vorgänge an der Poliklinik zu Göttingen im Jahre 1838 bis 39* (Hannover 1840).

Fuchs, Conrad Heinrich, 'Göttingen. Policlinic daselbst 1838/39', *Göttingische Gelehrte Anzeigen* 1 (1840b), 81–5.

Fuchs, Conrad Heinrich, *Lehrbuch der speciellen Nosologie und Therapie* (Göttingen 1845), vol. 1 (*Klassen und Familien*).

Fuchs, Conrad Heinrich, *Lehrbuch der speciellen Nosologie und Therapie* (Göttingen 1846), vol. 2, 1 (*Gattungen und Arten. Erste Abtheilung. Hämatonosen. Erste bis dritte Ordnung*).

Fuchs, Conrad Heinrich, Lehrbuch der speciellen Nosologie und Therapie (Göttingen 1847), vol. 2, 2 (*Gattungen und Arten. Zweite Abtheilung: Hämatonosen. Vierte bis sechste Ordnung*).

Fuchs, Conrad Heinrich, *Lehrbuch der speciellen Nosologie und Therapie* (Göttingen 1848), vol. 2, 3 (*Gattungen und Arten. Dritte Abtheilung: Neurosen, Morphonosen und Register*).

Fuchs, Conrad Heinrich, *Bericht über die medicinische Klinik zu Göttingen im Jahre 1853/54* (Göttingen 1855).

Fueter, Emanuel Eduard, *Achter Jahresbericht der poliklinischen Anstalt in Bern für das Jahr 1842* (Bern 1843).

Götz, Johann C., 'Observationes Physico-Medicae. Observatio CLXXXI-Observatio CCVII', *Nova Acta Physico-Medica Academiae Caesareae Leopoldino-Carolinae Naturae Curiosorum Exhibentia Ephemerides sive Observationes Historiae et Experimenta a Celeberrimis Germaniae et Exterrarum Regionum Viris Habita et Communicata Singulari Studio Collecta* 2 (1730), 423–63.

Glaser, Johann Friedrich, *Beschreibung einer neuerfundenen Blutwaage und Blutmeßgeschirrs: womit man beym Hand- und Fußaderlassen das Blut, ob es schon unter das dabey gebräuchlichen Wasser läuft, dennoch sobald Loth zu Loth richtig und sehr nützlich wägen und messen, auch sonst durch einen angewiesenen Vortheil unter Wasser abgesondert allein auffangen kann* (Hildburghausen 1758).

Glaser, Johann Friedrich, *Nützliches Verhalten bey der jetzo in Deutschland regierenden Fleckfieberseuche: um sie zu verhüten und wieder zu dämpfen* (Hildburghausen 1758).

Glaser, Johann Friedrich, *Nützliche und durch die Erfahrung bewährte Vorschläge, bey heftigen und geschwinden Feuersbrünsten Häuser und Mobilien sicher zu retten: nebst einer gründlichen Anweisung, grosse und gefährliche Feuersbrünste zu verhüten, und was man bey Auf- und Ausbauung der Häuser und Aufbehaltung der Mobilien, auf eine nutzbare Weise, zu beobachten habe* (Hildburghausen ^{3}1764).

Glaser, Johann Friedrich, *Physicalische Bewegungsgründe, die es wahrscheinlich und glaublich machen, daß bey der Hennebergischen Handelsstadt Suhla ein unterirdisch großer Schatz von Steinsalz verborgen liege, samt unmasgeblicher Vorschlägen, wie solcher Schatz schicklich aufzusuchen, und der gefundene sodann wohl genutzt werden kann* (Leipzig 1776).

Glaser, Johann Friedrich, *Auf richtige Erfahrung gegründete Abhandlung von der tödtlichen Knoten-Krankheit unterm Rindviehe und dem Rothwildprete in den Wäldern: wie solche abzuhalten und die angefallene gewiß zu curiren ist, und von der nicht tödtlichen Maul- und Fußrankheit unterm Rindviehe, wie solche leicht geheilet werden kann* (Leipzig 1780).

Glaser, Johann Friedrich, 'Observatio XLIX: De praeternaturali cranii depressione absque sugillationibus & fissuris per sectionem legalem detecta', *Nova Acta Physico-Medica Academiae Caesareae Leopoldino-Carolinae Naturae Curiosorum exhibentia Ephemerides sive Observationes Historiae et Experimenta a celeberrimis Germaniae et Exterrarum Regionum Viris Habita et Communicata singulari studio collecta* (1783) 230–2.

Glaser, Johann Friedrich, 'Beytrag zur Lebensgeschichte des vor kurzem verstorbenen Arztes D. Johann Friedrich Glaser (Aus einem eigenhändigen Brief desselben)', *Journal von und für Franken* 6, no. 1 (1793), 411–23.

Güldenklee, Balthasar Timaeus von, *Serenissimi Electoris Brandenburgici Archiatri Casus Medicinales Praxi Triginta Sex Annorum Observati* (Leipzig 1667).

Günther, Siegmund, 'Müller, Johannes', *Allgemeine Deutsche Biographie* 22 (1885), 564–81.

Grimm, Jacob, and Grimm, Wilhelm, *Das Deutsche Wörterbuch*, 16 vols (Leipzig 1854–1961) with register of sources (Leipzig 1971) [online access: http://woerterbuchnetz.de/DWB, accessed 21.1.2013].

Hahnemann, Samuel, *Materia Medica Pura*, ed. by Richard Hughes (Liverpool 1881), vol. II (Ledum–Verbascum).

Hahnemanns Krankenjournale Nr. 2 und 3, ed. by Heinz Henne (Stuttgart 1963).

Hahnemann, Samuel, *Organon of Medicine. Edition 6B*, ed. by Kurt Hochstetter (Santiago-Chile 1977).

Hahnemann, Samuel, *Krankenjournal D 5: (1803–1806)*, ed. by Helene Varady (Heidelberg 1991).

Hahnemann, Samuel, *Krankenjournal D F2: (1836–1842)*, ed. with transcription and commentary by Arnold Michalowski (Stuttgart, 2003).

Hahnemann, Samuel, *Krankenjournal D 16 (1817–1818)*, ed. with transcription and commentary by Ulrich Schuricht, 2 vols. (Stuttgart 2004).

Hahnemann, Samuel, *Die chronischen Krankheiten. Theoretische Grundlagen. Mit allen Änderungen von der 1. Auflage (1828) zur 2. Auflage (1835) auf einen Blick*, revised by Matthias Wischner (Stuttgart [3]2006).

Hahnemann, Samuel, *Krankenjournal D 38 (1833–1835)*, ed. with a commentary of the transcription by Monika Papsch (Stuttgart 2007).

Hahnemann, Samuel, *Krankenjournal D 22 (1821)*, ed. with transcription and commentary by Markus Mortsch, 2 vols. (Stuttgart 2008).

Hecker, August F., *Des achtzehenten Jahrhunderts Geschichte der Erfindungen, Theorien und Systeme in der Natur- und Arzneywissenschaft* (Gotha 1799).

Hensler, Philipp Gabriel, 'Ueber KrankenBesuchs-Anstalten', *Schlözers Stats-Anzeigen* 17 (1785), 273–96.

Hildebrand, Bruno (ed.), *Beiträge zur Statistik des Kantons Bern, im Auftrage der Berner Staatsregierung herausgegeben.* (Bern 1860), vol. 1, 1 (*Die Bevölkerung*).

Horsch, Philipp Joseph, *Versuch einer Topographie der Stadt Würzburg in Beziehung auf den allgemeinen Gesundheitszustand und die dahin zielenden Anstalten* (Arnstadt/Rudolstadt 1805).

Horsch, Philipp Joseph, *Über die Bildung des Arztes als Klinikers und als Staatsdieners. Ein Programm, womit seine Vorlesungen über Heilmittellehre, psychische Medizin und Klinik ankündigt Philipp Joseph Horsch* (Würzburg 1807).

Hufeland, Christoph Wilhelm, *Einrichtung und Gesetze der Herzoglich medicinisch-chirurgischen Krankenanstalt zu Jena* (Jena 1799).

Hufeland, Christoph Wilhelm, 'Ankündigung des Königl. Poliklinischen Instituts auf der Universität zu Berlin nebst den Gesetzen desselben', *Journal der practischen Arzneykunde u. Wundarzneykunst* 31, no. 10 (1810), 3–56.

Hufeland, Christoph Wilhelm, 'Erster Jahresbericht des Königl. Poliklinischen Instituts der Universität zu Berlin vom Jahre 1810', *Journal der practischen Heilkunde* 32, no. 6 (1811), 1–96.

Hufeland, Christoph Wilhelm, *Armen-Pharmakopöe. Zugleich eine Auswahl bewährter Arzneimittel und Arzneiformeln, Fünfte vermehrte Auflage* (Berlin 1828).

Hufeland, Christoph Wilhelm, *Enchiridion medicum oder Anleitung zur medicinischen Praxis. Vermächtnis einer fünfzehnjährigen Erfahrung* (Herisau 1837).

Husemann, Theodor, 'Fuchs, Konrad Heinrich', *Allgemeine Deutsche Biographie* 8 (1878), 168–9.

Küchler, Hermann, *Bemerkungen über die Erkenntniss der Krankheiten nebst einer Anleitung zur Führung einer geordneten Krankencontrole* (Darmstadt 1844).

Kirkland, Thomas, *Untersuchung über den gegenwärtigen Zustand der Medizinal-Chirurgie* (Leipzig 1785).

Knolz, Joseph, *Darstellung der Medicinal-Verfassung in den k.k. Staaten Oesterreichs, in Beziehung auf den Wirkungskreis der Kreiswundärzte, der Civil-, Stadt- und Landwundärzte und der Landesthierärzte* (Wien 1829).

Kräutermann, Valentin, *Der Thüringische Theophrastus Paracelsus, Wunder- und Kräuterdoctor oder Zauber-Arzt* (Arnstadt/Leipzig 1730).

Kraus, Ludwig August (ed.), *Preussische Pharmakopöe. Nach der neuesten Ausgabe aus dem Lateinischen übertragen und mit vielen Zusätzen versehen* (Braunschweig 1805).

Kraus, Ludwig August, *Dr. Justus Arnemann's chirurgische Arzneimittellehre* (Göttingen 1818).

Krünitz, Johann Georg, 'Hirschhorngeist', *Oekonomische Enzyklopädie, oder allgemeines System der Staats- Staat- Haus- u. Landwirtschaft* 23 (1781), 739–40 [online access: http://www.kruenitz1.uni-trier.de/, accessed 22.1.2013].

Lusitanus, Zacutus, *De medicorum principum historia liber tertius, in quo medicinales omnes medicorum principum historiae de uteri, genitalium, & inferiorum partium affectibus, describuntur & copendiose explanantur* (Amsterdam 1637).

Magirus, Johannes, *Prognosticon Astrologicum auff das Jahr nach der gnadenreichen Geburt unseres Herrn und Seligmachers MDCXLVI* (Altenburg [1645]).

Magirus, Johannes, *Epistolica declaratio collegii medici mathematici et practici* (Berlin 1646).

Magirus, Johannes, *Alter und Neuer Schreibkalender auf das Jahr nach Christi unseres Erlösers und Seligmachers Geburt MDCIL* (Leipzig [1648a]).

Magirus, Johannes, *Prognosticon, das ist: Astrologische Mutmaßung vom Gewitter, Frucht- und Unfruchtbarkeit, Gesund-oderKranckheiten, Krieg und Friedens des MDCXLIX. Jahres* nach Christi Geburt (Leipzig [1648b]).

Magirus, Johannes, *Alt und Neu Prognosticon Astrologicum, das ist Vernünfftige und aus dem Himmelslauff genommene Mutmaßung vom Gewitter, Frucht- und Unfruchtbarkeit, Gesund-oder Kranckheit, Krieg, Frieden und anderen Welthändeln des MDCLV. Jahres nach Christ unseres Herrn Geburt* (Nürnberg 1654).

Marx, Karl Friedrich, *Goettingen in medicinischer, physischer und historischer Hinsicht* (Göttingen 1824).

Niemeyer, Paul, 'Von ärztlicher Untersuchung', *Aerztliche Sprechstunden* 1, no. 3 (1879), 1–20.

Nushard, Franz Willibald, *Theoretische Medicin für Wundärzte als Leitfaden zu Vorlesungen entworfen* (Prag 1824), vol. 2 (*Die Grundzüge der allgemeinen Therapie, Arzneimittellehre, Krankendiätetis und Receptirkunde*).

Oesterley, Ferdinand, *Bemerkungen über die Bekanntmachung vom 23. December d.J. die ärztliche Behandlung und Aufnahme hiesiger kranker Gesellen. Lehrlinge und Dienstboten in das neue academische Hospital betreffend* (Göttingen 1851).

Oesterlen, Friedrich, *Handbuch der medicinischen Statistik* (Tübingen [2]1874).

Pfingsten, Johann Hermann, *Deutsches Dispensatorium oder allgemeines Apothekerbuch nach den neuesten und besten lateinischen Dispensatorien und Pharmacopoen zum gemeinnützlichen Gebrauch nach alphabetischer Ordnung eingerichtet* (Frankfurt/Leipzig 1795).

Platter, Felix, *Observationum in hominis affectibus plerisque, corpori et animo, functionum laesione, dolore, aliave molestia et vitio incommodantibus libri tres* (Basel 1614).

Platter, Felix, *Tagebuch (Lebensbeschreibung)* 1536–1567, ed. by Valentin Lötscher (Basel 1976).

Puchelt, Friedrich August B., *Das System der Medicin: in Umrissen dargestellt* (Heidelberg 1826).

Pupikofer, Johann Adam, *Der Kanton Thurgau historisch, geographisch, statistisch geschildert* (St. Gallen 1837).

Rahn, Johann Friedrich, *Fieberlehre. Ein nachgelassenes Bruchstück des Handbuchs der praktischen Heilkunde* (Zürich 1814).

Rambach, Johann Jacob, *Versuch einer physisch-medizinischen Beschreibung von Hamburg* (Hamburg 1801).

Rat der Stadt Nürnberg, *Verneuerte Gesetz und Ordnung Eines Hoch Edlen und Hochweisen Raths, des Heiligen Reichs Stadt Nürnberg, Dem Collegio Medico, den Apotheckern, und anderen Angehörigen daselbst gegeben* (Nürnberg 1700).

Ratier, Félix Séverin, *Coup-d'oeil sur les cliniques médicales de la faculté de médecine et des hôpitaux civils de Paris* (Paris 1830).

Razoux, Jean, *Tables nosologiques et météorologiques très-étendues dressées à l'Hôtel-Dieu des Nîmes depuis le premier juin 1757 jusques au premier janvier 1762* (Basel 1767).

Richter, Georg August (ed.), *Die specielle Therapie nach den hinterlassenen Papieren des verstorbenen August Gottlieb Richter* (Berlin 1813), vol. 1 (*Die acuten Krankheiten*).

Richter, Hermann E., '169. Bemerkungen über die Erkenntniss der Krankheiten, nebst einer Anleitung zur Führung einer geordneten Krankencontrole. Küchler, Arzt zu Darmstadt', *Carl Christian Schmidt's Jahrbücher der in- und ausländischen gesammten Medicin* 44 (1844), 128–9.

Rinecker, Franz von, *Medicinische Statistik der poliklinischen Anstalt an der Julius-Maximilians-Universität zu Würzburg in ihrem vierten Decenium 1837–1847* (Würzburg 1848).

Sennert, Daniel, *Opera omnia in quatuor tomos divisa* (Lyon 1656).

Schickert, Otto, *Die Militärärztlichen Bildungsanstalten von ihrer Gründung bis zur Gegenwart. Festschrift zur Feier des hundertjährigen Bestehens des medicinisch-chirurgischen Friedrich-Wilhelms-Instituts* (Berlin 1895).

Schmalz, Carl-Gustav, *Versuch einer medicinisch-chirurgischen Diagnostik in Tabellen oder Erkenntniß und Unterscheidung der innern und äußern Krankheiten mittelst Nebeneinanderstellung der ähnlichen Formen. Zweite, viel vermehrte und verbesserte Auflage* (Dresden 1812).

Schmidt, W., 'Vorschläge zur Verminderung der Arzneikosten', *C.W. Hufeland's Journal der practischen Heil-Kunde, fortgesetzt von Dr. E. Osann* 84 (1837), 24–48.

Schmidtmüller, Johann A., *Beiträge zur Vervollkommung der Staatsarzneikunde. Eine Beilage zum Handbuch der Staatsarzneikunde* (Landshut 1806).

Staffler, Johann Jakob, *Das deutsche Tirol und Vorarlberg, topographisch, mit geschichtlichen Bemerkungen II* (Innsbruck 1847).

Stahl, Georg E., and Götz, Johann C., *Scripta D. Georg. Ern. Stahlii Sacr. Reg. Maiest. Boruss. Consil. Aulici & Archiatri Primarii Aliorumque Ad Eius Mentem Disserentium Serie Chronologica Recenset D. Joh. Christoph Goetzivs* (Nürnberg 1729) [online: http://reader.digitale-sammlungen.de/de/fs1/object/display/bsb10054543_00005.html].

Statistisches Jahrbuch für den Kanton Bern. Herausgegeben vom kantonalen statistischen Bureau, 3. Jahrgang (Bern 1870).

Sydenham, Thomas, *Observationes Medicae Circa Morborum Acutorum Historiam Et Curationem* (London 1676).
Thomann, Joseph Nikolaus, *Über die klinische Anstalt an dem Julius-Spitale zu Würzburg* (Würzburg 1798).
Valleriola, François, *Observationum medicinalium libri sex* (Lyon 1573).
Wagemann, Ludwig Gerhard, *Nachricht über die Versorgung der Armen in Göttingen vom Jahr 1796* (Göttingen 1797).
Weber, Ferdinand, *Kurze Bemerkungen über die Section der Leiche zu pathologischen Zwecken: ein Gruss an seine Herren Collegen in den Herzogthumern Schleswig-Holstein und Lauenburg, von Dr. F. Weber, Prosector an der Universitat in Kiel* (Kiel 1847).
Weiz, Friedrich August, *Das gelehrte Sachsen oder Verzeichniß derer in den Churfürstl. Sächs. und incorporirten Ländern jetztlebenden Schriftsteller und ihrer Schriften* (Leipzig 1780).
Werther, Ferdinand, *Chronik der Stadt Suhl in der gefürsteten Graffschaft Henneberg* (Suhl 1846).
Wichmann, Ernst Johann, *Ideen zur Diagnostik, beobachtenden Ärzten mitgetheilet* (Hannover 1800).
Wittwer, Philipp L., *Entwurf einer Geschichte des Kollegiums der Aerzte in der Reichsstadt Nürnberg* (Nürnberg 1792).
Zedler, Johann Heinrich, *Grosses vollständiges Universal-Lexicon aller Wissenschafften und Künste, Welche bißhero durch menschlichen Verstand und Witz erfunden und verbessert worden* (Halle/ Leipzig 1732–1754) [online access: http://www.zedler-lexikon.de, accessed 31.3.2014].
Zimmermann, Johann Georg, *Von der Erfahrung in der Arzneykunst. 1. Theil* (Zürich 1763).

II Secondary Sources

Ackerknecht, Erwin H., 'A Plea for a "Behaviorist" Approach in Writing the History of Medicine', *Journal of the History of Medicine and Allied Sciences* 22 (1967), 211–4.
Antos, Gerd, and Weber, Tilo, Introduction to Tilo Weber (ed.), *Typen von Wissen. Begriffliche Unterscheidung und Ausprägungen in der Praxis des Wissenstransfers* (Frankfurt am Main 2009), 1–10.
Assion, Peter, and Telle, Joachim, 'Der Nürnberger Stadtarzt Johannes Magenbuch. Zu Leben und Werk eines Mediziners der Reformationszeit', *Sudhoffs Archiv* 56 (1972), 353–421.
Atzl, Isabel, Helms, Roland, Neuner, Stephanie, and Schilling, Ruth (eds), *Praxiswelten. Zur Geschichte der Begegnung zwischen Arzt und Patient* (Ingolstadt 2013).

Baal, Anne Hilde van, *In Search of a Cure. The Patients of the Ghent Homoeopathic Physician Gustave A. van den Berghe (1837–1902)* (PhD diss., Universiteet Amsterdam, 2004; published under the same title: Rotterdam 2008).

Bächtold-Stäubli, Hanns, and Hoffmann-Krayer, Eduard (eds), *Handwörterbuch des deutschen Aberglaubens,* vol. 4 (Berlin [3]2000).

Bähler, Arnold, *Biel vor hundert Jahren* (Biel 1916).

Balster, Wolfgang, *Medizinische Wissenschaft und ärztliche Praxis im Leben des Bochumer Arztes Karl Arnold Kortum (1745–1824). Medizinhistorische Analyse seines Patiententagebuches* (Med. diss., Universität Bochum, 1990).

Bannert, Uwe, *Diskussionen um ärztliche Gebührenordnungen im 19. Jahrhundert. Ein Beitrag zur Geschichte des ärztlichen Honorars* (Neumünster 1986).

Barthel, Christian, *Medizinische Polizey und medizinische Aufklärung. Aspekte des öffentlichen Gesundheitsdiskurses im 18. Jahrhundert* (Frankfurt am Main 1989).

Baschin, Marion, 'Sozial- und medizingeschichtliche Untersuchung einer württembergischen Oberamtsstadt im 19. Jahrhundert', in Marion Baschin and Andreas Kozlik (eds), *Studien zur südwestdeutschen Demographie. Die Sterblichkeit in Württemberg im 18./19. Jahrhundert und in Esslingen im 19. Jahrhundert* (Remshalden 2008), 11–139.

Baschin, Marion, 'Carl von Bönninghausen. Ein vergessener Homöopath und seine Lernzeit', *Medizin, Gesellschaft und Geschichte* 28 (2009), 237–80 (with English summary).

Baschin, Marion, *Wer lässt sich von einem Homöopathen behandeln? Die Patienten des Clemens Maria Franz von Bönninghausen, 1785–1864* (Stuttgart 2010).

Baschin, Marion, 'Die Medikation in der Praxis Friedrichs von Bönninghausen', *Zeitschrift für Klassische Homöopathie* 55 (2011), 116–21.

Baschin, Marion, 'How Patients Built the Practice of the Lay Homoeopath Clemens von Bönninghausen. Quantitative and Qualitative Aspects of Patient History', *Dynamis* 31, no. 2 (2011), 475–95.

Baschin, Marion, *Die Geschichte der Selbstmedikation in der Homöopathie* (Essen 2012).

Baschin, Marion, 'Warum besucht man einen Homöopathen? Eine Praxis in Münster im 19. Jahrhundert', in Lina Gafner, Iris Ritzmann and Katharina Weikl (eds), *Penning Patient's Histories – Doctors' Records and the Medical Market in the 18th and 19th Century* [special issue: *Gesnerus* 69, no. 1 (2012)], 126–40 (with English summary).

Baschin, Marion, 'The Patients' Choice. How and Why Sick People Used Homeopathy in 19th Century Münster', in Martin Dinges (ed.), *Medical Pluralism and Homoeopathy in India and Germany (1810–2010). A Comparison of Practices* (Stuttgart 2014), 149–65.

Baschin, Marion, *Ärztliche Praxis im letzten Drittel des 19. Jahrhunderts. Der Homöopath Dr. Friedrich Paul von Bönninghausen (1828–1910)* (Stuttgart 2014).

Bauer, Axel W., and Ho, Anthony D., *Nicht blos künstlich in einem Spitale. Zweihundert Jahre Medizinische Universitäts-Poliklinik Heidelberg und ihr Weg von der Stadtpraxis bis zur Blutstammzelltransplantation* (Heidelberg 2005).

Bernard, Anna, *Die Revokation des Edikts von Nantes und die Protestanten in Südostfrankreich (Provence und Dauphiné) 1685–1730* (München 2003).

Behrens, Rudolf, and Zelle, Carsten (eds), *Der ärztliche Fallbericht. Epistemische Grundlagen und textuelle Strukturen dargestellter Beobachtung* (Wiesbaden 2012).

Beier, Lucinda McCray, *Sufferers and Healers. The Experience of Illness in Seventeenth-Century England* (London 1987).

Berg, Marc, and Bowker, Geoffrey, 'The Multiple Bodies of the Medical Record. Toward a Sociology of an Artifact', *The Sociological Quarterly* 38 (1997), 513–37.

Berger, Eva, 'Frühmodernes Krankenhaus und "Pflegeversicherung" in der ersten Hälfte des 19. Jahrhunderts', in Alfons Labisch and Reinhard Spree (eds), *"Einem jeden Kranken in einem Hospitale sein eigenes Bett." Zur Sozialgeschichte des Allgemeinen Krankenhauses in Deutschland im 19. Jahrhundert* (Frankfurt am Main 1996), 208–35.

Bernet, Brigitta, '"Eintragen und Ausfüllen." Der Fall des psychiatrischen Formulars', in Sibylle Brändli, Barbara Lüthi and Gregor Spuhler (eds), *Zum Fall machen, zum Fall werden. Wissensproduktion und Patientenerfahrung in Medizin und Psychiatrie des 19. und 20. Jahrhunderts* (Frankfurt am Main 2009), 62–91.

Bieger, Alfons, *Prüfung der Ärzte und Wundärzte im Thurgau (1798–1867)* (Dietikon 1988).

Bieger, Alfons, *Schröpfende Heiler – schwitzende Kranke. Das Thurgauer Medizinalwesen im 18. und frühen 19. Jahrhundert* (Frauenfeld 2004).

Blair, Ann, 'The Practices of Erudition According to Morhof', in Françoise Waquet (ed.), *Mapping the World of Learning. The Polyhistor of Daniel Georg Morhof* (Wiesbaden 2000), 59–74.

Blanarsch, Matthias, 'Die Arzt-Patienten-Beziehung zu Beginn des 18. Jahrhunderts, untersucht anhand Johann Storchs Kasuistik zu Molenschwangerschaften', *Medizin, Gesellschaft und Geschichte* 28 (2009), 121–52 (with English summary).

Bleker, Johanna, Eva Brinkschulte and Pascal Grosse (eds), *Kranke und Krankheiten im Juliusspital zu Würzburg 1819–1829. Zur frühen Geschichte des Allgemeinen Krankenhauses in Deutschland* (Husum 1995).

Bleker, Johanna, 'Die medizinische Funktion des frühmodernen Krankenhauses. Zur Analyse des Diagnosespektrums der Inneren Abteilung des Würzburger Juliusspitals 1819–1829', in Alfons Labisch and Reinhard Spree (eds), *"Einem jeden Kranken in einem Hospitale sein eigenes Bett." Zur Sozialgeschichte des allgemeinen Krankenhauses in Deutschland im 19. Jahrhundert* (Frankfurt am Main 1996), 123–44.

Bleker, Johanna, 'Fuchs, Konrad Heinrich', in Wolfgang U. Eckart and Christoph Gradmann (eds), *Ärztelexikon* (München 1995), 146–7 (=Heidelberg ^{3}2006, 130).

Bödeker, Hans Erich, 'On the Origins of the "Statistical Gaze." Modes of Perception, Forms of Knowledge and Ways of Writing in the Early Social Sciences', in Peter Becker and William Clark (eds), *Little Tools of Knowledge. Historical Essays on Academic and Bureaucratic Practices* (Ann Arbor 2001), 169–95.

Bolognese-Leuchtenmüller, Birgit, *Bevölkerungsentwicklung und Berufsstruktur, Gesundheits- und Fürsorgewesen in Österreich 1750–1918* (Wien 1978).

Boschung, Urs, 'Albrecht Haller's Patient Records (Berne 1731–1736)', *Gesnerus* 53 (1996), 5–14.

Boschung, Urs, 'Franz Peter Siffert (1837–1881), Landarzt in Kleingurmels und Wallenbuch', in Jean P. Anderegg, Anton Scherer and Josef Vaucher (eds), *Deutschfreiburg im Aufbruch. Festschrift zum 40. Jahrestag der Gründung der Deutschfreiburgischen Arbeitsgemeinschaft am 15. Januar 1999. Ihrem Gründer und Ehrenobmann Peter Boschung in Dankbarkeit gewidmet* (Freiburg im Üechtland 1999), 36–60.

Boucherin, Nadine, 'Die Krankengeschichten von C.A. Bloesch. (1804–1863)', in Elisabeth Dietrich-Daum, Martin Dinges and Robert Jütte (eds), *Arztpraxen im Vergleich. 18.–20. Jahrhundert* (Innsbruck 2008), 147–65.

Bourdieu, Pierre, 'The Forms of Capital', in John G. Richardson (ed.), *Handbook of Theory and Research for the Sociology of Education* (New York 1986), 241–60.

Bourquin, Werner, and Bourquin, Marcus, *Biel. Stadtgeschichtliches Lexikon von der Römerzeit (Petinesca) bis Ende der 1930er Jahre, historisch, biographisch, topographisch, mit Ergänzungen für den Zeitraum bis 1999* (Biel 1999).

Boutier, Jean, 'Compétence internationale, émergence d'une "profession" et circulation des savoirs. Le tuteur aristocratique dans l'Angleterre du XVIIe siècle', in Maria Pia Paoli (ed.), *Saperi a confronto nell' Europa dei secoli XIII–XIX* (Pisa 2009), 151–80.

Brähler, Elmar, and Felder, Hildegard (eds), *Weiblichkeit, Männlichkeit und Gesundheit* (Opladen ²1999).

Brändli, Sebastian, *"Die Retter der leidenden Menschheit." Sozialgeschichte der Chirurgen und Ärzte auf der Zürcher Landschaft (1700–1850)* (Zürich 1990).

Brendecke, Arndt, 'Tabellen und Formulare als Regulative der Wissenserfassung und Wissenspräsentation', in Wulf Oesterreicher, Gerhard Regn and Winfried Schulze (eds), *Autorität der Form – Autorisierung – Institutionelle Autorität* (Münster 2003), 37–53.

Bresadola, Marco, 'A Physician and a Man of Science. Patients, Physicians, and Diseases in Marcello Malpighi's Medical Practice', *Bulletin of the History of Medicine* 85 (2011), 193–221.

Brinkschulte, Eva, 'Die Institutionalisierung des modernen Krankenhauses im Rahmen aufgeklärter Sozialpolitik. Die Beispiele Würzburg und Bamberg', in Alfons Labisch and Reinhard Spree (eds), *"Einem jeden Kranken in einem Hospitale sein eigenes Bett." Zur Sozialgeschichte des Allgemeinen Krankenhauses in Deutschland im 19. Jahrhundert* (Frankfurt am Main 1996), 189–207.

Brockbank, William, and Kay, Marjory Lilian, 'Extracts from the Diary of Richard Kay of Baldingstone, Bury, Surgeon (1737–50)', *Medical History* 3 (1959), 58–68.

Brockliss, Laurence, and Jones, Colin, *The Medical World of Early Modern France* (Oxford 1997).

Brockmeyer, Bettina, *Selbstverständnisse. Dialoge über Körper und Gemüt im frühen 19. Jahrhundert* (Göttingen 2009).

Broman, Thomas H., 'The Medical Sciences', in Roy Porter (ed.), *Eighteenth-Century Science* (Cambridge 2003), 463–84.

Broman, Thomas, 'Rethinking Professionalisation. Theory, Practice, and Professional Ideology in Eighteenth-Century German Medicine', *The Journal of Modern History* 67, no. 4 (1995), 835–72.

Brosseder, Claudia, *Im Bann der Sterne. Caspar Peucer, Philipp Melanchthon und andere Wittenberger Astrologen* (Berlin 2004).

Brückner, Burkhart, 'Psychiatriegeschichte und Patientengeschichte. Eine Literaturübersicht zum Stand der deutschsprachigen Forschung', *Sozialpsychiatrische Informationen* 35 (2006), 26–30.

Brügelmann, Jan, *Der Blick des Arztes auf die Krankheit im Alltag 1779–1850. Medizinische Topographien als Quelle für die Sozialgeschichte des Gesundheitswesens* (PhD diss., Freie Universität Berlin, 1982).

Bueltzingsloewen, Isabelle von, *Machines à instruire, machines à guérir. Les hôpitaux universitaires et la médicalisation de la société allemande 1730–1850* (Lyon 1997).

Bueltzingsloewen, Isabelle von, 'Die Entstehung des klinischen Unterrichts an den deutschen Universitäten des 18. Jahrhunderts und das Göttinger Accouchierhaus', in Jürgen Schlumbohm and Claudia Wiesemann (eds), *Die Entstehung der Geburtsklinik in Deutschland 1751–1850. Göttingen, Kassel, Braunschweig* (Göttingen 2004), 15–30.

Bünz, Enno, 'Leibärzte', in Werner Paravicini (ed.), *Höfe und Residenzen im spätmittelalterlichen Reich. Bilder und Begriffe* (Ostfildern 2005), 156–7.

Burke, Peter, *A Social History of Knowledge. From Gutenberg to Diderot* (Cambridge 2000).

Burnham, John, 'The Death of the Sick Role', *Social History of Medicine Advance Access* (2012), 1–16 [online http://shm.oxfordjournals.org/content/early/2012/04/18/shm.hks018, doi: 10.1093/shm/hks018].

Campe, Rüdiger, 'Die Schreibszene, Schreiben', in Sandro Zanetti (ed.), *Schreiben als Kulturtechnik* (Berlin 2012), 269–82.

Carpenter, Mary Wilson, *Health, Society and Medicine in Victorian England* (Santa Barbara [CA] 2010).

Castan, Joachim, *Hochschulwesen und reformierte Konfessionalisierung. Das Gymnasium Illustre des Fürstentums Anhalt in Zerbst* (Halle/Saale 1999).

Condrau, Flurin, *Lungenheilanstalt und Patientenschicksal. Sozialgeschichte der Tuberkulose in Deutschland und England im späten 19. und frühen 20. Jahrhundert* (Göttingen 2000).

Condrau, Flurin, 'The Patient's View Meets the Clinical Gaze', *Social History of Medicine* 20 (2007), 525–40.

Cook, Harold J., *The Decline of the Old Medical Regime in Stuart London* (Ithaca 1986).

Cook, Harold J., 'Medicine', in Katharine Park and Lorraine Daston (eds), *The Cambridge History of Science. 3. Early Modern Science* (Cambridge 2006), 407–34.

Cosmacini, Giorgio, *Storia della medicina e della sanità in Italia* (Roma ³1987).

Crellin, John, 'Theory and Clinical Experience in Eighteenth-Century Extemporaneous Prescriptions – A Reciprocal Relationship', *Pharmacy in History* 48 (2006), 3–13.

Croxson, Bronwyn, 'The Public and Private Faces of Eighteenth-Century London Dispensary Charity', *Medical History* 41 (1997), 127–49.

Daems, Willem F., *Johann Anton Grass von Portein, 1684–1770. Arzt, Chirurg, Zahnarzt, Harndiagnostiker, Pharmazeut, Viehdoktor und Dorfpolitiker. Ein Beitrag zur Kultur- und Medizingeschichte des Domleschgs und Heinzenbergs im 18. Jahrhundert* (Chur 1985).

Danninger, Gabriele, *"...daß sie auch vor den Krancken = Betten müsten das Maul halten..." Frauen zwischen "traditioneller Heiltätigkeit" und "gelehrter Medizin" um 1800 anhand Salzburger Quellen* (Wien 1998).

Daston, Lorraine, and Lunbeck, Elizabeth (eds), *Histories of Scientific Observation* (Chicago 2011).

Dauser, Regina, Hächler, Stefan, Kempe, Michael, Mauelshagen, Franz, and Stuber, Martin (eds), *Wissen im Netz. Botanik und Pflanzentransfer in europäischen Korrespondenznetzen des 18. Jahrhunderts* (Berlin 2008).

Delpiano, Patrizia, 'Precettori e allievi nel Settecento. La circolazione di saperi in Italia e in Europa', in Maria Pia Paoli (ed.), *Saperi a confronto nell' Europa dei secoli XIII–XIX* (Pisa 2009), 181–96.

Denzler, Alice, *Medizingeschichtliches aus der 1. Hälfte des 19. Jahrhunderts. Aus den Berichten der thurgauischen Bezirksärzte 1839–1862* (Basel 1938).

Diefenbacher, Michael, and Endres, Rudolf (eds), *Stadtlexikon Nürnberg* (Nürnberg 1999) [online: http://online-service.nuernberg.de/stadtarchiv/dok_start.fau?prj=biblio&dm=Stadtlexikon, accessed 14.09.2012].

Dietrich-Daum, Elisabeth, '"Historiae Morborum" – Die Krankengeschichten des Franz von Ottenthal. Ein Interreg IIIA Projekt. Österreich–Italien 2002–2007', *Virus. Zeitschrift für Sozialgeschichte der Medizin* 4 (2004) 83–9.

Dietrich-Daum, Elisabeth, 'Die Klage der Ärzte. Marktprobleme und Professionalisierung in der zweiten Hälfte des 19. Jahrhunderts in Österreich', in Alois Unterkircher, with Reinhard Bodner and Kathrin Sohm (eds), *Medikale Kulturen* [special issue: *Bricolage* 5 (2008)], 202–21.

Dietrich-Daum, Elisabeth, 'Arztpraxen im Vergleich. 18.–20. Jahrhundert. Der Vergleich des Unvergleichbaren?', in *eadem*, Martin Dinges, Robert Jütte and Christine Roilo (eds), *Arztpraxen im Vergleich. 18.–20. Jahrhundert* (Innsbruck 2008), 9–22.

Dietrich-Daum, Elisabeth, 'Reporting Death. The Case of Austrian Tuberculosis Mortality Registration. Critique and Consequences for Historical Epidemiology', *Prague Medical Report* 110 (2009), 146–158.

Dietrich-Daum, Elisabeth, and Oberhofer, Andreas, 'Die "Historiae Morborum" des Franz von Ottenthal. Ein Interreg IIIA-Projekt. Österreich–Italien 2002–2007', in Rainer Alsheimer and Roland Weibezahn (eds), *Körperlichkeit und Kultur 2004. Interdisziplinäre Medikalkulturforschung. Dokumentation des 7. Arbeitstreffens des "Netzwerk Gesundheit und Kultur in der volkskundlichen Forschung" Würzburg, 31. März–2. April 2004* (Bremen 2005), 203–14.

Dietrich-Daum, Elisabeth, and Oberhofer, Andreas, 'Medizingeschichte als Sozialgeschichte. Die Krankengeschichten des Südtiroler Landarztes Franz von Ottenthal (1818–1899), ein Interreg III A-Projekt. Österreich-Italien (2002–2007)', in Christoph Haidacher and Richard Schober (eds), *Bericht über den 24. Österreichischen Historikertag in Innsbruck* (Innsbruck 2006), 646–55.

Dietrich-Daum, Elisabeth, and Taddei, Elena, '"Historiae Morborum" – Le anamnesi del dottor Franz von Ottenthal. Un progetto Interreg IIIA. Austria–Italia 2002–2007', in Elisabeth Dietrich-Daum and Rodolfo Taiani (eds), *Medikalisierung auf dem Lande – Medicalizzazione in aree alpine* [special issue: *Geschichte und Region/Storia e regione* 14, no. 1 (2005)], 157–60.

Dietrich-Daum, Elisabeth, and Taddei, Elena, 'Curare – segregare – amministrare. L'assistenza e la gestione dei "mentecatti" in un contado del Tirolo. L'esempio del medico generico Franz von Ottenthal (1818–1899) di Campo Tures', in Elisabeth Dietrich-Daum and Rodolfo Taiani (eds), *Psychiatrielandschaft/Oltre il manicomio* [special issue: *Geschichte und Region/Storia e Regione* 17, no. 2 (2008)], 83–102.

Dietrich-Daum, Elisabeth, and Taddei, Elena, 'Psychiatrische Versorgung am Land. Die Rolle der praktischen Ärzte am Beispiel der Südtiroler Landarztpraxis des Franz von Ottenthal (1818–1899)', *Virus. Beiträge zur Sozialgeschichte der Medizin* 8 (2009), 27–42.

Dietrich-Daum, Elisabeth, and Taiani, Rodolfo, editorial to *Medikalisierung auf dem Lande/Medicalizzazione in aree alpine* [special issue: *Geschichte und Region/Storia e regione* 14, no.1 (2005)], 5–18.

Dietrich-Daum, Elisabeth, and Unterkircher, Alois, '"Medical History from Below" – Die Krankengeschichten eines Südtiroler Arztes als historische Quelle zur Geschichte der Medikalisierung am Land', in Sonia Horn, Gabriele Dorffner and Rosemarie Eichinger (eds), *Wissensaustausch in der Medizin des 15. bis 18. Jahrhunderts* (Wien 2007), 75–92.

Digby, Anne, *Making a Medical Living. Doctors and Patients in the English Market for Medicine 1720–1911* (Cambridge 1994).

Digby, Anne, 'The Patient's View', in Irvine Loudon (ed.), *Western Medicine* (Oxford 1997), 291–305.

Digby, Anne, *The Evolution of British General Practice, 1850–1948* (Oxford 1999).

Dijksterhuis, Eduard J., *The Mechanization of the World Picture. Pythagoras to Newton* (Princeton 1986).

Dinges, Martin, 'Medizinische Aufklärung bei Johann Georg Zimmermann. Zum Verhältnis von Macht und Wissen bei einem Arzt der Aufklärung', in Martin Fontius and Helmut Holzhey (eds), *Schweizer im Berlin des 18. Jahrhunderts* (Berlin 1996), 137–50.

Dinges, Martin, 'Soldatenkörper in der Frühen Neuzeit. Erfahrungen mit einem unzureichend geschützten, formierten und verletzten Körper in Selbstzeugnissen', in Richard van Dülmen (ed.), *Körpergeschichten* (Frankfurt am Main 1996), 71–98.

Dinges, Martin, 'Schmerzerfahrung und Männlichkeit. Der russische Gutsbesitzer und Offizier Andrej Bolotow (1738–1795)', *Medizin, Gesellschaft und Geschichte* 15 (1997), 55–78.

Dinges, Martin, '"Medicinische Policey" zwischen Heilkundigen und "Patienten" (1750–1830)', in Karl Härter (ed.), *Policey und frühneuzeitliche Gesellschaft* (Frankfurt am Main 2000), 263–95.

Dinges, Martin, 'Patients in the History of Homoeopathy', introduction to *idem* (ed.), *Patients in the History of Homoeopathy* (Sheffield 2002), 1–32.

Dinges, Martin, 'Social History of Medicine in Germany and France in the Late Twentieth Century. From History of Medicine toward a History of Health', in Frank Huisman and John Warner (eds), *Locating Medical History. The Stories and Their Meanings* (Baltimore 2004), 209–36.

Dinges, Martin, 'Immer schon 60 % Frauen in den Arztpraxen? Zur geschlechtsspezifischen Inanspruchnahme des medizinischen Angebotes (1600–2000)', in *idem* (ed.), *Männlichkeit und Gesundheit im historischen Wandel 1850–2000* (Stuttgart 2007), 295–322.

Dinges, Martin, 'Arztpraxen 1500–1900. Zum Stand der Forschung', in Elisabeth Dietrich-Daum, Martin Dinges, Robert Jütte and Christine Roilo (eds), *Arztpraxen im Vergleich. 18.–20. Jahrhundert* (Innsbruck 2008), 23–61.

Dinges, Martin, 'The Current State of Research on the History of Homeopathy', in Claudia Witt and Henning Albrecht (eds), *New Directions in Homeopathy Research. Advice from an Interdisciplinary Conference* (Essen 2009), 13–41.

Dinges, Martin, 'Hahnemanns Falldokumentation in historischer Perspektive', *Naturheilpraxis* 63 (2010), 1356–62.

Dinges, Martin, 'Medizin- und gesundheitsgeschichtliche Paradigmen zur geschlechterspezifischen Ungleichheit seit ca. 1750', in Martin Dinges and Andreas Weigl (eds), *Gesundheit und Geschlecht. Österreichische Zeitschrift für Geschichtswissenschaften* Sonderheft 22, no. 2 (2011), 8–49.

Dinges, Martin, and Barras, Vincent (eds), *Krankheit in Briefen im deutschen und französischen Sprachraum 17.–21. Jahrhundert* (Stuttgart 2007).

Dinges, Martin, and Holzapfel, Klaus, 'Von Fall zu Fall. Falldokumentation und Fallredaktion Clemens von Bönninghausen und Annette von Droste-Hülshoff', *Zeitschrift für klassische Homöopathie* 48 (2004), 149–67.

Dinges, Martin, and Weigl, Andreas (eds), *Gesundheit und Geschlecht* [special issue: *Österreichische Zeitschrift für Geschichtswissenschaften* 22, no. 2 (2011)].

Doebner, Eduard, *Bausteine zu einer Geschichte der Stadt Meiningen* (Meiningen 1902).

Dollinger, Heinz (ed.), *Die Universität Münster. 1780–1980* (Münster 1980).

Drees, Anette, *Die Ärzte auf dem Weg zu Prestige und Wohlstand. Sozialgeschichte der württembergischen Ärzte im 19. Jahrhundert* (Köln 1988).

Dross, Fritz, '"Der Kranke allein ist arm." Die Diskussion über den Zusammenhang von Krankheit und Armut um 1800', *Vierteljahrsschrift für Sozial- und Wirtschaftsgeschichte* 92, no. 4 (2005), 1–15.

Dross, Fritz, and Weyer-von Schoultz, Martin, 'Armenwesen und Krankenhäuser in der ersten Hälfte des 19. Jahrhunderts. Aspekte ihrer Funktion, Finanzierung und Klientel – das Düsseldorfer Beispiel', in Alfons Labisch and Reinhard Spree (eds), *Krankenhaus-Report 19. Jahrhundert. Krankenhausträger, Krankenhausfinanzierung, Krankenhauspatienten.* (Frankfurt am Main 2001), 295–337.

Duchhardt, Heinz (ed.), *Der Exodus der Hugenotten. Die Aufhebung des Edikts von Nantes 1685 als europäisches Ereignis* (Köln 1985).

Duden, Barbara, *The Woman Beneath the Skin. A Doctor's Patients in Eighteenth-Century Germany* (Cambridge 1991).

Duden, Barbara, *Geschichte des Ungeborenen. Zur Erfahrungs- und Wissenschaftsgeschichte der Schwangerschaft, 17.–20. Jahrhundert* (Göttingen 2002).

Duffin, Jacalyn, 'A Rural Pratice in Nineteenth-Century Ontario. The Continuing Medical Education of James Mile Langstaff', *Canadian Bulletin of Medical History* 5 (1988), 3–28.

Duffin, Jacalyn, *Langstaff. A Nineteenth-Century Medical Life* (Toronto 1993, repr. 1999).

Duffin, Jacalyn, *Lovers and Livers. Disease Concepts in History* (Toronto 2005).

Dumont, Franz, 'Nicht nur Hölderlin. Das ärztliche Besuchsbuch Soemmerings als Quelle für sein soziales Umfeld in Frankfurt am Main', *Medizinhistorisches Journal* 28 (1993), 123–54.

Eckart, Wolfgang, *Geschichte der Medizin* (Berlin ²1994).

Eckart, Wolfgang, and Jütte, Robert, *Medizingeschichte. Eine Einführung* (Köln 2007).

Efron, John M., *Medicine and the German Jews. A History* (New Haven 2001).

Ellerkamp, Marlene, *Industriearbeit, Krankheit und Geschlecht* (Göttingen 1991).

Endres, Rudolf, 'Bevölkerungsentwicklung', in Michael Diefenbacher and Rudolf Endres (eds), *Stadtlexikon Nürnberg* (Nürnberg 1999) [http://online-service2.nuernberg.de/stadtarchiv/doindex.FAU?sid=AEF99F2E12&dm=2&ind=1, 11 August 2014].

Engel, Regina, *Das Patientengut eines praktischen Arztes um die Mitte des 19. Jahrhunderts, gezeigt am Beispiel des Patientenjournals von Dr. Heinrich Grotjahn in Schladen am Harz* (Med. diss., Humboldt-Universität Berlin, 1978).

Engstrom, Eric J., Volker Hess und Ulrike Thoms (eds), *Figurationen des Experten. Ambivalenzen der wissenschaftlichen Expertise im ausgehenden 18. und frühen 19. Jahrhundert* (Frankfurt am Main 2005).

Ernst, Katharina, 'Patientengeschichte. Die kulturhistorische Wende in der Medizinhistoriographie', in Ralf Bröer (ed.), *Eine Wissenschaft emanzipiert sich. Medizinhistoriographie von der Aufklärung bis zur Postmoderne* (Pfaffenweiler 1999), 97–108.

Ernst, Katharina, *Krankheit und Heilung. Die medikale Kultur württembergischer Pietisten im 18. Jahrhundert* (Stuttgart 2003).

Ernst, Katharina, 'Fragments and Gaps. The Form of Illness Narratives in 18th Century Wurttemberg Pietist Diaries', *Medizin in Gesellschaft und Geschichte* 24 (2005), 33–45.

Faltin, Thomas, *Heil und Heilung. Geschichte der Laienheilkunde und Struktur antimodernistischer Weltanschauungen in Kaiserreich und Weimarer Republik am Beispiel von Eugen Wenz (1856–1945)* (Stuttgart 2000).

Farrell Brodie, Janet, 'Menstrual Interventions in the Nineteenth-Century United States', in Elisha P. Renne and Etienne van de Walle (eds), *Regulating Menstruation. Beliefs, Practices, Interpretations* (Chicago 2001), 39–63.

Faure, Olivier, 'La clientèle d'un homéopathe parisien au XXe siècle. Recherche sur les patients de L. Vannier 1928–1948', in *idem* (ed.), *Praticiens, patients et militants de l'homéopathie aux XIXe et XXe siècles (1800–1940)* (Lyon 1992), 175–96.

Faure, Olivier, *Les Français et leur médecine au XIXe siècle* (Paris 1993).

Fischer, Ulrich, *Die Chronischen Miasmen Hahnemanns. Grundgedanken zum Verständnis und zur Therapie chronischer Krankheiten aus homöopathischer Sicht* (Karlsruhe 1993).

Fissell, Mary E., *Patients, Power, and the Poor in Eighteenth-Century* (Cambridge 2002).

Flach, Hendrick, *Der Landarzt geht in Rente. Entwicklungstendenzen und Optionen zur Sicherstellung der ambulanten medizinischen Versorgung im ländlichen Raum* (Berlin 2012).

Föppl, August, *Lebenserinnerungen. Rückblick auf meine Lehr- und Aufstiegsjahre* (München 1925).

Foucault, Michel, 'La vie des hommes infâmes (1977)', in Daniel Defert and François Ewald, in collaboration with Jacques Lagrange (eds), *Dits et Ecrits*, vol. 3 (Paris 1994), 237–53.

Foucault, Michel, *The Birth of the Clinic. An Archaeology of Medical Perception* (London 2006).

Franke, Hans, and Schröder, Joachim, *Die Würzburger Medizinische Universitäts-Poliklinik 1807–1957* (Stuttgart 1957).

Frei, Daniel, 'Mediation', in Hanno Helbling (ed.), *Handbuch der Schweizer Geschichte* (Zürich 1977), vol. 2, 841–69.

French, Roger, 'Harvey in Holland. Circulation and the Calvinists', in Roger French and Andrew Wear (eds), *The Medical Revolution of the Seventeenth Century* (Cambridge 1989), 46–86.

Frevert, Ute, *Krankheit als politisches Problem 1770–1880. Soziale Unterschichten in Preußen zwischen medizinischer Polizei und staatlicher Sozialversicherung* (Göttingen 1984).

Frevert, Ute, 'Vertrauen. Eine historische Spurensuche', in Ute Frevert (ed.), *Vertrauen. Historische Annäherungen* (Göttingen 2003), 7–66.

Friedrich, Christoph, 'Pharmakopöen. Spiegel der pharmazeutischen Technologie', *Pharmazeutische Zeitung* 145 (2000), 2122–7.

Friedrich, Susanne, *Drehscheibe Regensburg. Das Informations- und Kommunikationssystem des Immerwährenden Reichstags um 1700* (Berlin 2007).

Fuchs, Peter, 'Metternich', in *Neue Deutsche Biographie*, vol. 17 (Berlin 1994), 232–35.

Füssel, Marian, *Gelehrtenkultur als symbolische Praxis. Rang, Ritual und Konflikt an der Universität der Frühen Neuzeit* (Darmstadt 2006).

Gaab, Hans, *Der Altdorfer Mathematik- und Physikdozent Abdias Trew (1597–1669). Astronom, Astrologe und Kalendermacher* (Frankfurt am Main 2011).

Gafner, Lina, Iris Ritzmann and Katharina Weikl (eds), *Penning Patient's Histories – Doctors' Records and the Medical Market in the 18th and 19th Century* [special issue: *Gesnerus* 69, no. 1 (2012)].

Gafner, Lina, Iris Ritzmann and Katharina Weikl, 'Gesundheitsmarkt in ärztlichen Praxisjournalen des 18. und 19. Jahrhunderts', introduction to *eaedem* (eds), *Penning Patient's Histories – Doctors' Records and the Medical Market in the 18th and 19th Century* [special issue: *Gesnerus* 69, no. 1 (2012)], 5–12.

Gafner, Lina, 'Ärztliche Zeugnisse auf dem medizinischen Markt. Die professionalisierte Zeugenschaft im wachsenden Verwaltungsapparat des frühen 19. Jahrhunderts, am Beispiel einer Bieler Arztpraxis', in *eadem*, Iris Ritzmann and Katharina Weikl (eds), *Penning Patient's Histories – Doctors' Records and the Medical Market in the 18th and 19th Century* [special issue: *Gesnerus* 69, no. 1 (2012c)], 95–109 (with English summary).

Gelfand, Toby, *Professionalizing Modern Medicine. Paris Surgeons and Medical Science and Institutions in the 18th Century* (Westport [Conn] 1980).

Genneper, Thomas, 'Die homöopathische Anamnese', in Thomas Genneper and Andreas Wegener (eds), *Lehrbuch Homöopathie. Grundlagen und Praxis der klassischen Homöopathie* (Stuttgart ^{3}2011), 80–106.

Genschorek, Wolfgang, *Wegbereiter der Chirurgie. Johann Friedrich Dieffenbach. Theodor Billroth* (Leipzig 1983).

Geyer-Kordesch, Johanna, 'Medizinische Fallbeschreibungen und ihre Bedeutung in der Wissensreform des 17. und 18. Jahrhunderts', *Medizin in Geschichte und Gesellschaft* 9 (1990), 7–19 (with English summary).

Geyer-Kordesch, Johanna, *Pietismus, Medizin und Aufklärung in Preußen im 18. Jahrhundert. Das Leben und Werk Georg Ernst Stahls* (Tübingen 2000).

Gillis, Jonathan, 'The History of the Patient History since 1850', *Bulletin of the History of Medicine* 80 (2006), 490–511.

Glenzdorf, Johann, and Treichel, Fritz, *Henker, Schinder und arme Sünder*, 2 vols. (Bad Münder am Deister 1970).

Göckenjan, Gerd, *Kurieren und Staat machen. Gesundheit und Medizin in der bürgerlichen Welt* (Frankfurt am Main 1985).

Gossmann, Heinz, *Das Collegium Pharmaceuticum Norimbergense und sein Einfluss auf das Nürnbergische Medizinalwesen* (Frankfurt am Main 1966).

Gotthard, Axel, *Der Augsburger Religionsfriede* (Münster 2006).

Goubert, Jean Pierre, 'Die Medikalisierung der französischen Gesellschaft am Ende des Ancien Régime. Die Bretagne als Beispiel', *Medizinhistorisches Journal* 17 (1982), 89–114.

Green, David, *Working-Class Patients and the Medical Establishment. Self-Help in Britain from the Mid-Nineteenth Century to 1948* (Aldershot 1985).

Griesel, Sabina Verena, *Medizinische und chirurgische Observationen, 1785–1787* (Med. diss., Universität Basel, 1984).

Gross, Dominik, *Die Aufhebung des Wundarztberufs – Ursachen, Begleitumstände und Auswirkungen am Beispiel des Königreichs Württemberg (1806–1918)* (Stuttgart 1999).

Gross, Dominik, 'Zur Rekonstruktion der Wundarztdichte im Königreich Württemberg. Ein Beitrag zur Sozialgeschichte der Handwerkschirurgen im 19. Jahrhundert', *Medizinhistorisches Journal* 32 (1997), 57–81.

Grumbach, Torsten, *Kurmainzer Medicinalpolicey 1650–1803. Eine Darstellung entlang der landesherrlichen Verordnungen* (Frankfurt am Main 2006).

Gundert, Beate, 'Humoralpathologie', in Karl-Heinz Leven (ed.), *Antike Medizin. Ein Lexikon* (München 2005), 436–41.

Günther, Martina, and Römermann, Hans, 'The Homoeopathic Patient in General Practice. Findings of a Comparative Poll of Patients in Conventional Medical Practices and Homoeopathic Private and Health Insurance Scheme Practices', in Martin Dinges (ed.), *Patients in the History of Homoeopathy* (Sheffield 2002), 281–99.

Günther, Siegmund, 'Müller, Johannes', in *Allgemeine Deutsche Biographie* 22 (1885), 625–8.

Habrich, Christa, 'Untersuchungen zur pietistischen Medizin und ihrer Ausprägung bei Johann Samuel Carl (1677–1757) und seinem Kreis' (Habilitationsschrift, University of Munich 1981).

Habrich, Christa, 'Johann Samuel Carl (1677–1757) und die Philadelphische Ärztegemeinschaft', in Hartmut Lehmann, Hans-Jürgen Schrader and Heinz Schilling (eds), *Jansenismus, Quietismus, Pietismus. Im Auftrag der Historischen Kommission zur Erforschung des Pietismus.* (Göttingen 2002), 272–89.

Haehl, Richard, *Samuel Hahnemann. Sein Leben und Schaffen*, 2 vols. (Leipzig 1922).

Hansch-Mock, Barbara, *Deutschschweizerische Kalender des 19. Jahrhunderts als Vermittler schul- und volksmedizinischer Vorstellungen* (Aarau 1976).

Hauser, Albert, *Schweizerische Wirtschafts- und Sozialgeschichte. Von den Anfängen bis zur Gegenwart* (Zürich 1961).

Heischkel, Edith, 'Die Poliklinik des 18. Jahrhunderts in Deutschland', *Deutsches Medizinisches Journal* 5, no. 1/2 (1954), 223–5.

Helm, Jürgen, and Wilson, Renate (eds), *Medical Theory and Therapeutic Practice in the Eighteenth Century. A Transatlantic Perspective* (Stuttgart 2008).

Herbst, Klaus-Dieter, and Kratochwil, Stefan (eds), *Kommunikation in der Frühen Neuzeit* (Frankfurt am Main 2009).

Hess, Volker, 'Samuel Hahnemann und die Semiotik', *Medizin, Gesellschaft und Geschichte* 12 (1993), 177–204 (with English summary).

Hess, Volker, *Von der semiotischen zur diagnostischen Medizin. Die Entstehung der klinischen Methode zwischen 1750 und 1850* (Husum 1993).

Hess, Volker, *Der wohltemperierte Mensch. Wissenschaft und Alltag des Fiebermessens (1850–1900)* (Frankfurt am Main 2000).

Hess, Volker, 'Ärztlicher Alltag in Thüringen um 1750. Auswertung eines Praxistagebuchs', in Elisabeth Dietrich-Daum, Martin Dinges, Robert Jütte and Christine Roilo (eds), *Arztpraxen im Vergleich. 18.–20. Jahrhundert* (Innsbruck 2008), 87–107.

Hess, Volker, 'Formalisierte Beobachtung. Die Genese der modernen Krankenakte am Beispiel der Berliner und Pariser Medizin (1725–1830)', *Medizinhistorisches Journal* 45 (2010), 293–340 (with English summary).

Hess, Volker, 'Medizin zwischen Sammeln und Experimentieren', in Rüdiger vom Bruch and Heinz-Elmar Tenorth (eds), *Geschichte der Universität zu Berlin, 1810–2010. Biographie einer Institution, Praxis ihrer Disziplinen*, vol. 4 (Berlin 2010), 489–566.

Hess, Volker, 'Das Material einer guten Geschichte. Register, Reglements und Formulare', in Sheila Dickson, Stefan Goldmann and Christof Wingertszahn (eds), *Fakta, und kein moralisches Geschwätz. Zu den Fallgeschichten im "Magazin zur Erfahrungsseel enkunde"(1783–1793)* (Göttingen 2011), 115–39.

Hess, Volker, and Mendelsohn, Andrew, 'Fallgeschichte, Historia, Klassifikation. François Boissier de Sauvages bei der Schreibarbeit', *NTM. Zeitschrift für Geschichte der Wissenschaften, Technik und Medizin* 21 (2013), 61–92 (with English summary).

Hilber, Marina, 'Der Landarzt als Geburtshelfer – Dr. Franz von Ottenthal und der medizinische Markt in Südtirol (1860–1869)', in Lina Gafner, Iris Ritzmann and Katharina Weikl (eds), *Penning Patient's Histories – Doctors' Records and the Medical Market in the 18th and 19th Century* [special issue: *Gesnerus* 69, no. 1 (2012)], 141–57 (with English summary).

Hinrichs, Hans Jürgen, *Geschichte des Gymnasium Gleichense in Ohrdruf 1564–2001* (Gotha 2001).

Hippel, Wolfgang von, Ute Mocker and Sylvia Schraut, 'Wohnen im Zeitalter der Industrialisierung. Esslingen am Neckar 1800–1914', *Esslinger Studien* 26 (1987), 47–180.

Hirsch, August (ed.), *Biographisches Lexikon der hervorragenden Ärzte aller Zeiten und Völker*, vol. 5 (München ³1962).

Hobhouse, Edmund (ed.), *The Diary of a West Country Physician, A.D. 1684–1726* (London 1934).

Hoffmann, Christoph, 'Festhalten, bereitstellen. Verfahren der Aufzeichnung', in *idem* (ed.), *Daten sichern. Schreiben und Zeichnen als Verfahren der Aufzeichnung* (Zürich 2008), 7–20.

Hoffmann, Christoph, 'Schreiben als Verfahren der Forschung', in Michael Gamper (ed.), *Experiment und Literatur. Themen, Methoden, Theorien* (Göttingen 2010), 181–207.

Hoffmann, Susanne, *Gesundheit und Krankheit bei Ulrich Bräker (1735–1798)* (Dietikon 2005).

Hoffmann, Susanne, *Gesunder Alltag im 20. Jahrhundert? Geschlechterspezifische Diskurse und gesundheitsrelevante Verhaltensstile in deutschsprachigen Ländern* (Stuttgart 2010).

Hoffmeyer-Zlotnik, Jürgen, 'Operationalisierung von "Beruf" als zentrale Variable zur Messung von sozio-ökonomischem Status', *ZUMA-Nachrichten* 32 (1993), 135–41.

Holenstein, André, *"Gute Policey" und lokale Gesellschaft im Staat des Ancien Régime. Das Fallbeispiel der Markgrafschaft Baden(-Durlach)* (Epfendorf 2003).

Holzapfel, Klaus, 'Repertorien und Repertorisation', in Thomas Genneper and Andreas Wegener (eds), *Lehrbuch Homöopathie. Grundlagen und Praxis der klassischen Homöopathie* (Stuttgart ³2011), 143–69.

Hörning, Karl H., and Reuter, Julia (eds), *Doing Culture. Neue Positionen zum Verhältnis von Kultur und sozialer Praxis* (Bielefeld 2004).

Hortzitz, Nicoline, *Der Judenarzt. Historische und sprachliche Untersuchungen zur Diskriminierung eines Berufsstands in der frühen Neuzeit* (Heidelberg 1994).

Huerkamp, Claudia, 'Ärzte und Professionalisierung in Deutschland. Überlegungen zum Wandel des Arztberufs im 19. Jahrhundert', *Geschichte und Gesellschaft* 6 (1980), 349–82.

Huerkamp, Claudia, *Der Aufstieg der Ärzte im 19. Jahrhundert. Vom gelehrten Stand zum professionellen Experten. Das Beispiel Preussens* (Göttingen 1985).

Huerkamp, Claudia, 'Ärzte und Patienten. Zum strukturellen Wandel der Arzt-Patient-Beziehung vom ausgehenden 18. bis zum frühen 20. Jahrhundert', in Reinhard Spree and Alfons Labisch (eds), *Medizinische Deutungsmacht im sozialen Wandel des 19. und 20. Jahrhunderts* (Bonn 1989), 57–73.

Huerkamp, Claudia, 'Das unterschiedliche Verhältnis von Arzt und Patient in der Krankenhauspraxis und der privaten ärztlichen Praxis im 19. Jahrhundert', in Peter

Schneck and Hans-Uwe Lamme (eds), *Die Medizin an der Berliner Universität und an der Charité zwischen 1810 und 1850* (Husum 1995), 254–68.

Huisman, Tim, *The Finger of God. Anatomical Practice in 17th-Century Leiden* (Leiden 2009).

Hurrelmann, Klaus, and Kolip, Petra (eds), *Geschlecht, Gesundheit, Krankheit. Männer und Frauen im Vergleich* (Bern 2002).

Husemann, Theodor, 'Fuchs, Konrad Heinrich', in *Allgemeine Deutsche Biographie* 8 (1878), 168–9.

Illich, Ivan, *Medical Nemesis. The Expropriation of Health* (London 1975).

Jacobi, Ursula, *Der Hochpotenzstreit. Von Hahnemann bis heute* (Stuttgart 1994).

Jankrift, Kay Peter, 'Eigenes und Fremdes. Zur sozialen Stellung jüdischer Gelehrter und ihrer Rolle für die Vermittlung von Wissen im Mittelalter', in Kay Peter Jankrift and Florian Steger (eds), *Gesundheit – Krankheit. Kulturtransfer medizinischen Wissens von der Spätantike bis in die Frühe Neuzeit* (Köln 2004), 139–54.

Jankrift, Kay-Peter, Annemarie Kinzelbach and Marion Ruisinger, 'Ernst von Metternich (1656–1727). Ein patientenzentrierter Einblick in den medizinischen Markt um 1720', in Lina Gafner, Iris Ritzmann and Katharina Weikl (eds), *Penning Patient's Histories – Doctors' Records and the Medical Market in the 18th and 19th Century* [special issue: *Gesnerus* 69, no. 1 (2012)], 12–35 (with English summary).

Jenner, Mark S.R., *Medicine and the Market in England and its Colonies, c.1450–c.1850* (Basingstoke 2007).

Jenny, Franziska, *"Sie habe 3024 Kinder gehoben." Das Thurgauische Hebammenwesen 1799–1869* (Dietikon 2005).

Jentzsch, Gunda, *Zur Geschichte der Klinischen Medizin in Göttingen. Das Ernst-August-Hospital 1850–1890* (Med. diss., Universität Göttingen, 1988).

Jewson, Nicolas, 'Medical Knowledge and the Patronage System in 18th Century England. A Sociological Analysis', *Sociology* 8 (1974), 369–85.

Jewson, Nicholas D., 'The Disappearance of the Sick Man from the Medical Cosmology', 1770–1870, *Sociology* 10 (1976), 225–44.

Jorland, Gérard, *Une société à soigner. Hygiène et salubrité publiques en France au XIXe siècle* (Paris 2010).

Jürgensen, Renate, *Utile cum dulci. Die Blütezeit des Pegnesischen Blumenordens in Nürnberg 1644–1744* (Wiesbaden 1994).

Jütte, Robert, 'A Seventeenth-Century German Barber-Surgeon and His Patients', *Medical History* 33 (1989), 184–98.

Jütte, Robert, 'Sozialgeschichte der Medizin. Inhalte – Methoden – Ziele', *Medizin, Gesellschaft und Geschichte* 9 (1991), 149–64 (with English summary).

Jütte, Robert, *Ärzte, Heiler und Patienten. Medizinischer Alltag in der frühen Neuzeit* (München 1991).

Jütte, Robert (ed.), *Die Krankenjournale von Samuel Hahnemann; D 2: (1801–1802)/nach der Ed. von Heinz Henne. Bearb. von Arnold Michalowski*, 2 (Heidelberg 1993).

Jütte, Robert, *Geschichte der Alternativen Medizin. Von der Volksmedizin zu den unkonventionellen Therapien von heute* (München 1996).

Jütte, Robert, 'Samuel Hahnemanns Patientenschaft', in Martin Dinges (ed.), *Homöopathie. Patienten, Heilkundige, Institutionen. Von den Anfängen bis heute* (Heidelberg 1996), 23–45.

Jütte, Robert, 'Case Taking in Homoeopathy in the 19th and 20th Centuries', *British Homoeopathic Journal* 87 (1998), 39–47.

Jütte, Robert, '"Und es sammelte sich ohne Verdruß von Seiten des Kranken in des Arztes Beutel." Samuel Hahnemann und die Honorarfrage', *Medizin, Gesellschaft und Geschichte* 18 (1999), 149–67 (with English summary).

Jütte, Robert, 'The Historiography of Homoeopathy in Germany', *Orvostörténeti Közlemények. Communicationes de Historia Artis Medicinae* 186/187 (2004), 123–30.

Jütte, Robert, 'Die Arzt-Patient-Beziehung im Spiegel der Krankenjournale Samuel Hahnemanns', in Elisabeth Dietrich-Daum, Martin Dinges and Robert Jütte (eds), *Arztpraxen im Vergleich. 18.–20. Jahrhundert* (Innsbruck 2008), 109–27.

Jütte, Robert, *The LM Potencies in Homoeopathy. From their Beginnings to the Present Day* (Stuttgart 2008), [online: http://www.igm-bosch.de/content/language2/downloads/lm_potencies(1).pdf, accessed 01.04.2014].

Jütte, Robert, 'Homöopathie. Das besondere Arzt-Patienten-Verhältnis', in Martin Momburg and Dietmar Schulte (eds), *Das Verhältnis von Arzt und Patient. Wie menschlich ist die Medizin?* (München 2010), 71–87.

Jütte, Robert, *Samuel Hahnemann. The Founder of Homeopathy*, 2012 [online: http://www.igm-bosch.de/content/language1/downloads/SamuelHahnemannTheFounderofHomeopathy.pdf, accessed 01.04.2014].

Jütte, Robert, and Dinges, Martin, *Homoeopathy. A Medical Approach and Its History* (Stuttgart 2009).

Karenberg, Axel, *Lernen am Bett der Kranken. Die frühen Universitätskliniken in Deutschland (1760–1840)* (Stuttgart 1997).

Kayne, Steven B., *Traditional Medicine. A Global Perspective* (London 2006).

Kass, Amalie M., 'The Obstetrical Casebook of Walter Channing, 1811–22', *Bulletin of the History of Medicine* 67 (1993), 494–523.

Keel, Othmar, *L'avènement de la médecine clinique moderne en Europe 1750–1815. Politiques, institutions et savoirs* (Montréal 2001).

Keller, Hagen, 'Vom "heiligen Buch" zur "Buchführung." Lebensfunktionen der Schrift im Mittelalter', *Frühmittelalterliche Studien* 26 (1992), 1–31.

Kinzelbach, Annemarie, *Gesundbleiben, Krankwerden, Armsein in der frühneuzeitlichen Gesellschaft. Gesunde und Kranke in den Reichsstädten Ueberlingen und Ulm, 1500–1700* (Stuttgart 1995).

Kinzelbach, Annemarie, 'Konstruktion und konkretes Handeln. Heilkundige Frauen im oberdeutschen Raum, 1450–1700', *Historische Anthropologie* 7 (1999), 165–90.

Kinzelbach, Annemarie, 'Infection, Contagion, and Public Health in Late Medieval and Early Modern German Imperial Towns', *Journal of the History of Medicine and Allied Sciences* 61 (2006), 369–89.

Kinzelbach, Annemarie, 'Women and Healthcare in Early Modern German Towns', *Renaissance Studies* 28 (2014), 619–38.

Kittler, Friedrich A., 'Das Subjekt als Beamter', in Manfred Frank, Gérard Raulet and Willem van Reijen (eds), *Die Frage nach dem Subjekt* (Frankfurt am Main 1988), 401–20.

Kittler, Friedrich A., *Discourse Networks 1800/1900* (Stanford 1990).

Klaas, Philipp, 'Cäsar Adolf Bloesch – ein Geburtshelfer? Einblicke in seine ärztliche Praxis zwischen 1832 und 1850', in Lina Gafner, Iris Ritzmann and Katharina Weikl (eds), *Penning Patient's Histories – Doctors' Records and the Medical Market in the 18th and 19th Century* [special issue: *Gesnerus* 69, no. 1 (2012)], 110–25 (with English summary).

Köllmann, Wolfgang, 'Die Bevölkerung des Regierungsbezirks Münster im 19. Jahrhundert', *Westfälische Forschungen* 40 (1990), 195–222.

König, Klaus G., *Der Nürnberger Stadtarzt Dr. Georg Palma (1543–1591)* (Stuttgart 1961).

Koselleck, Reinhart, '"Neuzeit." Zur Semantik moderner Bewegungsbegriffe', in *idem*, *Vergangene Zukunft* (Frankfurt am Main 1989), 300–47.

Kottwitz, Friedrich, *Bönninghausens Leben. Hahnemanns Lieblingsschüler* (Berg 1985).

Krieger, Martin, *Arme und Ärzte, Kranke und Kassen. Ländliche Gesundheitsversorgung und kranke Arme in der südlichen Rheinprovinz (1869 bis 1930)* (Stuttgart 2008).

Kümmel, Werner Friedrich, 'Der Puls und das Problem der Zeitmessung in der Geschichte der Medizin', *Medizinhistorisches Journal* 9 (1974), 1–22.

Kumsteller, Renate, *Die Anfänge der medizinischen Poliklinik zu Göttingen. Eine medizin- und kulturhistorische Studie zur zweiten Hälfte des 18. Jahrhunderts, gewonnen aus dem Archivmaterial der Universität* (Med. diss., Universität Göttingen, 1955).

Kunkle, Luise, 'Von Bönninghausens Verschreibungspraxis', *Zeitschrift für Klassische Homöopathie* 52 (2008), 172–78.

Labisch, Alfons, *Homo hygienicus. Gesundheit und Medizin in der Neuzeit* (Frankfurt am Main 1992).

Lachmund, Jens, *Der abgehorchte Körper. Zur historischen Soziologie der medizinischen Untersuchung* (Opladen 1997).

Lachmund, Jens, 'Between Scrutiny and Treatment. Physical Diagnosis and the Restructuring of 19th Century Medical Practice', *Sociology of Health & Illness* 20 (1998), 779–801.

Lachmund, Jens, and Stollberg, Gunnar, *Patientenwelten. Krankheit und Medizin vom späten 18. Jahrhundert bis zum frühen 20. Jahrhundert im Spiegel von Autobiographien* (Opladen 1995).

Lane, Joan, *John Hall and His Patients. The Medical Practice of Shakespeare's Son in Law. With Medical Commentary by Melvin Earles* (Stratford-upon-Avon 1996).

Lane, Joan, *A Social History of Medicine. Health, Healing and Disease in England, 1750–1950* (London 2001).

Latour, Bruno, and Woolgar, Steve, *Laboratory Life. The Social Construction of Scientific Facts* (Beverly Hills/London, 1979).

Lauer, Hans H., *Die Medizinische Poliklinik in Marburg und die Anfänge ihrer Selbständigkeit* (Marburg 1994).

Lee, Robert, Schneider, Michael, and Vögele, Jörg, 'Zur Entwicklung der Todesursachenstatistik in Preußen im 19. und frühen 20. Jahrhundert', in Josef Ehmer, Ursula Ferdinand and Jürgen Reulecke (eds), *Herausforderung Bevölkerung. Zu Entwicklungen des modernen Denkens über die Bevölkerung vor, im und nach dem "Dritten Reich"* (Wiesbaden 2007), 7–16.

Lehfeldt, Paul, and Voss, Georg, *Bau- und Kunstdenkmäler Thüringens. Vol. 2: Herzogtum Sachsen-Meiningen. Kreis Hildburghausen. Amtsgerichtsbezirke Hildburghausen, Eisfeld, Themar, Heldburg und Römhild* (Jena 1904).

Lenhardt, Friedrich, 'Zur Ikonographie der Blutschau', *Medizinhistorisches Journal* 17 (1982), 63–77.

Lenhardt, Friedrich, *Blutschau. Untersuchungen zur Entwicklung der Hämatoskopie* (Hannover 1986).

Léonard, Jacques, *Les médecins de l'Ouest au XIXe siècle*. 3 vols (Lille 1978).

Leven, Karl-Heinz, *Die Geschichte der Infektionskrankheiten. Von der Antike bis ins 20. Jahrhundert* (Landsberg 1997).

Leven, Karl-Heinz, 'Krankheiten. Historische Deutung versus retrospektive Diagnose', in Norbert Paul and Thomas Schlich (eds), *Medizingeschichte. Aufgaben, Probleme, Perspektiven* (Frankfurt am Main 1998), 153–85.

Lindemann, Mary, *Health and Healing in Eighteenth-Century Germany* (Baltimore 1996).

Lindemann, Mary, *Medicine and Society in Early Modern Europe* (Cambridge [2]2010).

Lindt, Wilhelm, *Zum Jubiläum der medizinisch-chirurgischen Gesellschaft des Kantons Bern* (Bern 1909).

Lobenwein, Elisabeth, *Höfe- und Häusergeschichte von St. Peter in Ahrn. Die Patientinnen und Patienten (aus St. Peter in Ahrn) des Franz von Ottenthal* (unpublished typescript). (Innsbruck 2012).

Loetz, Francisca, *Vom Kranken zum Patienten. "Medikalisierung" und medizinische Vergesellschaftung am Beispiel Badens 1750–1850* (Stuttgart 1993).

Loetz, Francisca, '"Medikalisierung" in Frankreich, Großbritannien und Deutschland 1750–1850. Ansätze, Ergebnisse und Perspektiven der Forschung', in Wolfgang Eckart and Robert Jütte (eds), *Das europäische Gesundheitswesen. Gemeinsamkeiten und Unterschiede in historischer Perspektive* (Stuttgart 1994), 123–61.

Loetz, Francisca, 'Andere Grenzen. Faktoren ärztlicher Inanspruchnahme in Deutschland 1780–1830. Empirische Ergebnisse und methodologische Überlegungen', in Thomas Schnalke and Claudia Wiesemann (eds), *Die Grenzen des Anderen. Medizingeschichte aus postmoderner Perspektive* (Köln 1998), 25–48.

Loudon, Irvine, 'A Doctor's Cash Book. The Economy of General Practice in the 1830s', *Medical History* 27 (1983), 249–68.

Loudon, Irvine, 'The Nature of Provincial Medical Practice in Eighteenth-Century England', *Medical History* 29 (1985), 1–32.

Loudon, Irvine, *Medical Care and the General Practitioner 1750–1850* (Oxford 1986).

Lüdtke, Alf, 'Alltagsgeschichte. Aneignung und Akteure. Oder – es hat doch kaum begonnen!', *WerkstattGeschichte* 17 (1997), 65–80.

Lüdtke, Alf, and Prass, Reiner, 'Einleitung. Gelehrtenleben. Wissenschaftspraxis in der Neuzeit', in Alf Lüdtke and Reiner Prass (eds), *Gelehrtenleben. Wissenschaftspraxis in der Neuzeit* (Köln 2008), 1–29.

Luhmann, Niklas, *Vertrauen. Ein Mechanismus der Reduktion sozialer Komplexität* (Stuttgart [3]1989).

Lütge, Friedrich, *Geschichte der deutschen Agrarverfassung vom frühen Mittelalter bis zum 19. Jahrhundert* (Stuttgart 1963).

Mader, Frank H., and Weißgerber, Herbert, *Allgemeinmedizin und Praxis. Anleitung in Diagnostik und Therapie* (Berlin [2]1995).

Mah, Harold, 'Phantasies of the Public Sphere. Rethinking the Habermas of Historians', *Journal of Modern History* 72 (2000), 153–82.

Mahlberg, Daniel Alexander, *Der ärztliche Bezirksverein Bern-Seeland. 1845–1895* (Med. diss., Universität Bern, 2005).

Marland, Hilary, *The Task of Healing. Medicine, Religion and Gender in England and the Netherlands, 1450–1800* (Rotterdam 1996).

Martinez, Matias, and Scheffel, Michael, *Einführung in die Erzähltheorie* (München [8]2009).

Marx-Jaskulski, Katrin, *Armut und Fürsorge auf dem Land. Vom Ende des 19. Jahrhunderts bis 1933* (Göttingen 2007).

Mattl, Siegfried, 'Formalisierung der Welt', *Österreichische Zeitschrift für Sozialwissenschaften* 7 (1996), 173–6.

Mauthe, Jürgen-Helmut (ed.), *Krankheit & Geschlecht. Konzepte & Kontroversen* (Sternenfels 1999).

Mauelshagen, Franz, 'Netzwerke des Vertrauens. Gelehrtenkorrespondenzen und wissenschaftlicher Austausch in der Frühen Neuzeit', in Ute Frevert (ed.), *Vertrauen. Historische Annäherungen* (Göttingen 2003), 119–51.

Meier-Oeser, Stephan, 'Wissen. (Frühe Neuzeit)', in Joachim Ritter (ed.), *Historisches Wörterbuch der Philosophie*, vol. 12 (Basel 2004), 881–4.

Meli, Domenico B., *Mechanism, Experiment, Disease. Marcello Malpighi and Seventeenth-Century Anatomy* (Baltimore 2011).

Messner, Wolfgang, 'Wirtschafts- und Sozialgeschichte des Tauferer Tales/Ahrntales im ausgehenden 18. und im Laufe des 19. Jahrhunderts' (Master thesis, University of Innsbruck, 2004).

Metzdorf, Jens, *Münster wird preußisch. Von der fürstbischöflichen Residenzstadt zur Provinzialhauptstadt* (Münster 1992).

Mixa, Elisabeth, and Malleier, Elisabeth (eds), *Körper – Geschlecht – Geschichte. Historische und aktuelle Debatten in der Medizin* (Innsbruck 1996).

Möller, Bernhard, 'Die Methodik Clemens von Bönninghausens. Dargestellt anhand seines Therapeutischen Taschenbuchs', in *Die Bönninghausen- und Bognermethodik* [special issue: *Homöopathie-Zeitschrift* (2002)], 6–25.

Mommertz, Monika, 'Das Wissen "auslocken." Eine Skizze zur Geschichte der epistemologischen Produktivität von Grenzüberschreitung, Transfer und Grenzziehung zwischen Universität und Gesellschaft', in Yuka Nakamura, Christine Böckelmann and Daniel Tröhler (eds), *Theorie versus Praxis? Perspektiven auf ein Missverständnis* (Zürich 2006), 19–51.

Moses, Simone, *Alt und krank. Ältere Patienten in der Medizinischen Klinik der Universität Tübingen zur Zeit der Entstehung der Geriatrie 1880 bis 1914* (Stuttgart 2005).

Moss, Ann, *Printed Commonplace-Books and the Structuring of Renaissance Thought* (Oxford 1996).

Mücke, Marion, and Schnalke, Thomas, *Briefnetz Leopoldina. Die Korrespondenz der Deutschen Akademie der Naturforscher um 1750* (Berlin 2009).

Müller-Jahncke, Wolf-Dieter, *Astrologisch-magische Theorie und Praxis in der Heilkunde der frühen Neuzeit* (Stuttgart 1985).

Nance, Brian K., *Turquet de Mayerne as Baroque Physician. The Art of Medical Portraiture* (Amsterdam 2001).

Nance, Brian K., 'Medical Casebooks in Early Modern Europe. A Survey of Recent Research and Strategies for Teaching', in Chester R. Burns *et al.* (eds), *The Proceedings of the 37th International Congress for the History of Medicine* (Galveston 2002), 214–7.

Nawrotzki, Bettina, *Zur Geschichte der Polikliniken im 19./20. Jahrhundert. Unter besonderer Berücksichtigung der Entwicklung in Hamburg* (Med. diss., Universität Aachen, 1987).

Nicolson, Malcolm, 'The Art of Diagnosis. Medicine and the Five Senses', in William F. Bynum and Roy Porter (eds), *Companion Encyclopedia of the History of Medicine*, Vol. 2 (London 1993), 801–25.

Niederhäusern, Raffael von, *Sozialtopografie der Stadt Biel Mitte 19. Jh.* (unpublished typescript). (Bern 2011).

Nolte, Karen, 'Die Erfahrung "zwischen den Zeilen." Eine patientenzentrierte Perspektive als eine andere Geschichte?', in Marguérite Bos, Bettina Vincenz and Tanja

Wirz (eds), *Erfahrung. Alles nur Diskurs? Zur Verwendung des Erfahrungsbegriffes in der Geschlechtergeschichte* (Zürich 2004), 273–81.

Nolte, Karen, 'Vom Umgang mit unheilbar Kranken und Sterbenden in "Kranken-Besuchs-Anstalten" zu Beginn des 19. Jahrhunderts', *Würzburger medizinhistorische Mitteilungen* 26 (2007), 28–52 (with English summary).

Nolte, Karen, 'Ärztliche Praxis am Sterbebett in der ersten Hälfte des 19. Jahrhunderts', in Walter Bruchhausen and Hans-Georg Hofer (eds), *Ärztliches Ethos im Kontext. Historische, phänomenologische und didaktische Analysen* (Göttingen 2010), 39–58.

Nolte, Karen, 'Die medizinische Versorgung der Göttinger Stadtarmen durch das Poliklinische Institut der Universität', *Würzburger medizinhistorische Mitteilungen* 29 (2010b), 131–57.

Nolte, Karen, 'Schwindsucht – Krankheit, Gesundheit und Moral im frühen 19. Jahrhundert', *Medizin, Gesellschaft und Geschichte* 29 (2010), 47–70 (with English summary).

Nolte, Karen, '"...wohltätige Hülfe für die notleidenden Armen und Bildung junger Ärzte." Polikliniken im frühen 19. Jahrhundert', *Historia Hospitalium* 27 (2011), 79–86.

Nowell-Smith, Harriet, 'Nineteenth-Century Narrative Case Histories. An Inquiry into Stylistics and History', *Canadian Bulletin of Medical History* 12 (1995), 47–67.

Nowosadtko, Jutta, *Scharfrichter und Abdecker. Der Alltag zweier "unehrlicher Berufe" in der Frühen Neuzeit* (Paderborn 1994).

Nowosadtko, Jutta, 'Rationale Heilbehandlung oder abergläubische Pfuscherei? Die medizinische Kompetenz von Scharfrichtern und ihre Ausgrenzung aus heilenden Tätigkeiten im 18. Jahrhundert', in Bettina Wahrig and Werner Sohn (eds), *Zwischen Aufklärung, Policey und Verwaltung. Zur Genese des Medizinalwesens, 1750–1850* (Wiesbaden 2003), 109–30.

Nutton, Vivian, 'Introduction' to *idem* (ed.), *Medicine at the Courts of Europe, 1500–1837* (London 1990), 1–14.

Oberhofer, Andreas, 'Häuer – Holzer – Schmelzer. Unfälle und Krankheiten im Bergbau Prettau. Eine Auswertung der Krankengeschichten des Südtiroler Landarztes Dr. Franz von Ottenthal (1818–1899)', in Wolfgang Ingenhaeff and Johann Bair (eds), *Bergvolk und Medizin. 3. Internationales Bergbausymposium Schwaz 2004* (Innsbruck 2005), 251–68.

Oberhofer, Andreas, 'Eine Landarztpraxis im 19. Jahrhundert am Beispiel der Ordination des Dr. Franz von Ottenthal (1818–1899)', in Elisabeth Dietrich-Daum, Martin Dinges, Robert Jütte and Christine Roilo (eds), *Arztpraxen im Vergleich. 18.–20. Jahrhundert* (Innsbruck 2008), 167–91 (with English summary).

Obrecht, Andreas, *Die Welt der Geistheiler. Zur Renaissance magischer Weltbilder* (Wien 1999).

Oesterlen, Friedrich, *Handbuch der medicinischen Statistik* (Tübingen ^{2}1874).

Oesterreicher, Wulf, Regn, Gerhard, and Schulze, Winfried (eds), *Autorität der Form, Autorisierung, institutionelle Autorität* (Münster 2003).

Oettli, Markus, *Das Amt des Bezirksarztes im Rahmen der Thurgauischen Sanitätsorganisation 1803–1869* (Med. diss., Universität Zürich, 1982).

Osten, Philipp (ed.), *Patientendokumente. Krankheit in Selbstzeugnissen* (Stuttgart 2010).

Oxfort, Judith, *Meine Nerven tanzen. Die Krankheiten der Madame de Graffigny (1695–1758)* (Köln 2010).

Papsch, Monika, *Krankenjournal D38 (1833–1835). Kommentarband zur Transkription* (Stuttgart 2007).

Papsch, Monika, 'Sozialstatistische Auswertung von Samuel Hahnemanns (1755–1843) homöopathischer Praxis von Dezember 1833 bis Mai 1835 anhand des Krankentagebuches "D38"', in Elisabeth Dietrich-Daum, Martin Dinges and Robert Jütte (eds), *Arztpraxen im Vergleich. 18.–20. Jahrhundert* (Innsbruck 2008), 129–45 (with English summary).

Park, Katharine, and Daston, Lorraine, 'The Age of the New', introduction to *eaedem* (eds), *The Cambridge History of Science, Vol. 3. Early Modern Science* (Cambridge 2006), 1–17.

Paul, Norbert W., 'Diagnose und Prognose', in Stefan Schulz, Klaus Steigleder, Heiner Fangerau and Norbert W. Paul (eds), *Geschichte, Theorie und Ethik der Medizin* (Frankfurt am Main 2006), 143–52.

Pechaček, Petra, *Scharfrichter und Wasenmeister in der Landgrafschaft Hessen-Kassel in der Frühen Neuzeit* (Frankfurt am Main 2003).

Perneel, Philip G., *Het Beroepsjournaal van Dr. J.F.Ph. Hers, Arts te Oud-Beijerland (1881–1915). Een reconstructie van een plattelandspraktijk omtreks 1900* (Rotterdam 2000).

Persaud, Rajendra, 'The Reporting of Psychiatric Symptoms in History. The Memorandum Book of Samuel Coates, 1785–1825', *History of Psychiatry* 4 (1993), 499–510.

Pfister, Christian, '"Uss gewüssen Ursachen." Hintergründe und Methoden statistischer Erhebungen im Kanton Bern 1528–1928', *Schweizerische Zeitschrift für Geschichte* 45 (1995), 29–50.

Pickering, Andrew, *The Mangle of Practice: Time, Agency and Science* (Chicago 1995).

Pilloud, Sévérine, Stefan Hächler and Vincent Barras, 'Consulter par lettre au XVIIIe siècle', *Gesnerus* 61 (2004), 232–53 (with English summary).

Pomata, Gianna, *Contracting a Cure. Patients, Healers, and the Law in Early Modern Bologna* (Baltimore 1998).

Pomata, Gianna, 'Medicina delle monache. Pratiche terapeutiche nei monasteri femminili di Bologna in età moderna', in *eadem* (ed.), *I monasteri femminili come centri di cultura fra Rinascimento e Barocco* (Roma 2005), 329–63.

Pomata, Gianna, 'Praxis Historialis. The Uses of Historia in Early Modern Medicine', in Gianna Pomata and Nancy G. Siraisi (eds), *Historia. Empiricism and Erudition in Early Modern Europe* (Cambridge 2005), 105–46.

Pomata, Gianna, 'Sharing Cases. The Observationes in Early Modern Medicine', *Early Science and Medicine* 15 (2010), 193–236.

Pomata, Gianna, 'Observation Rising. Birth of an Epistemic Genre, 1500–1650', in Lorraine Daston and Elizabeth Lunbeck (eds), *Histories of Scientific Observation* (Chicago 2011), 45–80.

Poovey, Mary, *A History of the Modern Fact. Problems of Knowledge in the Sciences of Wealth and Society* (Chicago 1998).

Porter, Roy, 'The Patient's View. Doing Medical History from Below', *Theory and Society* 14 (1985), 175–98.

Porter, Roy, *Health for Sale. Quackery in England, 1660–1850* (Manchester 1989).

Porter, Roy, *Disease, Medicine and Society in England* (Basingstoke 1989).

Porter, Roy, 'The Rise of Physical Examination', in William F. Bynum and Roy Porter (eds), *Medicine and the Five Senses* (Cambridge 2004), 179–97.

Porter, Roy, and Porter, Dorothy, *Patient's Progress. Doctors and Doctoring in 18th Century England*, (Cambridge 1989).

Powderly, Kathleen E., 'Patient Consent and Negotiation in the Brooklyn Gynaecological Practice of Alexander J.C. Skene. 1863–1900', *The Journal of Medicine and Philosophy* 25, no. 1 (2000), 12–27.

Probst, Christian, *Der Weg des ärztlichen Erkennens am Krankenbett. Herman Boerhaave und die ältere Wiener medizinische Schule, vol. 1, 1707–1787* (Wiesbaden 1972).

Probst, Christian, *Fahrende Heiler und Heilmittelhändler. Medizin von Marktplatz und Landstraße* (Rosenheim 1992).

Raffeiner, Wolfgang, *Der alte Doktor. Eine wahre Erzählung aus dem alten Tirol* (Innsbruck 2002).

Ramsey, Matthew, *Professional and Popular Medicine in France, 1770–1830. The Social World of Medical Practice* (Cambridge 1988).

Rau, Susanne, and Schwerhoff, Gerd, 'Öffentliche Räume in der Frühen Neuzeit. Überlegungen zu Leitbegriffen und Themen eines Forschungsfeldes', in Susanne Rau (ed.), *Zwischen Gotteshaus und Taverne. Öffentliche Räume in Spätmittelalter und Früher Neuzeit* (Köln 2004), 11–52.

Rau, Tilman T.R., *Das Commercium Litterarium. Die erste medizinische Wochenschrift in Deutschland und die Anfänge des medizinischen Journalismus* (Bremen 2009).

Reckwitz, Andreas, 'Grundelemente einer Theorie sozialer Praktiken. Eine sozialtheoretische Perspektive', *Zeitschrift für Soziologie* 32, no. 4 (2003), 282–301.

Reichardt, Sven, 'Praxeologische Geschichtswissenschaft. Eine Diskussionsanregung', *Sozial. Geschichte* 3 (2007), 43–65.

Reinhard, Wolfgang, *Freunde und Kreaturen. "Verflechtung" als Konzept zur Erforschung historischer Führungsgruppen. Römische Oligarchie um 1600* (München 1979).

Renzi, Silvia de, 'Medical Competence, Anatomy and the Polity in Seventeenth-Century Rome', in Sandra Cavallo and David Gentilcore (eds), *Spaces, Objects and Identities in Early Modern Italian Medicine* (Oxford 2007), 551–86.

Rieder, Philip, 'L'histoire du "patient." Aléa, moyen ou finalité de l'histoire médicale?', *Gesnerus* 60 (2003), 260–71.

Rieder, Philip, 'Médecins et patients à Genève: offre et consommations thérapeutiques à l'époque moderne', *Revue d'histoire moderne et contemporaine* 52 (2005), 39–63.

Rieder, Philip, *La figure du patient au XVIIIe siècle* (Genève 2010).

Rieder, Philip, 'The Physician Louis Odier and the Medical Market in Geneva (1774–1817)', *Gesnerus* 69 (2012), 54–75.

Risse, Guenter B., *Hospital Life in Enlightenment Scotland* (Cambridge 1986).

Risse, Guenter B., 'Clinical Instruction in Hospitals. The Boerhaavian Tradition in Leyden, Edinburgh, Vienna and Padua', in Harm Beukers and John Michael Henderson Molls (eds), *Clinical Teaching, Past and Present* (Amsterdam 1989), 1–19.

Ritzmann, Iris, 'Schweizerische Medizingeschichte in Zahlen. Zum Kapitel "Krankheiten und Todesursachen in der Historischen Statistik der Schweiz"', in Christoph Döbeli, Christian Pfister, Hannes Schüle and Remigius Wagner (eds), *Landesgeschichte und Informatik* (Muttenz 1996), 106–12.

Ritzmann, Iris, 'Der Verhaltenskodex des "Savoir faire" als Deckmantel ärztlicher Hilflosigkeit? Ein Beitrag zur Arzt-Patient-Beziehung im 18. Jahrhundert', *Gesnerus* 56 (1999), 197–219.

Ritzmann, Iris, 'Chirurgische Kinderheilkunde im alten Zürcher Spital', *Gesnerus* 58 (2001), 292–9.

Ritzmann, Iris, 'Leidenserfahrung in der historischen Betrachtung. Ein Seiltanz zwischen sozialem Konstrukt und humanbiologischer Konstanz', *Historische Zeitschrift Beiheft* (2001), 59–72.

Ritzmann, Iris, 'Der Faktor Nachfrage bei der Ausformung des modernen Medizinalwesens. Überlegungen am Beispiel der Kinderheilkunde', in Bettina Wahrig und Werner Sohn (eds), *Zwischen Aufklärung, Policey und Verwaltung. Zur Genese des Medizinalwesens 1750–1850* (Wiesbaden 2003), 163–78.

Ritzmann, Iris, '"Die der Welt und sich selbst zur Last sind." Behinderte Kinder und Jugendliche in der Frühen Neuzeit', *Traverse. Zeitschrift für Geschichte/Revue d'histoire* 3 (2006), 73–86.

Ritzmann, Iris, *Sorgenkinder. Kranke und behinderte Mädchen und Jungen im 18. Jahrhundert* (Köln 2008).

Ritzmann, Iris, 'Medikamentöse Kinderbehandlung im 18. Jahrhundert', *Arzneimittel-, Therapie-Kritik & Medizin und Umwelt* 41 (2009), 55–62 and 191–8.

Roelcke, Volker, '"Ars moriendi" und "euthanasia medica." Zur Neukonfiguration und ärztlichen Aneignung normativer Vorstellungen über den "guten Tod" um 1800', in Dietrich von Engelhardt, Jan Joerden and Lothar Jordan (eds), *Sterben und Tod bei Heinrich von Kleist und in seinem historischen Kontext* (Würzburg 2006), 29–44.

Roilo, Christine, '"Historiae Morborum" des Franz v. Ottenthal – ein Zwischenbericht', *Medizin, Geschichte und Gesellschaft* 18 (1999), 57–80 (with English summary).

Roilo, Christine, 'Medizin auf dem Lande. Die Historiae morborum des Franz von Ottenthal', in Elisabeth Dietrich-Daum, Werner Matt und Hanno Platzgummer (eds), *Geschichte und Medizin. Forschungsberichte – Fachgespräche. Dokumentation zur internationalen Tagung "Geschichte der Medizin", 5. Dornbirner Geschichtstage, 9. bis 12. Juni 1999* (Dornbirn 2001), 151–72.

Roland, Charles G., and Rubashewsky, Bohodar, 'The Economic Status of the Practice of Dr. Harmaunus Smith in Wentworth County, Ontario', *Canadian Bulletin of Medical History* 5 (1988), 29–49.

Rordorf, Hartmann, 'Ordinationsbücher und das Manual von Michael Schüppach, Chirurgus von Langnau (1707–1781)', *Pharmaceutica Acta Helvetiae* 5 (1934), 72–84.

Rosenberg, Charles, 'Framing Disease. Illness, Society and History', introduction to *idem* and Janet Golden (eds), *Framing Disease. Studies in Cultural History* (New Brunswick 1992), I–XXVI.

Rosenberg, Charles (ed.), *Right Living. An Anglo-American Tradition of Self-Help Medicine and Hygiene* (Baltimore 2003).

Ruisinger, Marion Maria, 'Briefpraxis versus Besuchspraxis. Das Beispiel Lorenz Heister (1683–1758)', in Elisabeth Dietrich-Daum, Martin Dinges, Robert Jütte and Christine Roilo (eds), *Arztpraxen im Vergleich. 18.–20. Jahrhundert* (Innsbruck 2008), 65–77.

Ruisinger, Marion Maria, *Patientenwege. Die Konsiliarkorrespondenz Lorenz Heisters (1683–1758) in der Trew-Sammlung Erlangen* (Stuttgart 2008).

Ruisinger, Marion Maria, 'The Circulation of the Blood and Venesection. On the Relation Between Medical Theory and Practice in the Early Eighteenth Century', in Jürgen Helm and Renate Wilson (eds), *Medical Theory and Therapeutic Practice in the Eighteenth Century. A Transatlantic Perspective* (Stuttgart 2008c), 37–59.

Ruisinger, Marion Maria, 'Die Medizinische Fakultät Altdorf im frühen 18. Jahrhundert – eine Wiege der wissenschaftlichen Chirurgie', in Hanns Christof Brennecke, Dirk Niefanger and Werner Wilhelm Schnabel (eds), *Akademie und Universität Altdorf. Studien zur Hochschulgeschichte Nürnbergs* (Köln 2011), 207–23.

Russell, Andrew W. (ed.), *The Town and State Physician in Europe from the Middle Ages to the Enlightenment* (Wolfenbüttel 1981).

Sachße, Christoph, and Tennstedt, Florian, *Geschichte der Armenfürsorge in Deutschland, vol. 1: Vom Spätmittelalter bis zum ersten Weltkrieg* (Stuttgart 1980).

Sander, Sabine, 'Die Bürokratisierung des Gesundheitswesens. Zur Problematik der "Modernisierung"', in Werner Kümmel (ed.), *Jahrbuch des Instituts für Geschichte der Medizin der Robert Bosch Stiftung* 6 (1987), 185–218.

Sander, Sabine, *Handwerkschirurgen. Sozialgeschichte einer verdrängten Berufsgruppe* (Göttingen 1989).

Sauerteig, Lutz, 'Vergleich. Ein Königsweg auch für die Medizingeschichte? Methodologische Fragen komparativen Forschens', in Norbert Paul and Thomas Schlich (eds), *Medizingeschichte. Aufgaben, Probleme, Perspektiven* (Frankfurt 1998), 266–91.

Scandola, Pietro (ed.), *Hochschulgeschichte Berns 1528–1984* (Bern 1984).

Schäfer, Daniel, *Alter und Krankheit in der Frühen Neuzeit* (Frankfurt am Main 2004).

Schäffner, Wolfgang, 'Nicht-Wissen um 1800. Buchführung und Statistik', in Joseph Vogl (ed.), *Poetologien des Wissens um 1800* (München 1999), 123–44.

Schatzki, Theodore R., Cetina, Karin Knorr, and Savigny, Eike von (eds), *The Practice Turn in Contemporary Theory* (London/New York 2001).

Schiebinger, Londa, 'Women of Natural Knowledge', in Katharine Park and Lorraine Daston (eds), *The Cambridge History of Science, vol. 3. Early Modern Science* (Cambridge 2006), 192–205.

Schilling, Heinz (ed.), *Der Augsburger Religionsfrieden 1555. Wissenschaftliches Symposium aus Anlass des 450. Jahrestages des Friedensschlusses. Augsburg 21. bis 25. September 2005* (Gütersloh 2007).

Schilling, Katharina, *"Ach gieb mir doch nur etwas Luft, Du hast der Luft so viel!" Palliativmedizin im frühen 19. Jahrhundert* (Duisburg 2011).

Schilling, Ruth, 'Amtsträger und Wissenschaftler – die Repräsentationsstrategien eines Scharfrichtersohns in der Mitte des 18. Jahrhunderts', *Kunstgeschichte. Open Peer Reviewed Journal* (2011) [online: http://www.kunstgeschichte-ejournal.net/129/, accessed 7.09.2012] (with English summary).

Schilling, Ruth, 'Stadt und Arzt im 18. Jahrhundert. Johann Friedrich Glaser, Stadtphysicus in Suhl', *Würzburger medizinhistorische Mitteilungen* 30 (2011), 310–33 (with English summary).

Schilling, Ruth, 'Raum und medizinischer Markt. Die Praxis des Stadt- und Amtsarztes Johann Friedrich Glaser (1707–1789) in Suhl', in Lina Gafner, Iris Ritzmann and Katharina Weikl (eds), *Penning Patient's Histories – Doctors' Records and the Medical Market in the 18th and 19th Century* [special issue: *Gesnerus* 69, no. 1 (2012)], 36–53 (with English summary).

Schilling, Ruth, Schlegelmilch, Sabine, and Splinter, Susan, 'Stadtarzt oder Arzt in der Stadt? Drei Ärzte der Frühen Neuzeit und ihr Verständnis des städtischen Amtes', *Medizinhistorisches Journal* 46 (2011), 99–133 (with English summary).

Schindler, Renate, 'Das Bürgerspital zum Heiligen Geist', in Ulrich Wagner (ed.), *Geschichte der Stadt Würzburg, vol. 3/1. Vom Übergang an Bayern bis zum 21. Jahrhundert.* (Stuttgart 2007), 780–5.

Schlegelmilch, Sabine, 'Johannes Magirus. Stadtarzt in Zerbst (1651–1656)', *Mitteilungen des Vereins für Anhaltische Geschichte* 20 (2011), 9–30.

Schlegelmilch, Sabine, 'Vom Nutzen des Nebensächlichen. Paratexte in den Kalendern des Johannes Magirus', in Klaus-Dieter Herbst (ed.), *Astronomie – Literatur – Volksaufklärung. Der Schreibkalender der Frühen Neuzeit mit seinen Text- und Bildbeigaben* (Bremen 2012), 393–411.

Schlegelmilch, Sabine, '"Hier sind Chymica, hier sind Chymica!" Die frühe Rezeption von Johann Baptist Helmonts "Ortus medicinae" (1648) in Berliner Ärztekreisen', *Morgen-Glantz* 23 (2013), 185–208.

Schlegelmilch, Sabine, *Die ärztliche Praxis des 17. Jahrhunderts - Johannes Magirus (1615–1697)* [scheduled for 2015].

Schlögl, Rudolf, 'Vergesellschaftung unter Anwesenden. Zur kommunikativen Form des Politischen in der vormodernen Stadt', in Rudolf Schlögl (ed.), *Interaktion und Herrschaft* (Konstanz 2004), 9–30.

Schlumbohm, Jürgen, *Lebendige Phantome. Ein Entbindungshospital und seine Patientinnen 1751–1830* (Göttingen 2012).

Schmidt-Herrling, Eleonore, *Die Briefsammlung des Nürnberger Arztes Christoph Jacob Trew (1695–1769) in der Universitätsbibliothek Erlangen* (Erlangen 1940).

Schnalke, Thomas, *Natur im Bild. Anatomie und Botanik in der Sammlung des Nürnberger Arztes Christoph Jacob Trew* (Erlangen 1995).

Schnalke, Thomas, 'Praxis, Theorie und Wissenschaft im Selbstverständnis des städtischen Arztes Christoph Jacob Trew (1695–1769)', *Gesnerus* 52 (1995), 40–53 (with English summary).

Schnalke, Thomas, *Medizin im Brief. Der städtische Arzt des 18. Jahrhunderts im Spiegel seiner Korrespondenz* (Stuttgart 1997).

Schnütgen, Hermann, 'Sanitätsrath Dr. Friedrich von Bönninghausen†', *Allgemeine Homöopathische Zeitung* 58 (1910), 351–2.

Schoop, Albert, *Geschichte der Thurgauer Miliz* (Frauenfeld 1948).

Schoop, Albert, *Geschichte des Kantons Thurgau, vol. 1. Chronologischer Teil* (Frauenfeld 1987).

Schraut, Sylvia, *Sozialer Wandel im Industrialisierungsprozeß. Esslingen 1800–1870* (Esslingen 1989).

Schreiber, Kathrin, *Samuel Hahnemann in Leipzig. Die Entwicklung der Homöopathie zwischen 1811 und 1821. Förderer, Gegner und Patienten* (Stuttgart 2002).

Schreiber, Kathrin, 'Was Hahnemann Driven out of Leipzig? The Leipzig Practice and Why Hahnemann Moved to Köthen in 1821. Patient Numbers and Polemics', in Martin Dinges (ed.), *Patients in the History of Homoeopathy* (Sheffield 2002), 53–63.

Schroers, Fritz, *Lexikon deutschsprachiger Homöopathen* (Stuttgart 2006).

Schuerer, Kevin, and Diederiks, Herman (eds), *The Use of Occupations in Historical Analysis* (Göttingen 1993).

Schweber, Libby, *Disciplining Statistics. Demography and Vital Statistics in France and England, 1830–1885* (Durham 2006).

Schweig, Nicole, *Gesundheitsverhalten von Männern. Gesundheit und Krankheit in Briefen 1800–1950* (Stuttgart 2009).

Seidel, Hans-Christoph, *Eine neue "Kultur des Gebärens." Die Medikalisierung von Geburt im 18. und 19. Jahrhundert in Deutschland* (Stuttgart 1998).

Seybold, Hans-Albrecht Rainer, *Conrad Heinrich Fuchs (1803–1855). Ein Lebens- und Zeitbild* (Göttingen 1963).

Sharma, Ursula, *Complementary Medicine Today. Practitioners and Patients* (London ²1995).

Shapin, Steven, *A Social History of Truth. Civility and Science in Seventeenth-Century England* (Chicago 2004).

Shephard, David, 'The Casebook, the Daybook, and the Diary as Sources in Medical Historiography', *Canadian Bulletin of Medical History* 17 (2000), 245–55.

Shorter, Edward, *Bedside Manners. The Troubled History of Doctors and Patients* (New York 1985).

Siegert, Bernhard, and Vogl, Joseph (eds), *Europa. Kultur der Sekretäre* (Zürich 2003).

Siegl, Gerhard, and Steiner, Guenther, *Ja, jetzt geht es mir gut...Entwicklung der bäuerlichen Sozialversicherung in Österreich* (Wien 2010).

Sigsworth, E.M., and Swan, P., 'An Eighteenth-Century Surgeon and Apothecary. William Elmhirst (1721–1773)', *Medical History* 26 (1982), 191–8.

Splinter, Susan, '"Der Aufrichtige Medicus." Eine Zeitschrift des Nürnberger Arztes Johann Christoph Götz (1688–1733) als Vorläufer des "Commercium Litterarium"', *Jahrbuch für Kommunikationsgeschichte* 13 (2011), 5–15 (with English summary).

Splinter, Susan, '"Ein jeder der Stadt Nürnberg bestelter Medicus und Physicus ordinarius soll geloben..." Medizinale Strukturen Nürnbergs zu Beginn des 18. Jahrhunderts', *Würzburger medizinhistorische Mitteilungen* 30 (2011b), 334–49 (with English summary).

Spree, Reinhard, *Soziale Ungleichheit vor Krankheit und Tod. Zur Sozialgeschichte des Gesundheitsbereichs im Deutschen Kaiserreich* (Göttingen 1981).

Spree, Reinhard, 'Kurpfuscherei-Bekämpfung und ihre sozialen Funktionen während des 19. und zu Beginn des 20. Jahrhunderts', in Alfons Labisch and Reinhard Spree (eds), *Medizinische Deutungsmacht im sozialen Wandel des 19. und frühen 20. Jahrhunderts* (Bonn 1989), 103–21.

Spree, Reinhard, 'Krankenhausentwicklung und Sozialpolitik in Deutschland während des 19. Jahrhunderts', *Historische Zeitschrift* 260 (1995), 75–105.

Spree, Reinhard, 'Quantitative Aspekte der Entwicklung des Krankenhauses im 19. und 20. Jahrhundert. Ein Bild innerer und äußerer Verhältnisse', in Alfons Labisch and Reinhard Spree (eds), *"Einem jeden Kranken in einem Hospitale sein eigenes Bett." Zur Sozialgeschichte des Allgemeinen Krankenhauses in Deutschland im 19. Jahrhundert* (Frankfurt am Main 1996), 51–90.

Stanislaw-Kemenah, Alexandra-Kathrin, 'Von der Hand Gottes berührt?! Krankheit, Alter und Armut im Spiegel von Bittgesuchen zur Aufnahme in Dresdner Hospitäler', in Cordula Nolte (ed.), *Homo debilis. Behinderte – Kranke – Versehrte in der Gesellschaft des Mittelalters* (Korb 2009), 225–44.

Steiger, Johann A., Richter Sandra, and Föcking Marc (eds), *Innovation durch Wissenstransfer in der Frühen Neuzeit. Kultur- und geistesgeschichtliche Studien zu Austauschprozessen in Mitteleuropa* (Amsterdam 2010).

Steinbrecher, Aline, *Verrückte Welten. Wahnsinn und Gesellschaft im barocken Zeitalter* (Zürich 2006).

Steinke, Hubert, 'Anatomie und Physiologie', in Hubert Steinke, Urs Boschung and Wolfgang Proß (eds), *Albrecht von Haller. Leben – Werk – Epoche* (Göttingen 2008), 226–54.

Steinke, Hubert, 'Der junge Arzt und seine Patienten. Albrecht von Hallers Praxis in Bern 1731–1736', in Elisabeth Dietrich-Daum, Martin Dinges, Robert Jütte and Christine Roilo (eds), *Arztpraxen im Vergleich. 18.–20. Jahrhundert* (Innsbruck 2008), 79–86.

Steinke, Hubert, 'Arzt und Patient um 1850', in Peter Martig (ed.), *Berns moderne Zeit. Das 19. und 20. Jahrhundert neu entdeckt* (Bern 2011), 181.

Steinke, Hubert, and Boschung, Urs, 'Nützliche Medizin. Theorie und Praxis bei Albrecht von Haller', in André Holenstein, Martin Stuber and Gerrendina Gerber-Visser (eds), *Nützliche Wissenschaft und Ökonomie im Ancien Régime. Akteure, Themen, Kommunikationsformen* (Heidelberg 2007), 133–47.

Steinke, Hubert, and Stuber, Martin, 'Haller und die Gelehrtenrepublik', in Hubert Steinke, Urs Boschung and Wolfgang Proß (eds), *Albrecht von Haller. Leben – Werk – Epoche* (Göttingen 2008), 381–414.

Steinke, Hubert, Boschung, Urs, and Proß, Wolfgang (eds), *Albrecht von Haller. Leben – Werk – Epoche* (Göttingen 2008).

Stenzel, Oliver, *Medikale Differenzierung. Der Konflikt zwischen akademischer Medizin und Laienheilkunde im 18. Jahrhundert* (Heidelberg 2005).

Stolberg, Michael, *Heilkunde zwischen Staat und Bevölkerung. Angebot und Annahme medizinischer Versorgung in Oberfranken im frühen 19. Jahrhundert* (Med. diss., Techn. Universität München, 1986).

Stolberg, Michael, 'Ärzte und ländliche Patienten. Soziologisch-historische Aspekte einer schwierigen Beziehung', *Die medizinische Welt* 43 (1992), 529–33.

Stolberg, Michael, 'Patientenschaft und Krankheitsspektrum in ländlichen Arztpraxen des 19. Jahrhunderts', *Medizinhistorisches Journal* 28 (1993), 3–27 (with English summary).

Stolberg, Michael, '"Mein äskulapisches Orakel!" Patientenbriefe als Quelle einer Kulturgeschichte der Krankheitserfahrung im 18. Jahrhundert', *Österreichische Zeitschrift für Geschichtswissenschaften* 7 (1996), 385–404 (with English summary).

Stolberg, Michael, 'Heilkundige. Professionalisierung und Medikalisierung', in Norbert Paul and Thomas Schlich (eds), *Medizingeschichte. Aufgaben, Probleme, Perspektiven* (Frankfurt am Main 1998), 69–86.

Stolberg, Michael, 'Probleme und Perspektiven einer Geschichte der "Volksmedizin"', in Thomas Schnalke and Claudia Wiesemann (eds), *Die Grenzen des Anderen. Medizingeschichte aus postmoderner Perspektive* (Köln 1998), 49–73.

Stolberg, Michael, *Die Geschichte der Homöopathie in Bayern (1800–1914)* (Heidelberg 1999a).

Stolberg, Michael, 'A Woman's Hell? Medical Perceptions of Menopause in Preindustrial Europe', *Bulletin of the History of Medicine* 73 (1999), 404–28.

Stolberg, Michael, 'Krankheitserfahrung und Arzt-Patienten-Beziehung in Samuel Hahnemanns Patientenkorrespondenz', *Medizin, Gesellschaft und Geschichte* 18 (1999), 169–88 (with English summary).

Stolberg, Michael, 'The Monthly Malady. A History of Premenstrual Suffering', *Medical History* 44 (2000), 301–22.

Stolberg, Michael, 'Formen und Strategien der Autorisierung in der frühneuzeitlichen Medizin', in Wulf Oesterreicher, Gerhard Regn and Winfried Schulze (eds), *Autorität der Form, Autorisierung, institutionelle Autorität* (Münster 2003), 205–18.

Stolberg, Michael, 'Menstruation and Sexual Difference in Early Modern Medicine', in Andrew Shail and Gillian Howie (eds), *Menstruation. A Cultural History* (New York 2005), 90–101.

Stolberg, Michael, '"Cura palliativa." Begriff und Diskussion der palliativen Krankheitsbehandlung in der vormodernen Medizin (ca. 1500–1850)', *Medizinhistorisches Journal* 42 (2007), 7–29 (with English summary).

Stolberg, Michael, 'Formen und Funktionen medizinischer Fallberichte in der Frühen Neuzeit (1500–1800)', in Johannes Süßmann, Susanne Scholz and Gisela Engel (eds), *Fallstudien. Theorie – Geschichte – Methode* (Berlin 2007), 81–95.

Stolberg, Michael, 'Medizin und Krankheit in der Frühen Neuzeit. Eine Einführung', *Geschichte in Wissenschaft und Unterricht* 2 (2008), 85–95.

Stolberg, Michael, *Die Geschichte der Palliativmedizin. Medizinische Sterbebegleitung von 1500 bis heute* (Frankfurt am Main 2011).

Stolberg, Michael, *Experiencing Illness and the Sick Body in Early Modern Europe* (Basingstoke 2011).

Stolberg, Michael, 'Möglichkeiten und Grenzen einer retrospektiven Diagnose', in Waltraud Pulz (ed.), *Zwischen Himmel und Erde. Körperliche Zeichen der Heiligkeit* (Stuttgart 2012), 209–27.

Stolberg, Michael, 'Examining the Body c. 1500–1750', in Sarah Toulalan and Kate Fisher (eds), *The Routledge History of Sex and the Body, 1500 to the Present* (London 2013), 91–120.

Stolberg, Michael, 'Medizinische *Loci communes*. Formen und Funktionen einer ärztlichen Aufzeichnungspraxis im 16. und 17. Jahrhundert', *NTM. Zeitschrift für Geschichte der Wissenschaften, Technik und Medizin* 21 (2013), 37–60 (with English summary).

Stolberg, Michael, *Die Harnschau. Eine Kultur- und Alltagsgeschichte* (Köln 2009); forthcoming English edn.: *Uroscopy in Early Modern Europe* (Farnham 2015).

Stolz, Silvia, *Ärztliche Theorie und Praxis im Spiegel von 61 Falldarstellungen eines Züricher Medizinschülers, 1785–1787* (Med. diss., Universität Basel, 1984).

Stowe, Steven M., *Doctoring the South. Southern Physicians and Everyday Medicine in the Mid-Nineteenth Century* (Chapel Hill 2004).

Stukenbrock, Karin, *"Der zerstückte Cörper." Zur Sozialgeschichte der anatomischen Sektionen in der frühen Neuzeit (1650–1800)* (Stuttgart 2001).

Stumm, Ingrid von, *Gesundheit, Arbeit und Geschlecht im Kaiserreich am Beispiel der Krankenstatistik der Leipziger Ortskrankenkasse 1887–1905* (Frankfurt am Main 1995).

Stürzbecher, Manfred, 'The Physici in German-Speaking Countries from the Middle Age to the Enlightenment', in Andrew W. Russell (ed.), *The Town and the State Physician in Europe from the Middle Ages to the Enlightenment* (Wolfenbüttel 1981), 123–30.

Szabo, Jason, *Incurable and Intolerable. Chronic Disease and Slow Death in Nineteenth-Century France* (New Brunswick 2009).

Taddei, Elena, 'Historiae morborum: Le anamnesi del medico generico Franz von Ottenthal', *Medicina e storia* 9 (2005), 89–98.

Taddei, Elena, 'Bestellungsverfahren von Ärzten in Tirol in der zweiten Hälfte des 19. Jahrhunderts. Qualifikationen, Auswahlkriterien und Erwartungen', in Elisabeth Dietrich-Daum, Martin Dinges, Robert Jütte and Christine Roilo (eds), *Arztpraxen im Vergleich. 18.–20. Jahrhundert* (Innsbruck 2008), 221–37.

Taddei, Elena, *Franz von Ottenthal. Arzt und Tiroler Landtagsabgeordneter (1818–1899)* (Wien 2010).

Teuteberg, Hans-Jürgen, 'Bevölkerungsentwicklung und Eingemeindungen (1816–1945)', in Franz-Josef Jakobi (ed.), *Geschichte der Stadt Münster*, vol. 2 (Münster 1993), 331–86.

Thomson, Donald M., 'General Practice and the Edinburgh Medical School. 200 Years of Teaching, Care and Research', *Journal of the Royal College of General Practitioners* 34 (1984), 9–12.

Thümmler, Andrea, *Rekonstruktion des Alltags eines thüringischen Arztes im 18. Jahrhundert anhand seines Praxistagebuchs 1750–1763* (Med. diss., Freie Universität Berlin, 2004).

Tomes, Nancy, 'Patients or Health-Care Consumers? Why the History of Contested Terms Matters', in Rosemary Stevens, Charles Rosenberg and Lawton Burns (eds), *History and Health Policy in the United States* (New Brunswick 2006), 83–110.

Trevor-Roper, Hugh, *Europe's Physician. The Various Life of Sir Theodore de Mayerne* (New Haven 2006).

Tröhler, Ulrich, 'The *Doctor* as *Naturalist*. The Idea and Practice of Clinical Teaching and Research in British Policlinics 1770–1850', in Harm Beukert (ed.), *Clinical Teaching, Past and Present* (Amsterdam 1989), 21–34.

Tutzke, Dietrich, and Engel, Regina, 'Tätigkeit und Einkommen eines Allgemeinpraktikers vor der Mitte des 19. Jahrhunderts. Ergebnisse einer historisch-statistischen Analyse', *Zeitschrift für die gesamte Hygiene* 24 (1978), 460–5.

Ulbricht, Otto, 'Supplikationen als Ego-Dokumente. Bittschriften von Leibeigenen aus der ersten Hälfte des 17. Jahrhunderts als Beispiel', in Winfried Schulze (ed.), *Ego-Dokumente. Annäherung an den Menschen in der Geschichte* (Berlin 1996), 149–74.

Ulbricht, Otto, 'Der Kampf des "Kurpfuschers." Friedrich Frantz Heinitz, 1786–1814', in *idem* (ed.), *Mikrogeschichte. Menschen und Konflikte in der frühen Neuzeit* (Frankfurt am Main 2009), 298–336.

Unterkircher, Alois, 'Die Praxis des Südtiroler Landarztes Franz von Ottenthal. Krankheitsspektrum und Gesundheitsverhalten seiner Patientenschaft um 1860', in Rainer Alsheimer and Roland Weibezahn (eds), *Körperlichkeit und Kultur 2004. Interdisziplinäre Medikalkulturforschung. Dokumentation des 7. Arbeitstreffens des "Netzwerk Gesundheit und Kultur in der volkskundlichen Forschung" Würzburg, 31. März–2. April 2004* (Bremen 2005), 215–37.

Unterkircher, Alois, 'Ausser Konkurrenz? Strategien der Inanspruchnahme einer Südtiroler Landarztpraxis um 1890 bei Erkrankungen von Kindern', in Lina Gafner, Iris Ritzmann and Katharina Weikl (eds), *Penning Patient's Histories – Doctors' Records and the Medical Market in the 18th and 19th Century* [special issue: *Gesnerus* 69, no. 1 (2012)], 158–77 (with English summary).

Unterkircher, Alois, Jungen und Männer als Patienten bei einem Südtiroler Landarzt (1860–1900) (Stuttgart 2014) [MedGG-Beiheft 51].

Vanja, Christina, 'Seele und Körper (Neuzeit)', in Peter Dinzelbach (ed.), *Europäische Mentalitätsgeschichte* (Stuttgart 1993), 200–7.

Vanja, Christina, 'Supplikationen als Quelle der Patientengeschichte', in Arnd Friedrich, Irmtraut Sahmland and Christina Vanja (eds), *An der Wende zur Moderne. Die hessischen Hohen Hospitäler im 18. und 19. Jahrhundert* (Petersberg 2008), 163–72.

Vieler, Ingrid, *Die deutsche Arztpraxis im 19. Jahrhundert* (Med. diss., Universität Mainz, 1958).

Vismann, Cornelia, *Files. Law and Media Technology* (Stanford 2008).

Vögele, Jürgen, *Urban Mortality Change in England and Germany 1870–1913* (Liverpool 1998).

Vögele, Jörg (ed.), *Stadt, Krankheit und Tod. Geschichte der städtischen Gesundheitsverhältnisse während der epidemiologischen Transition (vom 18. bis ins frühe 20. Jahrhundert)* (Berlin 2000).

Vögele, Jörg, Wolfgang Woelk and Bärbel Schürmann, 'Städtisches Armenwesen, Krankenkassen und Krankenhauspatienten während des späten 19. und frühen 20. Jahrhunderts in Düsseldorf', in Alfons Labisch and Reinhard Spree (eds), *Krankenhaus-Report 19. Jahrhundert. Krankenhausträger, Krankenhausfinanzierung, Krankenhauspatienten* (Frankfurt am Main 2001), 405–26.

Vogl, Michael, '"Nahe und entfernte Landpraxis." Untersuchungen zu Samuel Hahnemanns Eilenburger Patientenschaft 1801–1803', *Medizin, Gesellschaft und Geschichte* 9 (1990), 165–80 (with English summary).

Wahl, Volker, 'Die Mehliser Hirtenzeche. Über den historischen Ursprung des Hirtenfestes in Zella-Mehlis', in Bezirkskabinett für Kulturarbeit Suhl (ed.), *Zum Hirtenwesen im Thüringer Wald. Beiträge* (Suhl 1988), 46–52.

Wahrig, Bettina, and Sohn, Werner (eds), *Zwischen Aufklärung, Policey und Verwaltung. Zur Genese des Medizinalwesens 1750–1850* (Wiesbaden 2003).

Walter, Tilmann, 'Ärztehaushalte im 16. Jahrhundert. Einkünfte, Status und Praktiken der Repräsentation', *Medizin, Gesellschaft und Geschichte* 27 (2008), 31–74 (with English summary).

Warner, John H., *The Therapeutic Perspective. Medical Practice, Knowledge and Identity in America 1820–1885* (Cambridge 1986).

Warner, John Harley, *Against the Spirit of System. The French Impulse in Nineteenth-Century American Medicine* (Princeton 1998).

Warner, John Harley, 'The Uses of Patient Records by Historians – Patterns, Possibilities and Perplexities', *Health and History* 1 (1999), 101–11.

Wear, Andrew, 'Medical Practice in Late Seventeenth- and Early Eighteenth-Century England. Continuity and Union', in Roger French and Andrew Wear (eds), *The Medical Revolution of the Seventeenth Century* (Cambridge 1989), 294–320.

Wear, Andrew, *Knowledge and Practice in Early Modern English Medicine. 1550–1680* (New York 2000).

Weckbach, Hubert, 'Ein Scharfrichter wird Arzt. Zur Biographie des Dr. Johann Andreas Peter Bayer in Heilbronn', *Schwaben und Franken. Heimatgeschichtliche Beilage der Heilbronner Stimme* 33, No. 7 (1987), I–IV, No. 9 (1987), I–IV, No. 10 (1987), I–IV.

Weckbach, Hubert, 'Dr. med. Johann Christoph Paul Bayer. Der jüngste Sohn des Scharfrichters Johann Andreas Bayer in Heilbronn', *Schwaben und Franken. Heimatgeschichtliche Beilage der Heilbronner Stimme* 35 (1989), I–IV.

Wegener, Andreas, 'Homöopathische Symptomenlehre', in Thomas Genneper and Andreas Wegener (eds), *Lehrbuch Homöopathie. Grundlagen und Praxis der klassischen Homöopathie* (Stuttgart ^{3}2011), 50–78.

Wegener, Andreas, 'Mittelfindung nach dem Therapeutischen Taschenbuch von Bönninghausen', in Thomas Genneper and Andreas Wegener (eds), *Lehrbuch Homöopathie. Grundlagen und Praxis der klassischen Homöopathie* (Heidelberg 2001), 139–51 (=Stuttgart ^{3}2011), 118–30.

Wehren, Eugen, 'Das medizinische Werk des Wundarztes Michel Schüppach (1707–1781) an Hand seiner Rezept- und Ordinationsbücher', *Berner Zeitschrift für Geschichte und Heimatkunde* 47 (1985), 87–166.

Weikl, Katharina, and Ritzmann, Iris, 'Ein Auslaufmodell. Spuren einer ländlichen Heilerpraxis im Zeitalter der Akademisierung', in Lina Gafner, Iris Ritzmann and Katharina Weikl (eds), *Penning Patient's Histories – Doctors' Records and the Medical Market in the 18th and 19th Century* [special issue: *Gesnerus* 69, no. 1 (2012)], 76–94 (with English summary).

Weindling, Paul, 'Medical Practice in Imperial Germany. The Casebook of Alfred Grotjahn', *Bulletin of the History of Medicine* 61 (1987), 391–410.

Weller, Thomas, *Theatrum Praecedentiae. Zeremonieller Rang und gesellschaftliche Ordnung in der frühneuzeitlichen Stadt. Leipzig 1500–1800* (Darmstadt 2006).

White, Hayden, 'The Value of Narrativity in the Representation of Reality', *Critical Inquiry* 7 (1980), 5–27.

Whitteridge, Gweneth, *William Harvey and the Circulation of Blood* (London 1971).

Wiesemann, Claudia, and Tsouyopoulos, Nelly (eds), *Asklepios und die Philosophen. Paradigmawechsel in der Medizin im 19. Jahrhundert* (Stuttgart 2008).

Wiesing, Urban, *Kunst oder Wissenschaft? Konzeptionen der Medizin in der deutschen Romantik* (Stuttgart 1995).

Wilbertz, Gisela, *Scharfrichter und Abdecker im Hochstift Osnabrück. Untersuchungen zur Sozialgeschichte zweier "unehrlicher" Berufe im nordwestdeutschen Raum vom 16. bis zum 19. Jahrhundert* (Osnabrück 1979).

Wilbertz, Gisela, 'Zur sozialen und geographischen Mobilität einer Scharfrichterfamilie in der Frühen Neuzeit: Die Clauss (Clausen, Claessen, Clages) aus Lemgo', *Beiträge zur westfälischen Familienforschung* 53 (1995), 253–322; 54 (1997), 183–256; 55 (1998), 204–44.

Wilke, Gerhard, 'Die Sünden der Väter. Bedeutung und Wandel von Gesundheit und Krankheit im Dorfalltag', in Alfons Labisch and Reinhard Spree (eds), *Medizinische Deutungsmacht im sozialen Wandel des 19. und frühen 20. Jahrhunderts* (Bonn 1989), 123–40.

Winkelmann, Heike, *Das academische Hospital in Göttingen von 1781 bis 1850. Zur Geschichte der Göttinger Universitätsklinik* (Göttingen 1981).

Winker, Gabriele, and Degele, Nina, *Intersektionalität. Zur Analyse sozialer Ungleichheiten* (Bielefeld 2009).

Wolff, Eberhard, *Einschneidende Maßnahmen. Pockenschutzimpfung und traditionale Gesellschaft im Württemberg des frühen 19. Jahrhunderts* (Stuttgart 1998).

Wolff, Eberhard, 'Perspectives on Patients' History. Methodological Considerations on the Example of Recent German-Speaking Literature', *Canadian Bulletin of Medical History* 15 (1998), 207–28.

Wolff, Eberhard, '"Volksmedizin." Abschied auf Raten. Vom definitorischen zum heuristischen Begriffsverständnis', *Zeitschrift für Volkskunde* 94 (1998), 233–57.

Wolff, Jutta, and Wolff, Hans-Peter, 'Das Profil einer ärztlichen Allgemeinpraxis im Jahre 1862', *Das Deutsche Gesundheitswesen. Wochenschrift für die gesamte Medizin* 34 (1979), 568–71.

Wölfing, Günther, *Geschichte des Henneberger Landes zwischen Grabfeld, Rennsteig und Rhön. Ein Überblick* (Hildburghausen 1992).

Wulff, Henrik R., and Jungersen, Kirsten, 'A Danish Provincial Physician and His Patients. The Patient Records from the Pratice of Christopher Detlev Hahn in Aarhus around 1800', *Medizinhistorisches Journal* 40 (2005), 321–45.

Wüst, Wolfgang, Kreuzer, Georg, and Schümann, Nicola (eds), *Der Augsburger Religionsfriede 1555. Ein Epochenereignis und seine regionale Verankerung* (Augsburg 2005).

Zeber, Ulrike, *Die Geschichte des Pflasters. Von der traditionellen Arzneiform Pflaster zum Heftpflaster* (Stuttgart 2001).

Zedelmaier, Helmut, 'De ratione excerpendi. Daniel Georg Morhof und das Exzerpieren', in Françoise Waquet (ed.), *Mapping the World of Learning. The Polyhistor of Daniel Georg Morhof* (Wiesbaden 2000), 75–92.

Index